W9-AKS-517

STUDENT SOLUTIONS MANUAL

VOLUME ONE: CHAPTERS 1–20

SEARS & ZEMANSKY'S

UNIVERSITY PHYSICS

13TH EDITION

LEWIS FORD
WAYNE ANDERSON

PEARSON

Boston Columbus Indianapolis New York San Francisco Upper Saddle River
Amsterdam Cape Town Dubai London Madrid Milan Munich Paris Montreal Toronto
Delhi Mexico City São Paulo Sydney Hong Kong Seoul Singapore Taipei Tokyo

Publisher:	Jim Smith
Executive Editor:	Nancy Whilton
Project Editor:	Chandrika Madhavan
Editorial Manager:	Laura Kenney
Managing Editor:	Corinne Benson
Production Project Manager:	Beth Collins
Production Management and Compositor:	PreMediaGlobal
Manufacturing Buyer:	Jeffrey Sargent
Senior Marketing Manager:	Kerry Chapman
Cover Design:	Derek Bacchus and Seventeenth Street Design
Cover Photo Credits:	Getty Images/Mirko Cassanelli
Cover and Text Printer:	Bind-Rite Graphics

Copyright © 2012, 2008 Pearson Education, Inc., publishing as Addison-Wesley, 1301 Sansome Street, San Francisco, CA 94111. All rights reserved. Manufactured in the United States of America. This publication is protected by Copyright and permission should be obtained from the publisher prior to any prohibited reproduction, storage in a retrieval system, or transmission in any form or by any means, electronic, mechanical, photocopying, recording, or likewise. To obtain permission(s) to use material from this work, please submit a written request to Pearson Education, Inc., Permissions Department, 1900 E. Lake Ave., Glenview, IL 60025. For information regarding permissions, call (847) 486-2635.

Many of the designations used by manufacturers and sellers to distinguish their products are claimed as trademarks. Where those designations appear in this book, and the publisher was aware of a trademark claim, the designations have been printed in initial caps or all caps.

ISBN 10: 0-321-69668-9
ISBN 13: 978-0-321-69668-7

1 2 3 4 5 6 7 8 9 10—BR—14 13 12 11

CONTENTS

Thermodynamics

PREFACE

This Student Solutions Manual, Volume 1, contains detailed solutions for approximately one-third of the Exercises and Problems in Chapters 1 through 20 of the Thirteenth Edition of University Physics by Hugh Young and Roger Freedman. The Exercises and Problems included in this manual are selected solely from the odd-numbered Exercises and Problems in the text (for which the answers are tabulated at the back of the textbook). The Exercises and Problems included were not selected at random but rather were carefully chosen to include at least one representative example of each problem type. The remaining Exercises and Problems, for which solutions are not given here, constitute an ample set of problems for you to tackle on your own. In addition, there are the Challenge Problems in the text for which no solutions are given here.

This manual greatly expands the set of worked-out examples that accompanies the presentation of physics laws and concepts in the text. This manual was written to provide you with models to follow in working physics problems. The problems are worked out in the manner and style in which you should carry out your own problem solutions.

The Student Solutions Manual Volumes 2 and 3 companion volume is also available from your college bookstore.

Lewis Ford
Wayne Anderson
Sacramento, CA

UNITS, PHYSICAL QUANTITIES AND VECTORS

1.5.   **IDENTIFY:**   Convert volume units from in.^{3} to L.

SET UP: $1\text{ L} = 1000\text{ cm}^3$. $1\text{ in.} = 2.54\text{ cm}$.

EXECUTE: $(327\text{ in.}^3) \times (2.54\text{ cm/in.})^3 \times (1\text{ L}/1000\text{ cm}^3) = 5.36\text{ L}$

EVALUATE: The volume is 5360 cm^3. 1 cm^3 is less than 1 in.^3, so the volume in cm^3 is a larger number than the volume in in.^{3}.

1.9.   **IDENTIFY:**   Convert miles/gallon to km/L.

SET UP: $1\text{ mi} = 1.609\text{ km}$. $1\text{ gallon} = 3.788\text{ L}$.

EXECUTE: **(a)** $55.0\text{ miles/gallon} = (55.0\text{ miles/gallon})\left(\dfrac{1.609\text{ km}}{1\text{ mi}}\right)\left(\dfrac{1\text{ gallon}}{3.788\text{ L}}\right) = 23.4\text{ km/L}$.

(b) The volume of gas required is $\dfrac{1500\text{ km}}{23.4\text{ km/L}} = 64.1\text{ L}$. $\dfrac{64.1\text{ L}}{45\text{ L/tank}} = 1.4\text{ tanks}$.

EVALUATE: $1\text{ mi/gal} = 0.425\text{ km/L}$. A km is very roughly half a mile and there are roughly 4 liters in a gallon, so $1\text{ mi/gal} \sim \frac{2}{4}\text{ km/L}$, which is roughly our result.

1.11.   **IDENTIFY:**   We know the density and mass; thus we can find the volume using the relation density = mass/volume = m/V. The radius is then found from the volume equation for a sphere and the result for the volume.

SET UP: Density $= 19.5\text{ g/cm}^3$ and $m_{\text{critical}} = 60.0\text{ kg}$. For a sphere $V = \frac{4}{3}\pi r^3$.

EXECUTE: $V = m_{\text{critical}}/\text{density} = \left(\dfrac{60.0\text{ kg}}{19.5\text{ g/cm}^3}\right)\left(\dfrac{1000\text{ g}}{1.0\text{ kg}}\right) = 3080\text{ cm}^3$.

$r = \sqrt[3]{\dfrac{3V}{4\pi}} = \sqrt[3]{\dfrac{3}{4\pi}(3080\text{ cm}^3)} = 9.0\text{ cm}$.

EVALUATE: The density is very large, so the 130-pound sphere is small in size.

1.15.   **IDENTIFY:**   Use your calculator to display $\pi \times 10^7$. Compare that number to the number of seconds in a year.

SET UP: $1\text{ yr} = 365.24\text{ days}$, $1\text{ day} = 24\text{ h}$, and $1\text{ h} = 3600\text{ s}$.

EXECUTE: $(365.24\text{ days/1 yr})\left(\dfrac{24\text{ h}}{1\text{ day}}\right)\left(\dfrac{3600\text{ s}}{1\text{ h}}\right) = 3.15567\ldots \times 10^7\text{ s}$; $\pi \times 10^7\text{ s} = 3.14159\ldots \times 10^7\text{ s}$

The approximate expression is accurate to two significant figures. The percent error is 0.45%.

EVALUATE: The close agreement is a numerical accident.

1.19.   **IDENTIFY:**   Estimate the number of pages and the number of words per page.

SET UP: Assuming the two-volume edition, there are approximately a thousand pages, and each page has between 500 and a thousand words (counting captions and the smaller print, such as the end-of-chapter exercises and problems).

EXECUTE: An estimate for the number of words is about 10^6.

EVALUATE: We can expect that this estimate is accurate to within a factor of 10.

1.21.   **IDENTIFY:** Estimate the number of blinks per minute. Convert minutes to years. Estimate the typical lifetime in years.

SET UP: Estimate that we blink 10 times per minute. $1 \text{ y} = 365$ days. 1 day $= 24$ h, 1 h $= 60$ min. Use 80 years for the lifetime.

EXECUTE: The number of blinks is $(10 \text{ per min}) \left(\dfrac{60 \text{ min}}{1 \text{ h}} \right) \left(\dfrac{24 \text{ h}}{1 \text{ day}} \right) \left(\dfrac{365 \text{ days}}{1 \text{ y}} \right) (80 \text{ y/lifetime}) = 4 \times 10^8$

EVALUATE: Our estimate of the number of blinks per minute can be off by a factor of two but our calculation is surely accurate to a power of 10.

1.23.   **IDENTIFY:** Estimation problem

SET UP: Estimate that the pile is $18 \text{ in.} \times 18 \text{ in.} \times 5 \text{ ft } 8 \text{ in.}$. Use the density of gold to calculate the mass of gold in the pile and from this calculate the dollar value.

EXECUTE: The volume of gold in the pile is $V = 18 \text{ in.} \times 18 \text{ in.} \times 68 \text{ in.} = 22{,}000 \text{ in.}^3$. Convert to cm^3:

$$V = 22{,}000 \text{ in.}^3 (1000 \text{ cm}^3 / 61.02 \text{ in.}^3) = 3.6 \times 10^5 \text{ cm}^3.$$

The density of gold is 19.3 g/cm^3, so the mass of this volume of gold is

$$m = (19.3 \text{ g/cm}^3)(3.6 \times 10^5 \text{ cm}^3) = 7 \times 10^6 \text{ g}.$$

The monetary value of one gram is \$10, so the gold has a value of $(\$10/\text{gram})(7 \times 10^6 \text{ grams}) = \$7 \times 10^7$, or about $\$100 \times 10^6$ (one hundred million dollars).

EVALUATE: This is quite a large pile of gold, so such a large monetary value is reasonable.

1.25.   **IDENTIFY:** Estimate the number of students and the average number of pizzas eaten by each student in a school year.

SET UP: Assume a school of a thousand students, each of whom averages ten pizzas a year (perhaps an underestimate)

EXECUTE: They eat a total of 10^4 pizzas.

EVALUATE: The same answer applies to a school of 250 students averaging 40 pizzas a year each.

1.33.   **IDENTIFY:** Given the direction and one component of a vector, find the other component and the magnitude.

SET UP: Use the tangent of the given angle and the definition of vector magnitude.

EXECUTE: **(a)** $\tan 32.0° = \dfrac{|A_x|}{|A_y|}$

$|A_x| = (13.0 \text{ m}) \tan 32.0° = 8.12 \text{ m}.$ $A_x = -8.12 \text{ m}.$

(b) $A = \sqrt{A_x^2 + A_y^2} = 15.3 \text{ m}.$

EVALUATE: The magnitude is greater than either of the components.

1.35.   **IDENTIFY:** If $\vec{C} = \vec{A} + \vec{B}$, then $C_x = A_x + B_x$ and $C_y = A_y + B_y$. Use C_x and C_y to find the magnitude and direction of $\vec{C}$.

SET UP: From Figure E1.28 in the textbook, $A_x = 0$, $A_y = -8.00 \text{ m}$ and $B_x = +B \sin 30.0° = 7.50 \text{ m}$, $B_y = +B \cos 30.0° = 13.0 \text{ m}.$

EXECUTE: **(a)** $\vec{C} = \vec{A} + \vec{B}$ so $C_x = A_x + B_x = 7.50 \text{ m}$ and $C_y = A_y + B_y = +5.00 \text{ m}.$ $C = 9.01 \text{ m}.$

$\tan \theta = \dfrac{C_y}{C_x} = \dfrac{5.00 \text{ m}}{7.50 \text{ m}}$ and $\theta = 33.7°.$

(b) $\vec{B} + \vec{A} = \vec{A} + \vec{B}$, so $\vec{B} + \vec{A}$ has magnitude 9.01 m and direction specified by $33.7°.$

(c) $\vec{D} = \vec{A} - \vec{B}$ so $D_x = A_x - B_x = -7.50 \text{ m}$ and $D_y = A_y - B_y = -21.0 \text{ m}.$ $D = 22.3 \text{ m}.$

$\tan \phi = \dfrac{D_y}{D_x} = \dfrac{-21.0 \text{ m}}{-7.50 \text{ m}}$ and $\phi = 70.3°.$ $\vec{D}$ is in the 3rd quadrant and the angle θ counterclockwise from the $+x$ axis is $180° + 70.3° = 250.3°.$

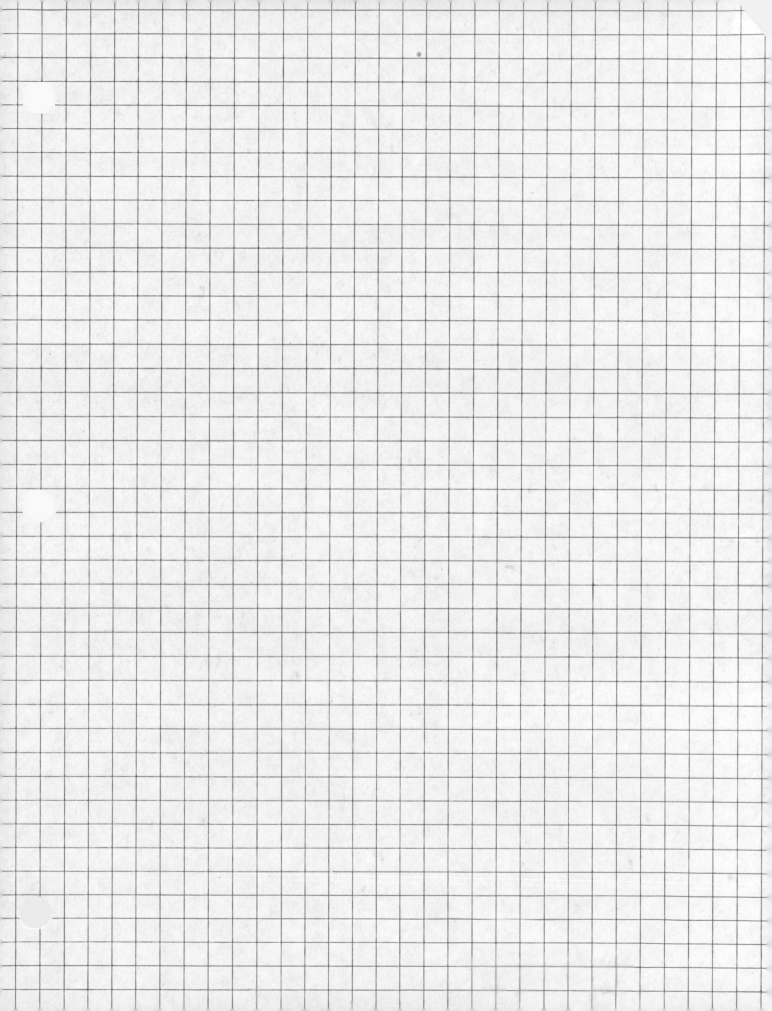

(d) $\vec{B} - \vec{A} = -(\vec{A} - \vec{B})$, so $\vec{B} - \vec{A}$ has magnitude 22.3 m and direction specified by $\theta = 70.3°$.

EVALUATE: These results agree with those calculated from a scale drawing in Problem 1.28.

1.43. **IDENTIFY:** Use trig to find the components of each vector. Use Eq. (1.11) to find the components of the vector sum. Eq. (1.14) expresses a vector in terms of its components.

SET UP: Use the coordinates in the figure that accompanies the problem.

EXECUTE: (a) $\vec{A} = (3.60 \text{ m})\cos 70.0° \hat{i} + (3.60 \text{ m})\sin 70.0° \hat{j} = (1.23 \text{ m})\hat{i} + (3.38 \text{ m})\hat{j}$

$\vec{B} = -(2.40 \text{ m})\cos 30.0° \hat{i} - (2.40 \text{ m})\sin 30.0° \hat{j} = (-2.08 \text{ m})\hat{i} + (-1.20 \text{ m})\hat{j}$

(b) $\vec{C} = (3.00)\,\vec{A} - (4.00)\,\vec{B} = (3.00)(1.23 \text{ m})\hat{i} + (3.00)(3.38 \text{ m})\hat{j} - (4.00)(-2.08 \text{ m})\hat{i} - (4.00)(-1.20 \text{ m})\hat{j}$

$$= (12.01 \text{ m})\hat{i} + (14.94)\hat{j}$$

(c) From Equations (1.7) and (1.8),

$$C = \sqrt{(12.01 \text{ m})^2 + (14.94 \text{ m})^2} = 19.17 \text{ m}, \ \arctan\left(\frac{14.94 \text{ m}}{12.01 \text{ m}}\right) = 51.2°$$

EVALUATE: C_x and C_y are both positive, so θ is in the first quadrant.

1.45. **IDENTIFY:** $\vec{A} \cdot \vec{B} = AB\cos\phi$

SET UP: For $\vec{A}$ and $\vec{B}$, $\phi = 150.0°$. For $\vec{B}$ and $\vec{C}$, $\phi = 145.0°$. For $\vec{A}$ and $\vec{C}$, $\phi = 65.0°$.

EXECUTE: (a) $\vec{A} \cdot \vec{B} = (8.00 \text{ m})(15.0 \text{ m})\cos 150.0° = -104 \text{ m}^2$

(b) $\vec{B} \cdot \vec{C} = (15.0 \text{ m})(12.0 \text{ m})\cos 145.0° = -148 \text{ m}^2$

(c) $\vec{A} \cdot \vec{C} = (8.00 \text{ m})(12.0 \text{ m})\cos 65.0° = 40.6 \text{ m}^2$

EVALUATE: When $\phi < 90°$ the scalar product is positive and when $\phi > 90°$ the scalar product is negative.

1.47. **IDENTIFY:** For all of these pairs of vectors, the angle is found from combining Eqs. (1.18) and (1.21), to give the angle ϕ as $\phi = \arccos\left(\dfrac{\vec{A} \cdot \vec{B}}{AB}\right) = \arccos\left(\dfrac{A_x B_x + A_y B_y}{AB}\right)$.

SET UP: Eq. (1.14) shows how to obtain the components for a vector written in terms of unit vectors.

EXECUTE: (a) $\vec{A} \cdot \vec{B} = -22$, $A = \sqrt{40}$, $B - \sqrt{13}$, and so $\phi - \arccos\left(\dfrac{-22}{\sqrt{40}\sqrt{13}}\right) = 165°$.

(b) $\vec{A} \cdot \vec{B} = 60$, $A = \sqrt{34}$, $B = \sqrt{136}$, $\phi = \arccos\left(\dfrac{60}{\sqrt{34}\sqrt{136}}\right) = 28°$.

(c) $\vec{A} \cdot \vec{B} = 0$ and $\phi = 90°$.

EVALUATE: If $\vec{A} \cdot \vec{B} > 0$, $0 \le \phi < 90°$. If $\vec{A} \cdot \vec{B} < 0$, $90° < \phi \le 180°$. If $\vec{A} \cdot \vec{B} = 0$, $\phi = 90°$ and the two vectors are perpendicular.

1.49. **IDENTIFY:** $\vec{A} \times \vec{D}$ has magnitude $AD\sin\phi$. Its direction is given by the right-hand rule.

SET UP: $\phi = 180° - 53° = 127°$

EXECUTE: (a) $|\vec{A} \times \vec{D}| = (8.00 \text{ m})(10.0 \text{ m})\sin 127° = 63.9 \text{ m}^2$. The right-hand rule says $\vec{A} \times \vec{D}$ is in the $-z$-direction (into the page).

(b) $\vec{D} \times \vec{A}$ has the same magnitude as $\vec{A} \times \vec{D}$ and is in the opposite direction.

EVALUATE: The component of $\vec{D}$ perpendicular to $\vec{A}$ is $D_{\perp} = D\sin 53.0° = 7.99 \text{ m}$.

$|\vec{A} \times \vec{D}| = AD_{\perp} = 63.9 \text{ m}^2$, which agrees with our previous result.

1.53. **IDENTIFY:** $\vec{A}$ and $\vec{B}$ are given in unit vector form. Find A, B and the vector difference $\vec{A} - \vec{B}$.

SET UP: $\vec{A} = -2.00\hat{i} + 3.00\hat{j} + 4.00\hat{k}$, $\vec{B} = 3.00\hat{i} + 1.00\hat{j} - 3.00\hat{k}$

Use Eq. (1.8) to find the magnitudes of the vectors.

EXECUTE: (a) $A = \sqrt{A_x^2 + A_y^2 + A_z^2} = \sqrt{(-2.00)^2 + (3.00)^2 + (4.00)^2} = 5.38$

$B = \sqrt{B_x^2 + B_y^2 + B_z^2} = \sqrt{(3.00)^2 + (1.00)^2 + (-3.00)^2} = 4.36$

(b) $\vec{A} - \vec{B} = (-2.00\hat{i} + 3.00\hat{j} + 4.00\hat{k}) - (3.00\hat{i} + 1.00\hat{j} - 3.00\hat{k})$

$\vec{A} - \vec{B} = (-2.00 - 3.00)\hat{i} + (3.00 - 1.00)\hat{j} + (4.00 - (-3.00))\hat{k} = -5.00\hat{i} + 2.00\hat{j} + 7.00\hat{k}$.

(c) Let $\vec{C} = \vec{A} - \vec{B}$, so $C_x = -5.00$, $C_y = +2.00$, $C_z = +7.00$

$$C = \sqrt{C_x^2 + C_y^2 + C_z^2} = \sqrt{(-5.00)^2 + (2.00)^2 + (7.00)^2} = 8.83$$

$\vec{B} - \vec{A} = -(\vec{A} - \vec{B})$, so $\vec{A} - \vec{B}$ and $\vec{B} - \vec{A}$ have the same magnitude but opposite directions.

EVALUATE: A, B and C are each larger than any of their components.

1.55. **IDENTIFY:** The density relates mass and volume. Use the given mass and density to find the volume and from this the radius.

SET UP: The earth has mass $m_E = 5.97 \times 10^{24}$ kg and radius $r_E = 6.38 \times 10^6$ m. The volume of a sphere is $V = \frac{4}{3}\pi r^3$. $\rho = 1.76$ g/cm^3 = 1760 km/m^3.

EXECUTE: **(a)** The planet has mass $m = 5.5m_E = 3.28 \times 10^{25}$ kg. $V = \dfrac{m}{\rho} = \dfrac{3.28 \times 10^{25} \text{ kg}}{1760 \text{ kg/m}^3} = 1.86 \times 10^{22}$ m^3.

$$r = \left(\frac{3V}{4\pi}\right)^{1/3} = \left(\frac{3[1.86 \times 10^{22} \text{ m}^3]}{4\pi}\right)^{1/3} = 1.64 \times 10^7 \text{ m} = 1.64 \times 10^4 \text{ km}$$

(b) $r = 2.57r_E$

EVALUATE: Volume V is proportional to mass and radius r is proportional to $V^{1/3}$, so r is proportional to $m^{1/3}$. If the planet and earth had the same density its radius would be $(5.5)^{1/3} r_E = 1.8r_E$. The radius of the planet is greater than this, so its density must be less than that of the earth.

1.57. **IDENTIFY:** Using the density of the oxygen and volume of a breath, we want the mass of oxygen (the target variable in part (a)) breathed in per day and the dimensions of the tank in which it is stored.

SET UP: The mass is the density times the volume. Estimate 12 breaths per minute. We know 1 day = 24 h, 1 h = 60 min and 1000 L = 1 m^3. The volume of a cube having faces of length l is $V = l^3$.

EXECUTE: **(a)** $(12 \text{ breaths/min})\left(\dfrac{60 \text{ min}}{1 \text{ h}}\right)\left(\dfrac{24 \text{ h}}{1 \text{ day}}\right) = 17,280$ breaths/day. The volume of air breathed in one day is $(\frac{1}{2}$ L/breath$)(17,280$ breaths/day$) = 8640$ L $= 8.64$ m^3. The mass of air breathed in one day is the density of air times the volume of air breathed: $m = (1.29 \text{ kg/m}^3)(8.64 \text{ m}^3) = 11.1$ kg. As 20% of this quantity is oxygen, the mass of oxygen breathed in 1 day is $(0.20)(11.1 \text{ kg}) = 2.2$ kg $= 2200$ g.

(b) $V = 8.64$ m^3 and $V = l^3$, so $l = V^{1/3} = 2.1$ m.

EVALUATE: A person could not survive one day in a closed tank of this size because the exhaled air is breathed back into the tank and thus reduces the percent of oxygen in the air in the tank. That is, a person cannot extract all of the oxygen from the air in an enclosed space.

1.59. **IDENTIFY:** Calculate the average volume and diameter and the uncertainty in these quantities.

SET UP: Using the extreme values of the input data gives us the largest and smallest values of the target variables and from these we get the uncertainty.

EXECUTE: **(a)** The volume of a disk of diameter d and thickness t is $V = \pi(d/2)^2 t$.

The average volume is $V = \pi(8.50 \text{ cm}/2)^2(0.50 \text{ cm}) = 2.837$ cm^3. But t is given to only two significant figures so the answer should be expressed to two significant figures: $V = 2.8$ cm^3.

We can find the uncertainty in the volume as follows. The volume could be as large as $V = \pi(8.52 \text{ cm}/2)^2(0.055 \text{ cm}) = 3.1$ cm^3, which is 0.3 cm^3 larger than the average value. The volume could be as small as $V = \pi(8.48 \text{ cm}/2)^2(0.045 \text{ cm}) = 2.5$ cm^3, which is 0.3 cm^3 smaller than the average value. The uncertainty is ± 0.3 cm^3, and we express the volume as $V = 2.8 \pm 0.3$ cm^3.

(b) The ratio of the average diameter to the average thickness is $8.50 \text{ cm}/0.050 \text{ cm} = 170$. By taking the largest possible value of the diameter and the smallest possible thickness we get the largest possible value for this ratio: $8.52 \text{ cm}/0.045 \text{ cm} = 190$. The smallest possible value of the ratio is $8.48/0.055 = 150$. Thus the uncertainty is ± 20 and we write the ratio as 170 ± 20.

EVALUATE: The thickness is uncertain by 10% and the percentage uncertainty in the diameter is much less, so the percentage uncertainty in the volume and in the ratio should be about 10%.

1.61. **IDENTIFY:** The number of atoms is your mass divided by the mass of one atom.

SET UP: Assume a 70-kg person and that the human body is mostly water. Use Appendix D to find the mass of one H_2O molecule: $18.015 \text{ u} \times 1.661 \times 10^{-27} \text{ kg/u} = 2.992 \times 10^{-26} \text{ kg/molecule}$.

EXECUTE: $(70 \text{ kg})/(2.992 \times 10^{-26} \text{ kg/molecule}) = 2.34 \times 10^{27}$ molecules. Each H_2O molecule has 3 atoms, so there are about 6×10^{27} atoms.

EVALUATE: Assuming carbon to be the most common atom gives 3×10^{27} molecules, which is a result of the same order of magnitude.

1.63. **IDENTIFY:** The cost would equal the number of dollar bills required; the surface area of the U.S. divided by the surface area of a single dollar bill.

SET UP: By drawing a rectangle on a map of the U.S., the approximate area is 2600 mi by 1300 mi or $3,380,000 \text{ mi}^2$. This estimate is within 10 percent of the actual area, $3,794,083 \text{ mi}^2$. The population is roughly 3.0×10^8 while the area of a dollar bill, as measured with a ruler, is approximately $6\frac{1}{8}$ in. by $2\frac{5}{8}$ in.

EXECUTE: $A_{\text{U.S.}} = (3,380,000 \text{ mi}^2)[(5280 \text{ ft})/(1 \text{ mi})]^2[(12 \text{ in.})/(1 \text{ ft})]^2 = 1.4 \times 10^{16} \text{ in.}^2$

$A_{\text{bill}} = (6.125 \text{ in.})(2.625 \text{ in.}) = 16.1 \text{ in.}^2$

Total cost $= N_{\text{bills}} = A_{\text{U.S.}}/A_{\text{bill}} = (1.4 \times 10^{16} \text{ in.}^2)/(16.1 \text{ in.}^2/\text{bill}) = 9 \times 10^{14}$ bills

Cost per person $= (9 \times 10^{14} \text{ dollars})/(3.0 \times 10^8 \text{ persons}) = 3 \times 10^6$ dollars/person

EVALUATE: The actual cost would be somewhat larger, because the land isn't flat.

1.67. **IDENTIFY:** $\vec{A} + \vec{B} = \vec{C}$ (or $\vec{B} + \vec{A} = \vec{C}$). The target variable is vector $\vec{A}$.

SET UP: Use components and Eq. (1.10) to solve for the components of $\vec{A}$. Find the magnitude and direction of $\vec{A}$ from its components.

EXECUTE: **(a)**

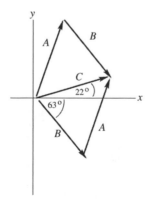

$C_x = A_x + B_x$, so $A_x = C_x - B_x$

$C_y = A_y + B_y$, so $A_y = C_y - B_y$

$C_x = C\cos 22.0° = (6.40 \text{ cm})\cos 22.0°$

$C_x = +5.934 \text{ cm}$

$C_y = C\sin 22.0° = (6.40 \text{ cm})\sin 22.0°$

$C_y = +2.397 \text{ cm}$

$B_x = B\cos(360° - 63.0°) = (6.40 \text{ cm})\cos 297.0°$

$B_x = +2.906 \text{ cm}$

$B_y = B\sin 297.0° = (6.40 \text{ cm})\sin 297.0°$

$B_y = -5.702 \text{ cm}$

Figure 1.67a

(b) $A_x = C_x - B_x = +5.934 \text{ cm} - 2.906 \text{ cm} = +3.03 \text{ cm}$

$A_y = C_y - B_y = +2.397 \text{ cm} - (-5.702) \text{ cm} = +8.10 \text{ cm}$

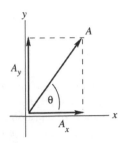

$$A = \sqrt{A_x^2 + A_y^2}$$

$$A = \sqrt{(3.03 \text{ cm})^2 + (8.10 \text{ cm})^2} = 8.65 \text{ cm}$$

$$\tan\theta = \frac{A_y}{A_x} = \frac{8.10 \text{ cm}}{3.03 \text{ cm}} = 2.67$$

$$\theta = 69.5°$$

Figure 1.67b

EVALUATE: The $\vec{A}$ we calculated agrees qualitatively with vector $\vec{A}$ in the vector addition diagram in part (a).

1.69.   **IDENTIFY:** Vector addition. Target variable is the 4th displacement.

SET UP: Use a coordinate system where east is in the $+x$-direction and north is in the $+y$-direction.

Let $\vec{A}$, $\vec{B}$, and $\vec{C}$ be the three displacements that are given and let $\vec{D}$ be the fourth unmeasured displacement. Then the resultant displacement is $\vec{R} = \vec{A} + \vec{B} + \vec{C} + \vec{D}$. And since she ends up back where she started, $\vec{R} = 0$.

$0 = \vec{A} + \vec{B} + \vec{C} + \vec{D}$, so $\vec{D} = -(\vec{A} + \vec{B} + \vec{C})$

$D_x = -(A_x + B_x + C_x)$ and $D_y = -(A_y + B_y + C_y)$

EXECUTE:

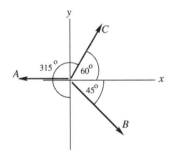

$A_x = -180 \text{ m}, \quad A_y = 0$

$B_x = B\cos 315° = (210 \text{ m})\cos 315° = +148.5 \text{ m}$

$B_y = B\sin 315° = (210 \text{ m})\sin 315° = -148.5 \text{ m}$

$C_x = C\cos 60° = (280 \text{ m})\cos 60° = +140 \text{ m}$

$C_y = C\sin 60° = (280 \text{ m})\sin 60° = +242.5 \text{ m}$

Figure 1.69a

$D_x = -(A_x + B_x + C_x) = -(-180 \text{ m} + 148.5 \text{ m} + 140 \text{ m}) = -108.5 \text{ m}$

$D_y = -(A_y + B_y + C_y) = -(0 - 148.5 \text{ m} + 242.5 \text{ m}) = -94.0 \text{ m}$

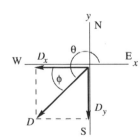

$$D = \sqrt{D_x^2 + D_y^2}$$

$$D = \sqrt{(-108.5 \text{ m})^2 + (-94.0 \text{ m})^2} = 144 \text{ m}$$

$$\tan\theta = \frac{D_y}{D_x} = \frac{-94.0 \text{ m}}{-108.5 \text{ m}} = 0.8664$$

$$\theta = 180° + 40.9° = 220.9°$$

($\vec{D}$ is in the third quadrant since both D_x and D_y are negative.)

Figure 1.69b

The direction of $\vec{D}$ can also be specified in terms of $\phi = \theta - 180° = 40.9°$; $\vec{D}$ is $41°$ south of west.

EVALUATE: The vector addition diagram, approximately to scale, is

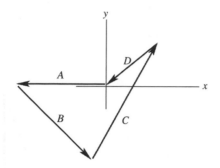

Vector $\vec{D}$ in this diagram agrees qualitatively with our calculation using components.

Figure 1.69c

1.73.   **IDENTIFY:**   We know the resultant of two forces of known equal magnitudes and want to find that magnitude (the target variable).

SET UP: Use coordinates having a horizontal $+x$ axis and an upward $+y$ axis. Then $A_x + B_x = R_x$ and $R_x = 5.60$ N.

SOLVE: $A_x + B_x = R_x$ and $A\cos 32° + B\sin 32° = R_x$. Since $A = B$,

$$2A\cos 32° = R_x, \text{ so } A = \frac{R_x}{(2)(\cos 32°)} = 3.30 \text{ N}.$$

EVALUATE: The magnitude of the x component of each pull is 2.80 N, so the magnitude of each pull (3.30 N) is greater than its x component, as it should be.

1.75.   **IDENTIFY:**   The sum of the vector forces on the beam sum to zero, so their x components and their y components sum to zero. Solve for the components of $\vec{F}$.

SET UP: The forces on the beam are sketched in Figure 1.75a. Choose coordinates as shown in the sketch. The 100-N pull makes an angle of $30.0° + 40.0° = 70.0°$ with the horizontal. $\vec{F}$ and the 100-N pull have been replaced by their x and y components.

EXECUTE: (a) The sum of the x-components is equal to zero gives $F_x + (100 \text{ N})\cos 70.0° = 0$ and $F_x = -34.2$ N. The sum of the y-components is equal to zero gives $F_y + (100 \text{ N})\sin 70.0° - 124 \text{ N} = 0$ and $F_y = +30.0$ N. $\vec{F}$ and its components are sketched in Figure 1.75b. $F = \sqrt{F_x^2 + F_y^2} = 45.5$ N.

$\tan\phi = \dfrac{|F_y|}{|F_x|} = \dfrac{30.0 \text{ N}}{34.2 \text{ N}}$ and $\phi = 41.3°$. $\vec{F}$ is directed at $41.3°$ above the $-x$-axis in Figure 1.75a.

(b) The vector addition diagram is given in Figure 1.75c. $\vec{F}$ determined from the diagram agrees with $\vec{F}$ calculated in part (a) using components.

EVALUATE: The vertical component of the 100 N pull is less than the 124 N weight so $\vec{F}$ must have an upward component if all three forces balance.

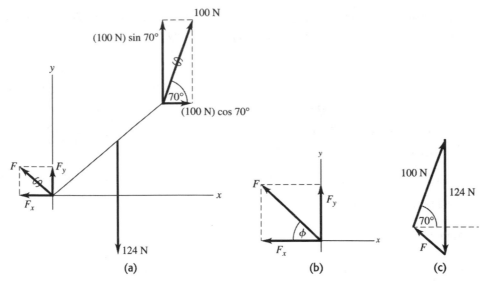

Figure 1.75

1.79.   **IDENTIFY:**   Vector addition. One force and the vector sum are given; find the second force.
SET UP: Use components. Let $+y$ be upward.

$\vec{B}$ is the force the biceps exerts.

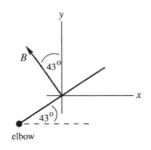

Figure 1.79a

$\vec{E}$ is the force the elbow exerts. $\vec{E} + \vec{B} = \vec{R}$, where $R = 132.5$ N and is upward.
$E_x = R_x - B_x$, $E_y = R_y - B_y$
EXECUTE: $B_x = -B\sin 43° = -158.2$ N, $B_y = +B\cos 43° = +169.7$ N, $R_x = 0$, $R_y = +132.5$ N
Then $E_x = +158.2$ N, $E_y = -37.2$ N.
$E = \sqrt{E_x^2 + E_y^2} = 160$ N;

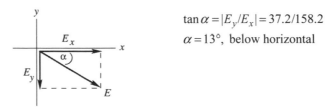

$\tan \alpha = |E_y/E_x| = 37.2/158.2$
$\alpha = 13°$, below horizontal

Figure 1.79b

EVALUATE: The x-component of $\vec{E}$ cancels the x-component of $\vec{B}$. The resultant upward force is less than the upward component of $\vec{B}$, so E_y must be downward.

1.83. **IDENTIFY:** We are given the resultant of three vectors, two of which we know, and want to find the magnitude and direction of the third vector.

SET UP: Calling $\vec{C}$ the unknown vector and $\vec{A}$ and $\vec{B}$ the known vectors, we have $\vec{A} + \vec{B} + \vec{C} = \vec{R}$. The components are $A_x + B_x + C_x = R_x$ and $A_y + B_y + C_y = R_y$.

EXECUTE: The components of the known vectors are $A_x = 12.0$ m, $A_y = 0$,

$B_x = -B\sin 50.0° = -21.45$ m, $B_y = B\cos 50.0° = +18.00$ m, $R_x = 0$, and $R_y = -10.0$ m. Therefore the

components of $\vec{C}$ are $C_x = R_x - A_x - B_x = 0 - 12.0$ m $- (-21.45$ m$) = 9.45$ m and

$C_y = R_y - A_y - B_y = -10.0$ m $- 0 - 18.0$ m $= -28.0$ m.

Using these components to find the magnitude and direction of $\vec{C}$ gives $C = 29.6$ m and $\tan\theta = \dfrac{9.45}{28.0}$ and

$\theta = 18.6°$ east of south

EVALUATE: A graphical sketch shows that this answer is reasonable.

1.85. **IDENTIFY:** Think of the displacements of the three people as vectors. We know two of them and want to find their resultant.

SET UP: Calling $\vec{A}$ the vector from John to Paul, $\vec{B}$ the vector from Paul to George, and $\vec{C}$ the vector from John to George, we have $\vec{A} + \vec{B} = \vec{C}$, which gives $A_x + B_x = C_x$ and $A_y + B_y = C_y$.

EXECUTE: The known components are $A_x = -14.0$ m, $A_y = 0$, $B_x = B\cos 37° = 28.75$ m, and

$B_y = -B\sin 37° = -21.67$ m. Therefore $C_x = -14.0$ m $+ 28.75$ m $= 14.75$ m, $C_y = 0 - 21.67$ m $= -21.67$ m.

These components give $C = 26.2$ m and $\tan\theta = \dfrac{14.75}{21.67}$, which gives $\theta = 34.2°$ east of south.

EVALUATE: A graphical sketch confirms that this answer is reasonable.

1.87. **IDENTIFY:** We know the scalar product and the magnitude of the vector product of two vectors and want to know the angle between them.

SET UP: The scalar product is $\vec{A} \cdot \vec{B} = AB\cos\theta$ and the vector product is $\left|\vec{A} \times \vec{B}\right| = AB\sin\theta$.

EXECUTE: $\vec{A} \cdot \vec{B} = AB\cos\theta = -6.00$ and $\left|\vec{A} \times \vec{B}\right| = AB\sin\theta = +9.00$. Taking the ratio gives $\tan\theta = \dfrac{9.00}{-6.00}$,

so $\theta = 124°$.

EVALUATE: Since the scalar product is negative, the angle must be between 90° and 180°.

1.91. **IDENTIFY:** Find the angle between specified pairs of vectors.

SET UP: Use $\cos\phi = \dfrac{\vec{A} \cdot \vec{B}}{AB}$

EXECUTE: **(a)** $\vec{A} = \hat{k}$ (along line ab)

$\vec{B} = \hat{i} + \hat{j} + \hat{k}$ (along line ad)

$A = 1$, $B = \sqrt{1^2 + 1^2 + 1^2} = \sqrt{3}$

$\vec{A} \cdot \vec{B} = \hat{k} \cdot (\hat{i} + \hat{j} + \hat{k}) = 1$

So $\cos\phi = \dfrac{\vec{A} \cdot \vec{B}}{AB} = 1/\sqrt{3}$; $\phi = 54.7°$

(b) $\vec{A} = \hat{i} + \hat{j} + \hat{k}$ (along line ad)

$\vec{B} = \hat{j} + \hat{k}$ (along line ac)

$A = \sqrt{1^2 + 1^2 + 1^2} = \sqrt{3}$; $B = \sqrt{1^2 + 1^2} = \sqrt{2}$

$\vec{A} \cdot \vec{B} = (\hat{i} + \hat{j} + \hat{k}) \cdot (\hat{i} + \hat{j}) = 1 + 1 = 2$

So $\cos\phi = \dfrac{\vec{A} \cdot \vec{B}}{AB} = \dfrac{2}{\sqrt{3}\sqrt{2}} = \dfrac{2}{\sqrt{6}}$; $\phi = 35.3°$

EVALUATE: Each angle is computed to be less than $90°$, in agreement with what is deduced from Figure P1.91 in the textbook.

1.95.   **IDENTIFY and SET UP:** The target variables are the components of $\vec{C}$. We are given $\vec{A}$ and $\vec{B}$. We also know $\vec{A} \cdot \vec{C}$ and $\vec{B} \cdot \vec{C}$, and this gives us two equations in the two unknowns C_x and C_y.

EXECUTE: $\vec{A}$ and $\vec{C}$ are perpendicular, so $\vec{A} \cdot \vec{C} = 0$. $A_x C_x + A_y C_y = 0$, which gives $5.0 C_x - 6.5 C_y = 0$. $\vec{B} \cdot \vec{C} = 15.0$, so $-3.5 C_x + 7.0 C_y = 15.0$

We have two equations in two unknowns C_x and C_y. Solving gives $C_x = 8.0$ and $C_y = 6.1$.

EVALUATE: We can check that our result does give us a vector $\vec{C}$ that satisfies the two equations $\vec{A} \cdot \vec{C} = 0$ and $\vec{B} \cdot \vec{C} = 15.0$.

MOTION ALONG A STRAIGHT LINE

2.3. **IDENTIFY:** Target variable is the time Δt it takes to make the trip in heavy traffic. Use Eq. (2.2) that relates the average velocity to the displacement and average time.

SET UP: $v_{\text{av-}x} = \dfrac{\Delta x}{\Delta t}$ so $\Delta x = v_{\text{av-}x}\Delta t$ and $\Delta t = \dfrac{\Delta x}{v_{\text{av-}x}}$.

EXECUTE: Use the information given for normal driving conditions to calculate the distance between the two cities:

$$\Delta x = v_{\text{av-}x}\Delta t = (105\ \text{km/h})(1\ \text{h}/60\ \text{min})(140\ \text{min}) = 245\ \text{km}.$$

Now use $v_{\text{av-}x}$ for heavy traffic to calculate Δt; Δx is the same as before:

$$\Delta t = \frac{\Delta x}{v_{\text{av-}x}} = \frac{245\ \text{km}}{70\ \text{km/h}} = 3.50\ \text{h} = 3\ \text{h}\ \text{and}\ 30\ \text{min}.$$

The trip takes an additional 1 hour and 10 minutes.

EVALUATE: The time is inversely proportional to the average speed, so the time in traffic is $(105/70)(140\ \text{min}) = 210\ \text{min}.$

2.5. **IDENTIFY:** Given two displacements, we want the average velocity and the average speed.

SET UP: The average velocity is $v_{\text{av-}x} = \dfrac{\Delta x}{\Delta t}$ and the average speed is just the total distance walked divided by the total time to walk this distance.

EXECUTE: **(a)** Let $+x$ be east. $\Delta x = 60.0\ \text{m} - 40.0\ \text{m} = 20.0\ \text{m}$ and $\Delta t = 28.0\ \text{s} + 36.0\ \text{s} = 64.0\ \text{s}.$ So

$$v_{\text{av-}x} = \frac{\Delta x}{\Delta t} = \frac{20.0\ \text{m}}{64.0\ \text{s}} = 0.312\ \text{m/s}.$$

(b) average speed $= \dfrac{60.0\ \text{m} + 40.0\ \text{m}}{64.0\ \text{s}} = 1.56\ \text{m/s}$

EVALUATE: The average speed is much greater than the average velocity because the total distance walked is much greater than the magnitude of the displacement vector.

2.7. **(a) IDENTIFY:** Calculate the average velocity using Eq. (2.2).

SET UP: $v_{\text{av-}x} = \dfrac{\Delta x}{\Delta t}$ so use $x(t)$ to find the displacement Δx for this time interval.

EXECUTE: $t = 0$: $x = 0$

$t = 10.0\ \text{s}$: $x = (2.40\ \text{m/s}^2)(10.0\ \text{s})^2 - (0.120\ \text{m/s}^3)(10.0\ \text{s})^3 = 240\ \text{m} - 120\ \text{m} = 120\ \text{m}.$

Then $v_{\text{av-}x} = \dfrac{\Delta x}{\Delta t} = \dfrac{120\ \text{m}}{10.0\ \text{s}} = 12.0\ \text{m/s}.$

(b) IDENTIFY: Use Eq. (2.3) to calculate $v_x(t)$ and evaluate this expression at each specified t.

SET UP: $v_x = \dfrac{dx}{dt} = 2bt - 3ct^2.$

EXECUTE: (i) $t = 0$: $v_x = 0$

(ii) $t = 5.0$ s: $v_x = 2(2.40 \text{ m/s}^2)(5.0 \text{ s}) - 3(0.120 \text{ m/s}^3)(5.0 \text{ s})^2 = 24.0 \text{ m/s} - 9.0 \text{ m/s} = 15.0 \text{ m/s}.$

(iii) $t = 10.0$ s: $v_x = 2(2.40 \text{ m/s}^2)(10.0 \text{ s}) - 3(0.120 \text{ m/s}^3)(10.0 \text{ s})^2 = 48.0 \text{ m/s} - 36.0 \text{ m/s} = 12.0 \text{ m/s}.$

(c) IDENTIFY: Find the value of t when $v_x(t)$ from part (b) is zero.

SET UP: $v_x = 2bt - 3ct^2$

$v_x = 0$ at $t = 0.$

$v_x = 0$ next when $2bt - 3ct^2 = 0$

EXECUTE: $2b = 3ct$ so $t = \dfrac{2b}{3c} = \dfrac{2(2.40 \text{ m/s}^2)}{3(0.120 \text{ m/s}^3)} = 13.3$ s

EVALUATE: $v_x(t)$ for this motion says the car starts from rest, speeds up, and then slows down again.

2.9.   **IDENTIFY:**   The average velocity is given by $v_{\text{av-}x} = \dfrac{\Delta x}{\Delta t}$. We can find the displacement Δt for each

constant velocity time interval. The average speed is the distance traveled divided by the time.
SET UP: For $t = 0$ to $t = 2.0$ s, $v_x = 2.0$ m/s. For $t = 2.0$ s to $t = 3.0$ s, $v_x = 3.0$ m/s. In part (b),

$v_x = -3.0$ m/s for $t = 2.0$ s to $t = 3.0$ s. When the velocity is constant, $\Delta x = v_x \Delta t.$
EXECUTE: **(a)** For $t = 0$ to $t = 2.0$ s, $\Delta x = (2.0 \text{ m/s})(2.0 \text{ s}) = 4.0$ m. For $t = 2.0$ s to $t = 3.0$ s,

$\Delta x = (3.0 \text{ m/s})(1.0 \text{ s}) = 3.0$ m. For the first 3.0 s, $\Delta x = 4.0 \text{ m} + 3.0 \text{ m} = 7.0$ m. The distance traveled is also

7.0 m. The average velocity is $v_{\text{av-}x} = \dfrac{\Delta x}{\Delta t} = \dfrac{7.0 \text{ m}}{3.0 \text{ s}} = 2.33$ m/s. The average speed is also 2.33 m/s.

(b) For $t = 2.0$ s to 3.0 s, $\Delta x = (-3.0 \text{ m/s})(1.0 \text{ s}) = -3.0$ m. For the first 3.0 s,

$\Delta x = 4.0 \text{ m} + (-3.0 \text{ m}) = +1.0$ m. The dog runs 4.0 m in the $+x$-direction and then 3.0 m in the

$-x$-direction, so the distance traveled is still 7.0 m. $v_{\text{av-}x} = \dfrac{\Delta x}{\Delta t} = \dfrac{1.0 \text{ m}}{3.0 \text{ s}} = 0.33$ m/s. The average speed is

$\dfrac{7.00 \text{ m}}{3.00 \text{ s}} = 2.33$ m/s.

EVALUATE: When the motion is always in the same direction, the displacement and the distance traveled are equal and the average velocity has the same magnitude as the average speed. When the motion changes direction during the time interval, those quantities are different.

2.11.   **IDENTIFY:**   Find the instantaneous velocity of a car using a graph of its position as a function of time.
SET UP: The instantaneous velocity at any point is the slope of the x versus t graph at that point. Estimate the slope from the graph.
EXECUTE: A: $v_x = 6.7$ m/s; B: $v_x = 6.7$ m/s; C: $v_x = 0$; D: $v_x = -40.0$ m/s; E: $v_x = -40.0$ m/s;

F: $v_x = -40.0$ m/s; G: $v_x = 0.$

EVALUATE: The sign of v_x shows the direction the car is moving. v_x is constant when x versus t is a

straight line.

2.13.   **IDENTIFY:**   The average acceleration for a time interval Δt is given by $a_{\text{av-}x} = \dfrac{\Delta v_x}{\Delta t}.$

SET UP: Assume the car is moving in the $+x$ direction. 1 mi/h $= 0.447$ m/s, so 60 mi/h $= 26.82$ m/s, 200 mi/h $= 89.40$ m/s and 253 mi/h $= 113.1$ m/s.
EXECUTE: **(a)** The graph of v_x versus t is sketched in Figure 2.13. The graph is not a straight line, so the acceleration is not constant.

(b) (i) $a_{\text{av-}x} = \dfrac{26.82 \text{ m/s} - 0}{2.1 \text{ s}} = 12.8 \text{ m/s}^2$ **(ii)** $a_{\text{av-}x} = \dfrac{89.40 \text{ m/s} - 26.82 \text{ m/s}}{20.0 \text{ s} - 2.1 \text{ s}} = 3.50 \text{ m/s}^2$

(iii) $a_{\text{av-}x} = \dfrac{113.1 \text{ m/s} - 89.40 \text{ m/s}}{53 \text{ s} - 20.0 \text{ s}} = 0.718 \text{ m/s}^2$. The slope of the graph of v_x versus t decreases as t

increases. This is consistent with an average acceleration that decreases in magnitude during each successive time interval.

EVALUATE: The average acceleration depends on the chosen time interval. For the interval between 0 and

53 s, $a_{\text{av-}x} = \dfrac{113.1\,\text{m/s} - 0}{53\,\text{s}} = 2.13\,\text{m/s}^2$.

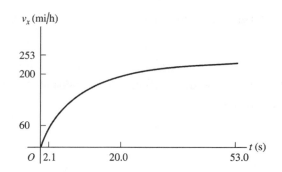

Figure 2.13

2.15.   IDENTIFY and SET UP: Use $v_x = \dfrac{dx}{dt}$ and $a_x = \dfrac{dv_x}{dt}$ to calculate $v_x(t)$ and $a_x(t)$.

EXECUTE: $v_x = \dfrac{dx}{dt} = 2.00\,\text{cm/s} - (0.125\,\text{cm/s}^2)t$

$a_x = \dfrac{dv_x}{dt} = -0.125\,\text{cm/s}^2$

(a) At $t = 0$, $x = 50.0$ cm, $v_x = 2.00$ cm/s, $a_x = -0.125$ cm/s^2.

(b) Set $v_x = 0$ and solve for t: $t = 16.0$ s.

(c) Set $x = 50.0$ cm and solve for t. This gives $t = 0$ and $t = 32.0$ s. The turtle returns to the starting point after 32.0 s.

(d) The turtle is 10.0 cm from starting point when $x = 60.0$ cm or $x = 40.0$ cm.
Set $x = 60.0$ cm and solve for t: $t = 6.20$ s and $t = 25.8$ s.
At $t = 6.20$ s, $v_x = +1.23$ cm/s.
At $t = 25.8$ s, $v_x = -1.23$ cm/s.
Set $x = 40.0$ cm and solve for t: $t = 36.4$ s (other root to the quadratic equation is negative and hence nonphysical).
At $t = 36.4$ s, $v_x = -2.55$ cm/s.

(e) The graphs are sketched in Figure 2.15.

Figure 2.15

EVALUATE: The acceleration is constant and negative. v_x is linear in time. It is initially positive, decreases to zero, and then becomes negative with increasing magnitude. The turtle initially moves farther away from the origin but then stops and moves in the $-x$-direction.

2.17.   IDENTIFY: The average acceleration is $a_{\text{av-}x} = \dfrac{\Delta v_x}{\Delta t}$. Use $v_x(t)$ to find v_x at each t. The instantaneous

acceleration is $a_x = \dfrac{dv_x}{dt}$.

SET UP: $v_x(0) = 3.00$ m/s and $v_x(5.00 \text{ s}) = 5.50$ m/s.

EXECUTE: **(a)** $a_{\text{av-}x} = \dfrac{\Delta v_x}{\Delta t} = \dfrac{5.50 \text{ m/s} - 3.00 \text{ m/s}}{5.00 \text{ s}} = 0.500 \text{ m/s}^2$

(b) $a_x = \dfrac{dv_x}{dt} = (0.100 \text{ m/s}^3)(2t) = (0.200 \text{ m/s}^3)t$. At $t = 0$, $a_x = 0$. At $t = 5.00$ s, $a_x = 1.00 \text{ m/s}^2$.

(c) Graphs of $v_x(t)$ and $a_x(t)$ are given in Figure 2.17.

EVALUATE: $a_x(t)$ is the slope of $v_x(t)$ and increases as t increases. The average acceleration for $t = 0$ to $t = 5.00$ s equals the instantaneous acceleration at the midpoint of the time interval, $t = 2.50$ s, since $a_x(t)$ is a linear function of t.

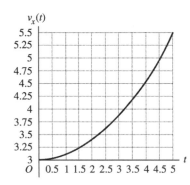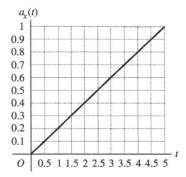

Figure 2.17

2.19. **IDENTIFY:** Use the constant acceleration equations to find v_{0x} and a_x.

(a) SET UP: The situation is sketched in Figure 2.19.

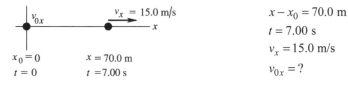

Figure 2.19

EXECUTE: Use $x - x_0 = \left(\dfrac{v_{0x} + v_x}{2}\right)t$, so $v_{0x} = \dfrac{2(x - x_0)}{t} - v_x = \dfrac{2(70.0 \text{ m})}{7.00 \text{ s}} - 15.0 \text{ m/s} = 5.0 \text{ m/s}$.

(b) Use $v_x = v_{0x} + a_x t$, so $a_x = \dfrac{v_x - v_{0x}}{t} = \dfrac{15.0 \text{ m/s} - 5.0 \text{ m/s}}{7.00 \text{ s}} = 1.43 \text{ m/s}^2$.

EVALUATE: The average velocity is $(70.0 \text{ m})/(7.00 \text{ s}) = 10.0$ m/s. The final velocity is larger than this, so the antelope must be speeding up during the time interval; $v_{0x} < v_x$ and $a_x > 0$.

2.23. **IDENTIFY:** Assume that the acceleration is constant and apply the constant acceleration kinematic equations. Set $|a_x|$ equal to its maximum allowed value.

SET UP: Let $+x$ be the direction of the initial velocity of the car. $a_x = -250 \text{ m/s}^2$. 105 km/h = 29.17 m/s.

EXECUTE: $v_{0x} = +29.17$ m/s. $v_x = 0$. $v_x^2 = v_{0x}^2 + 2a_x(x - x_0)$ gives

$$x - x_0 = \dfrac{v_x^2 - v_{0x}^2}{2a_x} = \dfrac{0 - (29.17 \text{ m/s})^2}{2(-250 \text{ m/s}^2)} = 1.70 \text{ m}.$$

EVALUATE: The car frame stops over a shorter distance and has a larger magnitude of acceleration. Part of your 1.70 m stopping distance is the stopping distance of the car and part is how far you move relative to the car while stopping.

2.27.   **IDENTIFY:** We know the initial and final velocities of the object, and the distance over which the velocity change occurs. From this we want to find the magnitude and duration of the acceleration of the object.

SET UP: The constant-acceleration kinematics formulas apply. $v_x^2 = v_{0x}^2 + 2a_x(x - x_0)$, where $v_{0x} = 0$, $v_x = 5.0 \times 10^3$ m/s, and $x - x_0 = 4.0$ m.

EXECUTE: (a) $v_x^2 = v_{0x}^2 + 2a_x(x - x_0)$ gives $a_x = \dfrac{v_x^2 - v_{0x}^2}{2(x - x_0)} = \dfrac{(5.0 \times 10^3 \text{ m/s})^2}{2(4.0 \text{ m})} = 3.1 \times 10^6 \text{ m/s}^2 = 3.2 \times 10^5 \ g$.

(b) $v_x = v_{0x} + a_x t$ gives $t = \dfrac{v_x - v_{0x}}{a_x} = \dfrac{5.0 \times 10^3 \text{ m/s}}{3.1 \times 10^6 \text{ m/s}^2} = 1.6$ ms.

EVALUATE: (c) The calculated a is less than 450,000 g so the acceleration required doesn't rule out this hypothesis.

2.29.   **IDENTIFY:** The average acceleration is $a_{\text{av-}x} = \dfrac{\Delta v_x}{\Delta t}$. For constant acceleration, Eqs. (2.8), (2.12), (2.13) and (2.14) apply.

SET UP: Assume the shuttle travels in the $+x$ direction. 161 km/h = 44.72 m/s and 1610 km/h = 447.2 m/s. 1.00 min = 60.0 s

EXECUTE: (a) (i) $a_{\text{av-}x} = \dfrac{\Delta v_x}{\Delta t} = \dfrac{44.72 \text{ m/s} - 0}{8.00 \text{ s}} = 5.59 \text{ m/s}^2$

(ii) $a_{\text{av-}x} = \dfrac{447.2 \text{ m/s} - 44.72 \text{ m/s}}{60.0 \text{ s} - 8.00 \text{ s}} = 7.74 \text{ m/s}^2$

(b) (i) $t = 8.00$ s, $v_{0x} = 0$, and $v_x = 44.72$ m/s. $x - x_0 = \left(\dfrac{v_{0x} + v_x}{2}\right)t = \left(\dfrac{0 + 44.72 \text{ m/s}}{2}\right)(8.00 \text{ s}) = 179$ m.

(ii) $\Delta t = 60.0$ s $- 8.00$ s $= 52.0$ s, $v_{0x} = 44.72$ m/s, and $v_x = 447.2$ m/s.

$x - x_0 = \left(\dfrac{v_{0x} + v_x}{2}\right)t = \left(\dfrac{44.72 \text{ m/s} + 447.2 \text{ m/s}}{2}\right)(52.0 \text{ s}) = 1.28 \times 10^4$ m.

EVALUATE: When the acceleration is constant the instantaneous acceleration throughout the time interval equals the average acceleration for that time interval. We could have calculated the distance in part (a) as $x - x_0 = v_{0x}t + \frac{1}{2}a_x t^2 = \frac{1}{2}(5.59 \text{ m/s}^2)(8.00 \text{ s})^2 = 179$ m, which agrees with our previous calculation.

2.31.   (a) **IDENTIFY** and **SET UP:** The acceleration a_x at time t is the slope of the tangent to the v_x versus t curve at time t.

EXECUTE: At $t = 3$ s, the v_x versus t curve is a horizontal straight line, with zero slope. Thus $a_x = 0$.

At $t = 7$ s, the v_x versus t curve is a straight-line segment with slope $\dfrac{45 \text{ m/s} - 20 \text{ m/s}}{9 \text{ s} - 5 \text{ s}} = 6.3 \text{ m/s}^2$.

Thus $a_x = 6.3 \text{ m/s}^2$.

At $t = 11$ s the curve is again a straight-line segment, now with slope $\dfrac{-0 - 45 \text{ m/s}}{13 \text{ s} - 9 \text{ s}} = -11.2 \text{ m/s}^2$.

Thus $a_x = -11.2 \text{ m/s}^2$.

EVALUATE: $a_x = 0$ when v_x is constant, $a_x > 0$ when v_x is positive and the speed is increasing, and $a_x < 0$ when v_x is positive and the speed is decreasing.

(b) **IDENTIFY:** Calculate the displacement during the specified time interval.

SET UP: We can use the constant acceleration equations only for time intervals during which the acceleration is constant. If necessary, break the motion up into constant acceleration segments and apply the constant acceleration equations for each segment. For the time interval $t = 0$ to $t = 5$ s the acceleration is constant and equal to zero. For the time interval $t = 5$ s to $t = 9$ s the acceleration is constant and equal to 6.25 m/s^2. For the interval $t = 9$ s to $t = 13$ s the acceleration is constant and equal to -11.2 m/s^2.

EXECUTE: During the first 5 seconds the acceleration is constant, so the constant acceleration kinematic formulas can be used.

$v_{0x} = 20$ m/s $a_x = 0$ $t = 5$ s $x - x_0 = ?$

$x - x_0 = v_{0x}t$ ($a_x = 0$ so no $\frac{1}{2}a_x t^2$ term)

$x - x_0 = (20$ m/s$)(5$ s$) = 100$ m; this is the distance the officer travels in the first 5 seconds.

During the interval $t = 5$ s to 9 s the acceleration is again constant. The constant acceleration formulas can be applied to this 4-second interval. It is convenient to restart our clock so the interval starts at time $t = 0$ and ends at time $t = 4$ s. (Note that the acceleration is *not* constant over the entire $t = 0$ to $t = 9$ s interval.)

$v_{0x} = 20$ m/s $a_x = 6.25$ m/s^2 $t = 4$ s $x_0 = 100$ m $x - x_0 = ?$

$x - x_0 = v_{0x}t + \frac{1}{2}a_x t^2$

$x - x_0 = (20$ m/s$)(4$ s$) + \frac{1}{2}(6.25$ m/s$^2)(4$ s$)^2 = 80$ m$ + 50$ m$ = 130$ m.

Thus $x - x_0 + 130$ m $= 100$ m $+ 130$ m $= 230$ m.

At $t = 9$ s the officer is at $x = 230$ m, so she has traveled 230 m in the first 9 seconds.

During the interval $t = 9$ s to $t = 13$ s the acceleration is again constant. The constant acceleration formulas can be applied for this 4-second interval but *not* for the whole $t = 0$ to $t = 13$ s interval. To use the equations restart our clock so this interval begins at time $t = 0$ and ends at time $t = 4$ s.

$v_{0x} = 45$ m/s (at the start of this time interval)

$a_x = -11.2$ m/s^2 $t = 4$ s $x_0 = 230$ m $x - x_0 = ?$

$x - x_0 = v_{0x}t + \frac{1}{2}a_x t^2$

$x - x_0 = (45$ m/s$)(4$ s$) + \frac{1}{2}(-11.2$ m/s$^2)(4$ s$)^2 = 180$ m$ - 89.6$ m$ = 90.4$ m.

Thus $x = x_0 + 90.4$ m $= 230$ m $+ 90.4$ m $= 320$ m.

At $t = 13$ s the officer is at $x = 320$ m, so she has traveled 320 m in the first 13 seconds.

EVALUATE: The velocity v_x is always positive so the displacement is always positive and displacement and distance traveled are the same. The average velocity for time interval Δt is $v_{\text{av-}x} = \Delta x / \Delta t$. For $t = 0$ to 5 s, $v_{\text{av-}x} = 20$ m/s. For $t = 0$ to 9 s, $v_{\text{av-}x} = 26$ m/s. For $t = 0$ to 13 s, $v_{\text{av-}x} = 25$ m/s. These results are consistent with Figure 2.37 in the textbook.

2.39.   **IDENTIFY:**   A ball on Mars that is hit directly upward returns to the same level in 8.5 s with a constant downward acceleration of 0.379g. How high did it go and how fast was it initially traveling upward?

SET UP: Take $+y$ upward. $v_y = 0$ at the maximum height. $a_y = -0.379g = -3.71$ m/s^2. The constant-acceleration formulas $v_y = v_{0y} + a_y t$ and $y = y_0 + v_{0y}t + \frac{1}{2}a_y t^2$ both apply.

EXECUTE: Consider the motion from the maximum height back to the initial level. For this motion $v_{0y} = 0$ and $t = 4.25$ s. $y = y_0 + v_{0y}t + \frac{1}{2}a_y t^2 = \frac{1}{2}(-3.71$ m/s$^2)(4.25$ s$)^2 = -33.5$ m. The ball went 33.5 m above its original position.

(b) Consider the motion from just after it was hit to the maximum height. For this motion $v_y = 0$ and $t = 4.25$ s. $v_y = v_{0y} + a_y t$ gives $v_{0y} = -a_y t = -(-3.71$ m/s$^2)(4.25$ s$) = 15.8$ m/s.

(c) The graphs are sketched in Figure 2.39.

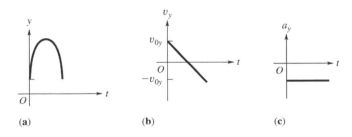

(a)   (b)   (c)

Figure 2.39

EVALUATE: The answers can be checked several ways. For example, $v_y = 0$, $v_{0y} = 15.8$ m/s, and

$a_y = -3.7$ m/s^2 in $v_y^2 = v_{0y}^2 + 2a_y(y - y_0)$ gives $y - y_0 = \dfrac{v_y^2 - v_{0y}^2}{2a_y} = \dfrac{0 - (15.8 \text{ m/s})^2}{2(-3.71 \text{ m/s}^2)} = 33.6$ m,

which agrees with the height calculated in (a).

2.41. **IDENTIFY:** Apply constant acceleration equations to the motion of the meterstick. The time the meterstick falls is your reaction time.

SET UP: Let $+y$ be downward. The meter stick has $v_{0y} = 0$ and $a_y = 9.80$ m/s^2. Let d be the distance the meterstick falls.

EXECUTE: **(a)** $y - y_0 = v_{0y}t + \frac{1}{2}a_y t^2$ gives $d = (4.90 \text{ m/s}^2)t^2$ and $t = \sqrt{\dfrac{d}{4.90 \text{ m/s}^2}}$.

(b) $t = \sqrt{\dfrac{0.176 \text{ m}}{4.90 \text{ m/s}^2}} = 0.190$ s

EVALUATE: The reaction time is proportional to the square of the distance the stick falls.

2.43. **IDENTIFY:** When the only force is gravity the acceleration is 9.80 m/s^2, downward. There are two intervals of constant acceleration and the constant acceleration equations apply during each of these intervals.

SET UP: Let $+y$ be upward. Let $y = 0$ at the launch pad. The final velocity for the first phase of the motion is the initial velocity for the free-fall phase.

EXECUTE: **(a)** Find the velocity when the engines cut off. $y - y_0 = 525$ m, $a_y = +2.25$ m/s^2, $v_{0y} = 0$.

$v_y^2 = v_{0y}^2 + 2a_y(y - y_0)$ gives $v_y = \sqrt{2(2.25 \text{ m/s}^2)(525 \text{ m})} = 48.6$ m/s.

Now consider the motion from engine cut-off to maximum height: $y_0 = 525$ m, $v_{0y} = +48.6$ m/s, $v_y = 0$ (at the maximum height), $a_y = -9.80$ m/s^2. $v_y^2 = v_{0y}^2 + 2a_y(y - y_0)$ gives

$y - y_0 = \dfrac{v_y^2 - v_{0y}^2}{2a_y} = \dfrac{0 - (48.6 \text{ m/s})^2}{2(-9.80 \text{ m/s}^2)} = 121$ m and $y = 121$ m $+ 525$ m $= 646$ m.

(b) Consider the motion from engine failure until just before the rocket strikes the ground:

$y - y_0 = -525$ m, $a_y = -9.80$ m/s^2, $v_{0y} = +48.6$ m/s. $v_y^2 = v_{0y}^2 + 2a_y(y - y_0)$ gives

$v_y = -\sqrt{(48.6 \text{ m/s})^2 + 2(-9.80 \text{ m/s}^2)(-525 \text{ m})} = -112$ m/s. Then $v_y = v_{0y} + a_y t$ gives

$t = \dfrac{v_y - v_{0y}}{a_y} = \dfrac{-112 \text{ m/s} - 48.6 \text{ m/s}}{-9.80 \text{ m/s}^2} = 16.4$ s.

(c) Find the time from blast-off until engine failure: $y - y_0 = 525$ m, $v_{0y} = 0$, $a_y = +2.25$ m/s^2.

$y - y_0 = v_{0y}t + \frac{1}{2}a_y t^2$ gives $t = \sqrt{\dfrac{2(y - y_0)}{a_y}} = \sqrt{\dfrac{2(525 \text{ m})}{2.25 \text{ m/s}^2}} = 21.6$ s. The rocket strikes the launch pad

21.6 s $+ 16.4$ s $= 38.0$ s after blast-off. The acceleration a_y is $+2.25$ m/s^2 from $t = 0$ to $t = 21.6$ s. It is -9.80 m/s^2 from $t = 21.6$ s to 38.0 s. $v_y = v_{0y} + a_y t$ applies during each constant acceleration segment, so the graph of v_y versus t is a straight line with positive slope of 2.25 m/s^2 during the blast-off phase and with negative slope of -9.80 m/s^2 after engine failure. During each phase $y - y_0 = v_{0y}t + \frac{1}{2}a_y t^2$. The sign of a_y determines the curvature of $y(t)$. At $t = 38.0$ s the rocket has returned to $y = 0$. The graphs are sketched in Figure 2.43.

EVALUATE: In part (b) we could have found the time from $y - y_0 = v_{0y}t + \frac{1}{2}a_y t^2$, finding v_y first allows us to avoid solving for t from a quadratic equation.

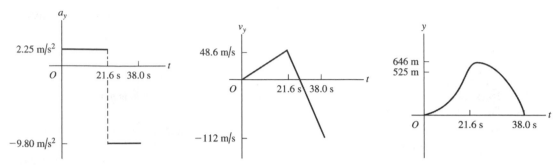

Figure 2.43

2.45.   **IDENTIFY:**   Use the constant acceleration equations to calculate a_x and $x - x_0$.

(a) SET UP: $v_x = 224$ m/s, $v_{0x} = 0$, $t = 0.900$ s, $a_x = ?$

$v_x = v_{0x} + a_x t$

EXECUTE: $a_x = \dfrac{v_x - v_{0x}}{t} = \dfrac{224 \text{ m/s} - 0}{0.900 \text{ s}} = 249$ m/s^2

(b) $a_x/g = (249 \text{ m/s}^2)/(9.80 \text{ m/s}^2) = 25.4$

(c) $x - x_0 = v_{0x}t + \frac{1}{2}a_x t^2 = 0 + \frac{1}{2}(249 \text{ m/s}^2)(0.900 \text{ s})^2 = 101$ m

(d) SET UP: Calculate the acceleration, assuming it is constant:

$t = 1.40$ s, $v_{0x} = 283$ m/s, $v_x = 0$ (stops), $a_x = ?$

$v_x = v_{0x} + a_x t$

EXECUTE: $a_x = \dfrac{v_x - v_{0x}}{t} = \dfrac{0 - 283 \text{ m/s}}{1.40 \text{ s}} = -202$ m/s^2

$a_x/g = (-202 \text{ m/s}^2)/(9.80 \text{ m/s}^2) = -20.6$; $a_x = -20.6g$

If the acceleration while the sled is stopping is constant then the magnitude of the acceleration is only 20.6g. But if the acceleration is not constant it is certainly possible that at some point the instantaneous acceleration could be as large as 40g.

EVALUATE: It is reasonable that for this motion the acceleration is much larger than g.

2.47.   **IDENTIFY:**   We can avoid solving for the common height by considering the relation between height, time of fall and acceleration due to gravity and setting up a ratio involving time of fall and acceleration due to gravity.

SET UP: Let g_{En} be the acceleration due to gravity on Enceladus and let g be this quantity on earth. Let h be the common height from which the object is dropped. Let $+y$ be downward, so $y - y_0 = h$. $v_{0y} = 0$

EXECUTE: $y - y_0 = v_{0y}t + \frac{1}{2}a_y t^2$ gives $h = \frac{1}{2}gt_E^2$ and $h = \frac{1}{2}g_{En}t_{En}^2$. Combining these two equations gives

$gt_E^2 = g_{En}t_{En}^2$ and $g_{En} = g\left(\dfrac{t_E}{t_{En}}\right)^2 = (9.80 \text{ m/s}^2)\left(\dfrac{1.75 \text{ s}}{18.6 \text{ s}}\right)^2 = 0.0868$ m/s^2.

EVALUATE: The acceleration due to gravity is inversely proportional to the square of the time of fall.

2.49.   **IDENTIFY:**   Two stones are thrown up with different speeds. (a) Knowing how soon the faster one returns to the ground, how long it will take the slow one to return? (b) Knowing how high the slower stone went, how high did the faster stone go?

SET UP: Use subscripts f and s to refer to the faster and slower stones, respectively. Take $+y$ to be upward and $y_0 = 0$ for both stones. $v_{0f} = 3v_{0s}$. When a stone reaches the ground, $y = 0$. The constant-acceleration formulas $y = y_0 + v_{0y}t + \frac{1}{2}a_y t^2$ and $v_y^2 = v_{0y}^2 + 2a_y(y - y_0)$ both apply.

EXECUTE: **(a)** $y = y_0 + v_{0y}t + \frac{1}{2}a_y t^2$ gives $a_y = -\dfrac{2v_{0y}}{t}$. Since both stones have the same a_y, $\dfrac{v_{0f}}{t_f} = \dfrac{v_{0s}}{t_s}$

and $t_s = t_f\left(\dfrac{v_{0s}}{v_{0f}}\right) = \left(\frac{1}{3}\right)(10 \text{ s}) = 3.3$ s.

(b) Since $v_y = 0$ at the maximum height, then $v_y^2 = v_{0y}^2 + 2a_y(y - y_0)$ gives $a_y = -\dfrac{v_{0y}^2}{2y}$. Since both

have the same a_y, $\dfrac{v_{0f}^2}{y_f} = \dfrac{v_{0s}^2}{y_s}$ and $y_f = y_s \left(\dfrac{v_{0f}}{v_{0s}}\right)^2 = 9H$.

EVALUATE: The faster stone reaches a greater height so it travels a greater distance than the slower stone and takes more time to return to the ground.

2.51. **IDENTIFY:** The acceleration is not constant, but we know how it varies with time. We can use the definitions of instantaneous velocity and position to find the rocket's position and speed.

SET UP: The basic definitions of velocity and position are $v_y(t) = \int_0^t a_y\,dt$ and $y - y_0 = \int_0^t v_y\,dt$.

EXECUTE: **(a)** $v_y(t) = \int_0^t a_y\,dt = \int_0^t (2.80 \text{ m/s}^3)t\,dt = (1.40 \text{ m/s}^3)t^2$

$y - y_0 = \int_0^t v_y\,dt = \int_0^t (1.40 \text{ m/s}^3)t^2\,dt = (0.4667 \text{ m/s}^3)t^3$. For $t = 10.0$ s, $y - y_0 = 467$ m.

(b) $y - y_0 = 325$ m so $(0.4667 \text{ m/s}^3)t^3 = 325$ m and $t = 8.864$ s. At this time

$v_y = (1.40 \text{ m/s}^3)(8.864 \text{ s})^2 = 110$ m/s.

EVALUATE: The time in part (b) is less than 10.0 s, so the given formulas are valid.

2.59. **IDENTIFY:** In time t_S the S-waves travel a distance $d = v_S t_S$ and in time t_P the P-waves travel a distance $d = v_P t_P$.

SET UP: $t_S = t_P + 33$ s

EXECUTE: $\dfrac{d}{v_S} = \dfrac{d}{v_P} + 33$ s. $d\left(\dfrac{1}{3.5 \text{ km/s}} - \dfrac{1}{6.5 \text{ km/s}}\right) = 33$ s and $d = 250$ km.

EVALUATE: The times of travel for each wave are $t_S = 71$ s and $t_P = 38$ s.

2.61. **IDENTIFY:** The average velocity is $v_{\text{av-}x} = \dfrac{\Delta x}{\Delta t}$.

SET UP: Let $+x$ be upward.

EXECUTE: **(a)** $v_{\text{av-}x} = \dfrac{1000 \text{ m} - 63 \text{ m}}{4.75 \text{ s}} = 197$ m/s

(b) $v_{\text{av-}x} = \dfrac{1000 \text{ m} - 0}{5.90 \text{ s}} = 169$ m/s

EVALUATE: For the first 1.15 s of the flight, $v_{\text{av-}x} = \dfrac{63 \text{ m} - 0}{1.15 \text{ s}} = 54.8$ m/s. When the velocity isn't constant the average velocity depends on the time interval chosen. In this motion the velocity is increasing.

2.67. **IDENTIFY:** When the graph of v_x versus t is a straight line the acceleration is constant, so this motion consists of two constant acceleration segments and the constant acceleration equations can be used for each segment. Since v_x is always positive the motion is always in the $+x$ direction and the total distance moved equals the magnitude of the displacement. The acceleration a_x is the slope of the v_x versus t graph.

SET UP: For the $t = 0$ to $t = 10.0$ s segment, $v_{0x} = 4.00$ m/s and $v_x = 12.0$ m/s. For the $t = 10.0$ s to 12.0 s segment, $v_{0x} = 12.0$ m/s and $v_x = 0$.

EXECUTE: **(a)** For $t = 0$ to $t = 10.0$ s, $x - x_0 = \left(\dfrac{v_{0x} + v_x}{2}\right)t = \left(\dfrac{4.00 \text{ m/s} + 12.0 \text{ m/s}}{2}\right)(10.0 \text{ s}) = 80.0$ m.

For $t = 10.0$ s to $t = 12.0$ s, $x - x_0 = \left(\dfrac{12.0 \text{ m/s} + 0}{2}\right)(2.00 \text{ s}) = 12.0$ m. The total distance traveled is 92.0 m.

(b) $x - x_0 = 80.0 \text{ m} + 12.0 \text{ m} = 92.0$ m

(c) For $t = 0$ to 10.0 s, $a_x = \dfrac{12.0 \text{ m/s} - 4.00 \text{ m/s}}{10.0 \text{ s}} = 0.800 \text{ m/s}^2$. For $t = 10.0$ s to 12.0 s,

$a_x = \dfrac{0 - 12.0 \text{ m/s}}{2.00 \text{ s}} = -6.00 \text{ m/s}^2$. The graph of a_x versus t is given in Figure 2.67.

EVALUATE: When v_x and a_x are both positive, the speed increases. When v_x is positive and a_x is negative, the speed decreases.

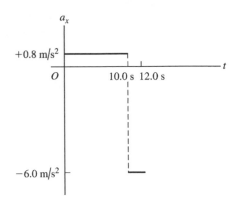

Figure 2.67

2.69.   **IDENTIFY** and **SET UP:** Apply constant acceleration equations.
Find the velocity at the start of the second 5.0 s; this is the velocity at the end of the first 5.0 s. Then find $x - x_0$ for the first 5.0 s.

EXECUTE: For the first 5.0 s of the motion, $v_{0x} = 0$, $t = 5.0$ s.

$v_x = v_{0x} + a_x t$ gives $v_x = a_x(5.0 \text{ s})$.

This is the initial speed for the second 5.0 s of the motion. For the second 5.0 s:

$v_{0x} = a_x(5.0 \text{ s})$, $t = 5.0$ s, $x - x_0 = 150$ m.

$x - x_0 = v_{0x}t + \frac{1}{2}a_x t^2$ gives $150 \text{ m} = (25 \text{ s}^2)a_x + (12.5 \text{ s}^2)a_x$ and $a_x = 4.0 \text{ m/s}^2$

Use this a_x and consider the first 5.0 s of the motion:

$x - x_0 = v_{0x}t + \frac{1}{2}a_x t^2 = 0 + \frac{1}{2}(4.0 \text{ m/s}^2)(5.0 \text{ s})^2 = 50.0 \text{ m}.$

EVALUATE: The ball is speeding up so it travels farther in the second 5.0 s interval than in the first. In fact, $x - x_0$ is proportional to t^2 since it starts from rest. If it goes 50.0 m in 5.0 s, in twice the time (10.0 s) it should go four times as far. In 10.0 s we calculated it went $50 \text{ m} + 150 \text{ m} = 200 \text{ m}$, which is four times 50 m.

2.71.   **IDENTIFY:** Apply constant acceleration equations to the motion of the two objects, you and the cockroach. You catch up with the roach when both objects are at the same place at the same time. Let T be the time when you catch up with the cockroach.
SET UP: Take $x = 0$ to be at the $t = 0$ location of the roach and positive x to be in the direction of motion of the two objects.
<u>roach:</u>
$v_{0x} = 1.50 \text{ m/s}$, $a_x = 0$, $x_0 = 0$, $x = 1.20 \text{ m}$, $t = T$
<u>you:</u>
$v_{0x} = 0.80 \text{ m/s}$, $x_0 = -0.90 \text{ m}$, $x = 1.20 \text{ m}$, $t = T$, $a_x = ?$

Apply $x - x_0 = v_{0x}t + \frac{1}{2}a_x t^2$ to both objects:

EXECUTE: <u>roach:</u> $1.20 \text{ m} = (1.50 \text{ m/s})T$, so $T = 0.800$ s.

<u>you:</u> $1.20 \text{ m} - (-0.90 \text{ m}) = (0.80 \text{ m/s})T + \frac{1}{2}a_x T^2$

$2.10 \text{ m} = (0.80 \text{ m/s})(0.800 \text{ s}) + \frac{1}{2}a_x(0.800 \text{ s})^2$

$2.10 \text{ m} = 0.64 \text{ m} + (0.320 \text{ s}^2)a_x$

$a_x = 4.6 \text{ m/s}^2.$

EVALUATE: Your final velocity is $v_x = v_{0x} + a_x t = 4.48$ m/s. Then $x - x_0 = \left(\dfrac{v_{0x} + v_x}{2}\right)t = 2.10$ m, which

checks. You have to accelerate to a speed greater than that of the roach so you will travel the extra 0.90 m you are initially behind.

2.73.   **IDENTIFY:** Apply constant acceleration equations to each object.

Take the origin of coordinates to be at the initial position of the truck, as shown in Figure 2.73a.

Let d be the distance that the auto initially is behind the truck, so $x_0(\text{auto}) = -d$ and $x_0(\text{truck}) = 0$. Let

T be the time it takes the auto to catch the truck. Thus at time T the truck has undergone a displacement

$x - x_0 = 40.0$ m, so is at $x = x_0 + 40.0 \text{ m} = 40.0$ m. The auto has caught the truck so at time T is also at

$x = 40.0$ m.

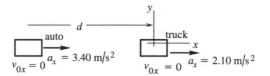

Figure 2.73a

(a) SET UP: Use the motion of the truck to calculate T:

$x - x_0 = 40.0$ m, $v_{0x} = 0$ (starts from rest), $a_x = 2.10 \text{ m/s}^2$, $t = T$

$x - x_0 = v_{0x}t + \frac{1}{2}a_x t^2$

Since $v_{0x} = 0$, this gives $t = \sqrt{\dfrac{2(x - x_0)}{a_x}}$

EXECUTE: $T = \sqrt{\dfrac{2(40.0 \text{ m})}{2.10 \text{ m/s}^2}} = 6.17$ s

(b) SET UP: Use the motion of the auto to calculate d:

$x - x_0 = 40.0 \text{ m} + d$, $v_{0x} = 0$, $a_x = 3.40 \text{ m/s}^2$, $t = 6.17$ s

$x - x_0 = v_{0x}t + \frac{1}{2}a_x t^2$

EXECUTE: $d + 40.0 \text{ m} = \frac{1}{2}(3.40 \text{ m/s}^2)(6.17 \text{ s})^2$

$d = 64.8 \text{ m} - 40.0 \text{ m} = 24.8$ m

(c) auto: $v_x = v_{0x} + a_x t = 0 + (3.40 \text{ m/s}^2)(6.17 \text{ s}) = 21.0$ m/s

truck: $v_x = v_{0x} + a_x t = 0 + (2.10 \text{ m/s}^2)(6.17 \text{ s}) = 13.0$ m/s

(d) The graph is sketched in Figure 2.73b.

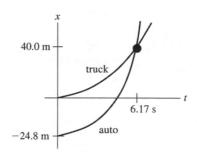

Figure 2.73b

EVALUATE: In part (c) we found that the auto was traveling faster than the truck when they came abreast. The graph in part (d) agrees with this: at the intersection of the two curves the slope of the x-t curve for the auto is greater than that of the truck. The auto must have an average velocity greater than that of the truck since it must travel farther in the same time interval.

2.75. **IDENTIFY:** The average speed is the distance traveled divided by the time. The average velocity is $v_{\text{av-}x} = \dfrac{\Delta x}{\Delta t}$.

SET UP: The distance the ball travels is half the circumference of a circle of diameter 50.0 cm so is $\frac{1}{2}\pi d = \frac{1}{2}\pi(50.0 \text{ cm}) = 78.5 \text{ cm}$. Let $+x$ be horizontally from the starting point toward the ending point, so Δx equals the diameter of the bowl.

EXECUTE: **(a)** The average speed is $\dfrac{\frac{1}{2}\pi d}{t} = \dfrac{78.5 \text{ cm}}{10.0 \text{ s}} = 7.85 \text{ cm/s}$.

(b) The average velocity is $v_{\text{av-}x} = \dfrac{\Delta x}{\Delta t} = \dfrac{50.0 \text{ cm}}{10.0 \text{ s}} = 5.00 \text{ cm/s}$.

EVALUATE: The average speed is greater than the magnitude of the average velocity, since the distance traveled is greater than the magnitude of the displacement.

2.85. **(a) IDENTIFY:** Consider the motion from when he applies the acceleration to when the shot leaves his hand.

SET UP: Take positive y to be upward. $v_{0y} = 0$, $v_y = ?$, $a_y = 35.0 \text{ m/s}^2$, $y - y_0 = 0.640 \text{ m}$,

$v_y^2 = v_{0y}^2 + 2a_y(y - y_0)$

EXECUTE: $v_y = \sqrt{2a_y(y - y_0)} = \sqrt{2(35.0 \text{ m/s}^2)(0.640 \text{ m})} = 6.69 \text{ m/s}$

(b) IDENTIFY: Consider the motion of the shot from the point where he releases it to its maximum height, where $v = 0$. Take $y = 0$ at the ground.

SET UP: $y_0 = 2.20 \text{ m}$, $y = ?$, $a_y = -9.80 \text{ m/s}^2$ (free fall), $v_{0y} = 6.69 \text{ m/s}$ (from part (a), $v_y = 0$ at maximum height), $v_y^2 = v_{0y}^2 + 2a_y(y - y_0)$

EXECUTE: $y - y_0 = \dfrac{v_y^2 - v_{0y}^2}{2a_y} = \dfrac{0 - (6.69 \text{ m/s})^2}{2(-9.80 \text{ m/s}^2)} = 2.29 \text{ m}$, $y = 2.20 \text{ m} + 2.29 \text{ m} = 4.49 \text{ m}$.

(c) IDENTIFY: Consider the motion of the shot from the point where he releases it to when it returns to the height of his head. Take $y = 0$ at the ground.

SET UP: $y_0 = 2.20 \text{ m}$, $y = 1.83 \text{ m}$, $a_y = -9.80 \text{ m/s}^2$, $v_{0y} = +6.69 \text{ m/s}$, $t = ?$ $y - y_0 = v_{0y}t + \frac{1}{2}a_y t^2$

EXECUTE: $1.83 \text{ m} - 2.20 \text{ m} = (6.69 \text{ m/s})t + \frac{1}{2}(-9.80 \text{ m/s}^2)t^2 = (6.69 \text{ m/s})t - (4.90 \text{ m/s}^2)t^2$,

$4.90t^2 - 6.69t - 0.37 = 0$, with t in seconds. Use the quadratic formula to solve for t:

$t = \dfrac{1}{9.80}\left(6.69 \pm \sqrt{(6.69)^2 - 4(4.90)(-0.37)}\right) = 0.6830 \pm 0.7362$. Since t must be positive,

$t = 0.6830 \text{ s} + 0.7362 \text{ s} = 1.42 \text{ s}$.

EVALUATE: Calculate the time to the maximum height: $v_y = v_{0y} + a_y t$, so $t = (v_y - v_{0y})/a_y =$

$-(6.69 \text{ m/s})/(-9.80 \text{ m/s}^2) = 0.68 \text{ s}$. It also takes 0.68 s to return to 2.2 m above the ground, for a total time of 1.36 s. His head is a little lower than 2.20 m, so it is reasonable for the shot to reach the level of his head a little later than 1.36 s after being thrown; the answer of 1.42 s in part (c) makes sense.

2.93. **IDENTIFY:** Apply constant acceleration equations to the motion of the rocket and to the motion of the canister after it is released. Find the time it takes the canister to reach the ground after it is released and find the height of the rocket after this time has elapsed. The canister travels up to its maximum height and then returns to the ground.

SET UP: Let $+y$ be upward. At the instant that the canister is released, it has the same velocity as the rocket. After it is released, the canister has $a_y = -9.80 \text{ m/s}^2$. At its maximum height the canister has $v_y = 0$.

EXECUTE: **(a)** Find the speed of the rocket when the canister is released: $v_{0y} = 0$, $a_y = 3.30 \text{ m/s}^2$,

$y - y_0 = 235 \text{ m}$. $v_y^2 = v_{0y}^2 + 2a_y(y - y_0)$ gives $v_y = \sqrt{2a_y(y - y_0)} = \sqrt{2(3.30 \text{ m/s}^2)(235 \text{ m})} = 39.4 \text{ m/s}$.

For the motion of the canister after it is released, $v_{0y} = +39.4 \text{ m/s}$, $a_y = -9.80 \text{ m/s}^2$, $y - y_0 = -235 \text{ m}$.

$y - y_0 = v_{0y}t + \frac{1}{2}a_y t^2$ gives $-235 \text{ m} = (39.4 \text{ m/s})t - (4.90 \text{ m/s}^2)t^2$. The quadratic formula gives $t = 12.0 \text{ s}$

as the positive solution. Then for the motion of the rocket during this 12.0 s,

$y - y_0 = v_{0y}t + \frac{1}{2}a_y t^2 = 235 \text{ m} + (39.4 \text{ m/s})(12.0 \text{ s}) + \frac{1}{2}(3.30 \text{ m/s}^2)(12.0 \text{ s})^2 = 945 \text{ m}$.

(b) Find the maximum height of the canister above its release point: $v_{0y} = +39.4 \text{ m/s}$, $v_y = 0$,

$a_y = -9.80 \text{ m/s}^2$. $v_y^2 = v_{0y}^2 + 2a_y(y - y_0)$ gives $y - y_0 = \dfrac{v_y^2 - v_{0y}^2}{2a_y} = \dfrac{0 - (39.4 \text{ m/s})^2}{2(-9.80 \text{ m/s}^2)} = 79.2 \text{ m}$. After its

release the canister travels upward 79.2 m to its maximum height and then back down $79.2 \text{ m} + 235 \text{ m}$ to the ground. The total distance it travels is 393 m.

EVALUATE: The speed of the rocket at the instant that the canister returns to the launch pad is

$v_y = v_{0y} + a_y t = 39.4 \text{ m/s} + (3.30 \text{ m/s}^2)(12.0 \text{ s}) = 79.0 \text{ m/s}$. We can calculate its height at this instant by

$v_y^2 = v_{0y}^2 + 2a_y(y - y_0)$ with $v_{0y} = 0$ and $v_y = 79.0 \text{ m/s}$. $y - y_0 = \dfrac{v_y^2 - v_{0y}^2}{2a_y} = \dfrac{(79.0 \text{ m/s})^2}{2(3.30 \text{ m/s}^2)} = 946 \text{ m}$, which

agrees with our previous calculation.

2.95.   **IDENTIFY** and **SET UP:**   Use $v_x = dx/dt$ and $a_x = dv_x/dt$ to calculate $v_x(t)$ and $a_x(t)$ for each car. Use these equations to answer the questions about the motion.

EXECUTE: $x_A = \alpha t + \beta t^2$, $v_{Ax} = \dfrac{dx_A}{dt} = \alpha + 2\beta t$, $a_{Ax} = \dfrac{dv_{Ax}}{dt} = 2\beta$

$x_B = \gamma t^2 - \delta t^3$, $v_{Bx} = \dfrac{dx_B}{dt} = 2\gamma t - 3\delta t^2$, $a_{Bx} = \dfrac{dv_{Bx}}{dt} - 2\gamma - 6\delta t$

(a) IDENTIFY and **SET UP:**   The car that initially moves ahead is the one that has the larger v_{0x}.

EXECUTE: At $t = 0$, $v_{Ax} = \alpha$ and $v_{Bx} = 0$. So initially car A moves ahead.

(b) IDENTIFY and **SET UP:**   Cars at the same point implies $x_A = x_B$.

$\alpha t + \beta t^2 = \gamma t^2 - \delta t^3$

EXECUTE: One solution is $t = 0$, which says that they start from the same point. To find the other solutions, divide by t: $\alpha + \beta t = \gamma t - \delta t^2$

$\delta t^2 + (\beta - \gamma)t + \alpha = 0$

$t = \dfrac{1}{2\delta}\left(-(\beta - \gamma) \pm \sqrt{(\beta - \gamma)^2 - 4\delta\alpha}\right) = \dfrac{1}{0.40}\left(+1.60 \pm \sqrt{(1.60)^2 - 4(0.20)(2.60)}\right) = 4.00 \text{ s} \pm 1.73 \text{ s}$

So $x_A = x_B$ for $t = 0$, $t = 2.27 \text{ s}$ and $t = 5.73 \text{ s}$.

EVALUATE: Car A has constant, positive a_x. Its v_x is positive and increasing. Car B has $v_{0x} = 0$ and a_x that is initially positive but then becomes negative. Car B initially moves in the $+x$-direction but then slows down and finally reverses direction. At $t = 2.27 \text{ s}$ car B has overtaken car A and then passes it. At $t = 5.73 \text{ s}$, car B is moving in the $-x$-direction as it passes car A again.

(c) IDENTIFY: The distance from A to B is $x_B - x_A$. The rate of change of this distance is $\dfrac{d(x_B - x_A)}{dt}$. If

this distance is not changing, $\dfrac{d(x_B - x_A)}{dt} = 0$. But this says $v_{Bx} - v_{Ax} = 0$. (The distance between A and B

is neither decreasing nor increasing at the instant when they have the same velocity.)

SET UP: $v_{Ax} = v_{Bx}$ requires $\alpha + 2\beta t = 2\gamma t - 3\delta t^2$

EXECUTE: $3\delta t^2 + 2(\beta - \gamma)t + \alpha = 0$

$$t = \frac{1}{6\delta}\left(-2(\beta-\gamma) \pm \sqrt{4(\beta-\gamma)^2 - 12\delta\alpha}\right) = \frac{1}{1.20}\left(3.20 \pm \sqrt{4(-1.60)^2 - 12(0.20)(2.60)}\right)$$

$t = 2.667 \text{ s} \pm 1.667 \text{ s}$, so $v_{Ax} = v_{Bx}$ for $t = 1.00$ s and $t = 4.33$ s.

EVALUATE: At $t = 1.00$ s, $v_{Ax} = v_{Bx} = 5.00$ m/s. At $t = 4.33$ s, $v_{Ax} = v_{Bx} = 13.0$ m/s. Now car B is slowing down while A continues to speed up, so their velocities aren't ever equal again.

(d) IDENTIFY and SET UP: $a_{Ax} = a_{Bx}$ requires $2\beta = 2\gamma - 6\delta t$

EXECUTE: $t = \dfrac{\gamma - \beta}{3\delta} = \dfrac{2.80 \text{ m/s}^2 - 1.20 \text{ m/s}^2}{3(0.20 \text{ m/s}^3)} = 2.67$ s.

EVALUATE: At $t = 0$, $a_{Bx} > a_{Ax}$, but a_{Bx} is decreasing while a_{Ax} is constant. They are equal at $t = 2.67$ s but for all times after that $a_{Bx} < a_{Ax}$.

MOTION IN TWO OR THREE DIMENSIONS

3.3. **(a) IDENTIFY** and **SET UP:** From $\vec{r}$ we can calculate x and y for any t. Then use Eq. (3.2), in component form.

EXECUTE: $\vec{r} = [4.0 \text{ cm} + (2.5 \text{ cm/s}^2)t^2]\hat{i} + (5.0 \text{ cm/s})t\hat{j}$

At $t = 0$, $\vec{r} = (4.0 \text{ cm})\hat{i}$.

At $t = 2.0 \text{ s}$, $\vec{r} = (14.0 \text{ cm})\hat{i} + (10.0 \text{ cm})\hat{j}$.

$(v_{av})_x = \dfrac{\Delta x}{\Delta t} = \dfrac{10.0 \text{ cm}}{2.0 \text{ s}} = 5.0 \text{ cm/s}.$

$(v_{av})_y = \dfrac{\Delta y}{\Delta t} = \dfrac{10.0 \text{ cm}}{2.0 \text{ s}} = 5.0 \text{ cm/s}.$

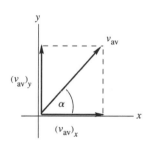

$v_{av} = \sqrt{(v_{av})_x^2 + (v_{av})_y^2} = 7.1 \text{ cm/s}$

$\tan \alpha = \dfrac{(v_{av})_y}{(v_{av})_x} = 1.00$

$\theta = 45°.$

Figure 3.3a

EVALUATE: Both x and y increase, so $\vec{v}_{av}$ is in the 1st quadrant.

(b) IDENTIFY and **SET UP:** Calculate $\vec{r}$ by taking the time derivative of $\vec{r}(t)$.

EXECUTE: $\vec{v} = \dfrac{d\vec{r}}{dt} = ([5.0 \text{ cm/s}^2]t)\hat{i} + (5.0 \text{ cm/s})\hat{j}$

$\underline{t = 0}$: $v_x = 0$, $v_y = 5.0 \text{ cm/s}$; $v = 5.0 \text{ cm/s}$ and $\theta = 90°$

$\underline{t = 1.0 \text{ s}}$: $v_x = 5.0 \text{ cm/s}$, $v_y = 5.0 \text{ cm/s}$; $v = 7.1 \text{ cm/s}$ and $\theta = 45°$

$\underline{t = 2.0 \text{ s}}$: $v_x = 10.0 \text{ cm/s}$, $v_y = 5.0 \text{ cm/s}$; $v = 11 \text{ cm/s}$ and $\theta = 27°$

(c) The trajectory is a graph of y versus x.

$x = 4.0 \text{ cm} + (2.5 \text{ cm/s}^2)t^2$, $y = (5.0 \text{ cm/s})t$

For values of t between 0 and 2.0 s, calculate x and y and plot y versus x.

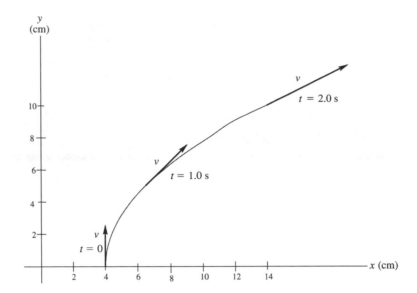

Figure 3.3b

EVALUATE: The sketch shows that the instantaneous velocity at any t is tangent to the trajectory.

3.7.   IDENTIFY and SET UP: Use Eqs. (3.4) and (3.12) to find v_x, v_y, a_x, and a_y as functions of time. The magnitude and direction of $\vec{r}$ and $\vec{a}$ can be found once we know their components.

EXECUTE: (a) Calculate x and y for t values in the range 0 to 2.0 s and plot y versus x. The results are given in Figure 3.7a.

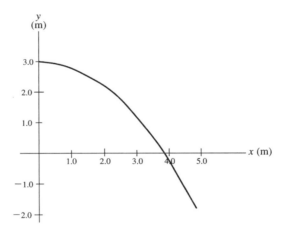

Figure 3.7a

(b) $v_x = \dfrac{dx}{dt} = \alpha$ $v_y = \dfrac{dy}{dt} = -2\beta t$

$a_y = \dfrac{dv_x}{dt} = 0$ $a_y = \dfrac{dv_y}{dt} = -2\beta$

Thus $\vec{v} = \alpha\hat{i} - 2\beta t\hat{j}$ $\vec{a} = -2\beta\hat{j}$

(c) velocity: At $t = 2.0$ s, $v_x = 2.4$ m/s, $v_y = -2(1.2 \text{ m/s}^2)(2.0 \text{ s}) = -4.8$ m/s

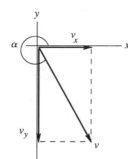

$$v = \sqrt{v_x^2 + v_y^2} = 5.4 \text{ m/s}$$

$$\tan \alpha = \frac{v_y}{v_x} = \frac{-4.8 \text{ m/s}}{2.4 \text{ m/s}} = -2.00$$

$$\alpha = -63.4° + 360° = 297°$$

Figure 3.7b

<u>acceleration:</u>   At $t = 2.0$ s, $a_x = 0$, $a_y = -2 (1.2 \text{ m/s}^2) = -2.4 \text{ m/s}^2$

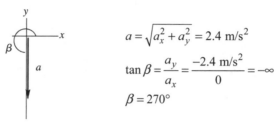

$$a = \sqrt{a_x^2 + a_y^2} = 2.4 \text{ m/s}^2$$

$$\tan \beta = \frac{a_y}{a_x} = \frac{-2.4 \text{ m/s}^2}{0} = -\infty$$

$$\beta = 270°$$

Figure 3.7c

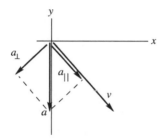

EVALUATE: (d) $\vec{a}$ has a component $a_\parallel$ in the same direction as $\vec{v}$, so we know that v is increasing (the bird is speeding up.) $\vec{a}$ also has a component $a_\perp$ perpendicular to $\vec{v}$, so that the direction of $\vec{v}$ is changing; the bird is turning toward the $-y$-direction (toward the right)

Figure 3.7d

$\vec{v}$ is always tangent to the path; $\vec{v}$ at $t = 2.0$ s shown in part (c) is tangent to the path at this t, conforming to this general rule. $\vec{a}$ is constant and in the $-y$-direction; the direction of $\vec{v}$ is turning toward the $-y$-direction.

3.11.   **IDENTIFY:** Each object moves in projectile motion.

SET UP: Take $+y$ to be downward. For each cricket, $a_x = 0$ and $a_y = +9.80 \text{ m/s}^2$. For Chirpy, $v_{0x} = v_{0y} = 0$. For Milada, $v_{0x} = 0.950 \text{ m/s}$, $v_{0y} = 0$.

EXECUTE: Milada's horizontal component of velocity has no effect on her vertical motion. She also reaches the ground in 3.50 s. $x - x_0 = v_{0x}t + \frac{1}{2}a_x t^2 = (0.950 \text{ m/s})(3.50 \text{ s}) = 3.32 \text{ m}$

EVALUATE: The x and y components of motion are totally separate and are connected only by the fact that the time is the same for both.

3.13.   **IDENTIFY:** The car moves in projectile motion. The car travels $21.3 \text{ m} - 1.80 \text{ m} = 19.5 \text{ m}$ downward during the time it travels 61.0 m horizontally.

SET UP: Take $+y$ to be downward. $a_x = 0$, $a_y = +9.80 \text{ m/s}^2$. $v_{0x} = v_0$, $v_{0y} = 0$.

EXECUTE: (a) Use the vertical motion to find the time in the air:

$y - y_0 = v_{0y}t + \frac{1}{2}a_y t^2$ gives $t = \sqrt{\dfrac{2(y - y_0)}{a_y}} = \sqrt{\dfrac{2(19.5 \text{ m})}{9.80 \text{ m/s}^2}} = 1.995 \text{ s}$

Then $x - x_0 = v_{0x}t + \frac{1}{2}a_x t^2$ gives $v_0 = v_{0x} = \dfrac{x - x_0}{t} = \dfrac{61.0 \text{ m}}{1.995 \text{ s}} = 30.6 \text{ m/s}.$

(b) $v_x = 30.6 \text{ m/s}$ since $a_x = 0$. $v_y = v_{0y} + a_y t = -19.6 \text{ m/s}.$ $v = \sqrt{v_x^2 + v_y^2} = 36.3 \text{ m/s}.$

EVALUATE: We calculate the final velocity by calculating its x and y components.

3.15. **IDENTIFY:** The ball moves with projectile motion with an initial velocity that is horizontal and has magnitude v_0. The height h of the table and v_0 are the same; the acceleration due to gravity changes from $g_E = 9.80 \text{ m/s}^2$ on earth to g_X on planet X.

SET UP: Let $+x$ be horizontal and in the direction of the initial velocity of the marble and let $+y$ be upward. $v_{0x} = v_0$, $v_{0y} = 0$, $a_x = 0$, $a_y = -g$, where g is either g_E or g_X.

EXECUTE: Use the vertical motion to find the time in the air: $y - y_0 = -h$. $y - y_0 = v_{0y}t + \frac{1}{2}a_y t^2$ gives $t = \sqrt{\dfrac{2h}{g}}$. Then $x - x_0 = v_{0x}t + \frac{1}{2}a_x t^2$ gives $x - x_0 = v_{0x}t = v_0 \sqrt{\dfrac{2h}{g}}$. $x - x_0 = D$ on earth and $2.76D$ on Planet X. $(x - x_0)\sqrt{g} = v_0\sqrt{2h}$, which is constant, so $D\sqrt{g_E} = 2.76D\sqrt{g_X}$.

$g_X = \dfrac{g_E}{(2.76)^2} = 0.131 g_E = 1.28 \text{ m/s}^2.$

EVALUATE: On Planet X the acceleration due to gravity is less, it takes the ball longer to reach the floor and it travels farther horizontally.

3.19. **IDENTIFY:** Take the origin of coordinates at the point where the quarter leaves your hand and take positive y to be upward. The quarter moves in projectile motion, with $a_x = 0$, and $a_y = -g$. It travels vertically for the time it takes it to travel horizontally 2.1 m.

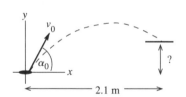

$v_{0x} = v_0 \cos\alpha_0 = (6.4 \text{ m/s}) \cos 60°$

$v_{0x} = 3.20 \text{ m/s}$

$v_{0y} = v_0 \sin\alpha_0 = (6.4 \text{ m/s}) \sin 60°$

$v_{0y} = 5.54 \text{ m/s}$

Figure 3.19

(a) SET UP: Use the horizontal (x-component) of motion to solve for t, the time the quarter travels through the air:

$t = ?$, $x - x_0 = 2.1 \text{ m}$, $v_{0x} = 3.2 \text{ m/s}$, $a_x = 0$

$x - x_0 = v_{0x}t + \frac{1}{2}a_x t^2 = v_{0x}t$, since $a_x = 0$

EXECUTE: $t = \dfrac{x - x_0}{v_{0x}} = \dfrac{2.1 \text{ m}}{3.2 \text{ m/s}} = 0.656 \text{ s}$

SET UP: Now find the vertical displacement of the quarter after this time:

$y - y_0 = ?$, $a_y = -9.80 \text{ m/s}^2$, $v_{0y} = +5.54 \text{ m/s}$, $t = 0.656 \text{ s}$

$y - y_0 + v_{0y}t + \frac{1}{2}a_y t^2$

EXECUTE: $y - y_0 = (5.54 \text{ m/s})(0.656 \text{ s}) + \frac{1}{2}(-9.80 \text{ m/s}^2)(0.656 \text{ s})^2 = 3.63 \text{ m} - 2.11 \text{ m} = 1.5 \text{ m}.$

(b) SET UP: $v_y = ?$, $t = 0.656 \text{ s}$, $a_y = -9.80 \text{ m/s}^2$, $v_{0y} = +5.54 \text{ m/s}$ $v_y = v_{0y} + a_y t$

EXECUTE: $v_y = 5.54 \text{ m/s} + (-9.80 \text{ m/s}^2)(0.656 \text{ s}) = -0.89 \text{ m/s}.$

EVALUATE: The minus sign for v_y indicates that the y-component of $\vec{v}$ is downward. At this point the quarter has passed through the highest point in its path and is on its way down. The horizontal range if it returned to its original height (it doesn't!) would be 3.6 m. It reaches its maximum height after traveling horizontally 1.8 m, so at $x - x_0 = 2.1$ m it is on its way down.

3.21.   **IDENTIFY:** Take the origin of coordinates at the roof and let the $+y$-direction be upward. The rock moves in projectile motion, with $a_x = 0$ and $a_y = -g$. Apply constant acceleration equations for the x and y components of the motion.

SET UP:

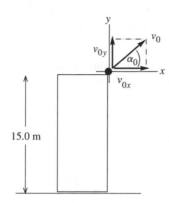

$v_{0x} = v_0 \cos \alpha_0 = 25.2$ m/s

$v_{0y} = v_0 \sin \alpha_0 = 16.3$ m/s

15.0 m

Figure 3.21a

(a) At the maximum height $v_y = 0$.

$a_y = -9.80$ m/s^2, $v_y = 0$, $v_{0y} = +16.3$ m/s, $y - y_0 = ?$

$v_y^2 = v_{0y}^2 + 2a_y (y - y_0)$

EXECUTE: $y - y_0 = \dfrac{v_y^2 - v_{0y}^2}{2a_y} = \dfrac{0 - (16.3 \text{ m/s})^2}{2(-9.80 \text{ m/s}^2)} = +13.6$ m

(b) SET UP: Find the velocity by solving for its x and y components.

$v_x = v_{0x} = 25.2$ m/s (since $a_x = 0$)

$v_y = ?$, $a_y = -9.80$ m/s^2, $y - y_0 = -15.0$ m (negative because at the ground the rock is below its initial position), $v_{0y} = 16.3$ m/s

$v_y^2 = v_{0y}^2 + 2a_y (y - y_0)$

$v_y = -\sqrt{v_{0y}^2 + 2a_y (y - y_0)}$ (v_y is negative because at the ground the rock is traveling downward.)

EXECUTE: $v_y = -\sqrt{(16.3 \text{ m/s})^2 + 2(-9.80 \text{ m/s}^2)(-15.0 \text{ m})} = -23.7$ m/s

Then $v = \sqrt{v_x^2 + v_y^2} = \sqrt{(25.2 \text{ m/s})^2 + (-23.7 \text{ m/s})^2} = 34.6$ m/s.

(c) SET UP: Use the vertical motion (y-component) to find the time the rock is in the air:

$t = ?$, $v_y = -23.7$ m/s (from part (b)), $a_y = -9.80$ m/s^2, $v_{0y} = +16.3$ m/s

EXECUTE: $t = \dfrac{v_y - v_{0y}}{a_y} = \dfrac{-23.7 \text{ m/s} - 16.3 \text{ m/s}}{-9.80 \text{ m/s}^2} = +4.08$ s

SET UP: Can use this t to calculate the horizontal range:

$t = 4.08$ s, $v_{0x} = 25.2$ m/s, $a_x = 0$, $x - x_0 = ?$

EXECUTE: $x - x_0 = v_{0x}t + \frac{1}{2}a_x t^2 = (25.2 \text{ m/s})(4.08 \text{ s}) + 0 = 103$ m

(d) Graphs of x versus t, y versus t, v_x versus t and v_y versus t:

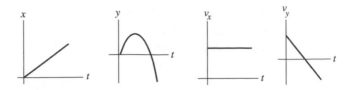

Figure 3.21b

EVALUATE: The time it takes the rock to travel vertically to the ground is the time it has to travel horizontally. With $v_{0y} = +16.3$ m/s the time it takes the rock to return to the level of the roof $(y = 0)$ is $t = 2v_{0y}/g = 3.33$ s. The time in the air is greater than this because the rock travels an additional 15.0 m to the ground.

3.23.   **IDENTIFY and SET UP:**   The stone moves in projectile motion. Its initial velocity is the same as that of the balloon. Use constant acceleration equations for the x and y components of its motion. Take $+y$ to be downward.
EXECUTE: **(a)** Use the vertical motion of the rock to find the initial height.

$t = 6.00$ s, $v_{0y} = +20.0$ m/s, $a_y = +9.80$ m/s^2, $y - y_0 = ?$

$y - y_0 = v_{0y}t + \frac{1}{2}a_y t^2$ gives $y - y_0 = 296$ m

(b) In 6.00 s the balloon travels downward a distance $y - y_0 = (20.0 \text{ m/s})(6.00 \text{ s}) = 120$ m. So, its height above ground when the rock hits is $296 \text{ m} - 120 \text{ m} = 176$ m.

(c) The horizontal distance the rock travels in 6.00 s is 90.0 m. The vertical component of the distance between the rock and the basket is 176 m, so the rock is $\sqrt{(176 \text{ m})^2 + (90 \text{ m})^2} = 198$ m from the basket when it hits the ground.

(d) (i) The basket has no horizontal velocity, so the rock has horizontal velocity 15.0 m/s relative to the basket. Just before the rock hits the ground, its vertical component of velocity is

$v_y = v_{0y} + a_y t = 20.0 \text{ m/s} + (9.80 \text{ m/s}^2)(6.00 \text{ s}) = 78.8$ m/s, downward, relative to the ground. The basket is moving downward at 20.0 m/s, so relative to the basket the rock has a downward component of velocity 58.8 m/s.
(ii) horizontal: 15.0 m/s; vertical: 78.8 m/s

EVALUATE: The rock has a constant horizontal velocity and accelerates downward

3.25.   **IDENTIFY:**   Apply Eq. (3.30).
SET UP: $T = 24$ h.

EXECUTE: **(a)** $a_{\text{rad}} = \dfrac{4\pi^2 (6.38 \times 10^6 \text{ m})}{((24 \text{ h})(3600 \text{ s/h}))^2} = 0.034$ m/s^2 $= 3.4 \times 10^{-3} g$.

(b) Solving Eq. (3.30) for the period T with $a_{\text{rad}} = g$, $T = \sqrt{\dfrac{4\pi^2 (6.38 \times 10^6 \text{ m})}{9.80 \text{ m/s}^2}} = 5070$ s $= 1.4$ h.

EVALUATE: a_{rad} is proportional to $1/T^2$, so to increase a_{rad} by a factor of $\dfrac{1}{3.4 \times 10^{-3}} = 294$ requires

that T be multiplied by a factor of $\dfrac{1}{\sqrt{294}}$. $\dfrac{24 \text{ h}}{\sqrt{294}} = 1.4$ h.

3.33.   **IDENTIFY:**   Apply the relative velocity relation.
SET UP: The relative velocities are $\vec{v}_{C/E}$, the canoe relative to the earth, $\vec{v}_{R/E}$, the velocity of the river relative to the earth and $\vec{v}_{C/R}$, the velocity of the canoe relative to the river.
EXECUTE: $\vec{v}_{C/E} = \vec{v}_{C/R} + \vec{v}_{R/E}$ and therefore $\vec{v}_{C/R} = \vec{v}_{C/E} - \vec{v}_{R/E}$. The velocity components of $\vec{v}_{C/R}$ are

-0.50 m/s $+ (0.40 \text{ m/s})/\sqrt{2}$, east and $(0.40 \text{ m/s})/\sqrt{2}$, south, for a velocity relative to the river of 0.36 m/s, at 52.5° south of west.

EVALUATE: The velocity of the canoe relative to the river has a smaller magnitude than the velocity of the canoe relative to the earth.

3.35. **IDENTIFY:** Relative velocity problem in two dimensions. His motion relative to the earth (time displacement) depends on his velocity relative to the earth so we must solve for this velocity.

(a) SET UP: View the motion from above.

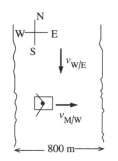

The velocity vectors in the problem are:

$\vec{v}_{M/E}$, the velocity of the man relative to the earth

$\vec{v}_{W/E}$, the velocity of the water relative to the earth

$\vec{v}_{M/W}$, the velocity of the man relative to the water

The rule for adding these velocities is

$\vec{v}_{M/E} = \vec{v}_{M/W} + \vec{v}_{W/E}$

Figure 3.35a

The problem tells us that $\vec{v}_{W/E}$ has magnitude 2.0 m/s and direction due south. It also tells us that $\vec{v}_{M/W}$ has magnitude 4.2 m/s and direction due east. The vector addition diagram is then as shown in Figure 3.35b.

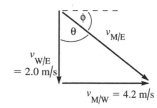

This diagram shows the vector addition

$\vec{v}_{M/E} = \vec{v}_{M/W} + \vec{v}_{W/E}$

and also has $\vec{v}_{M/W}$ and $\vec{v}_{W/E}$ in their specified directions. Note that the vector diagram forms a right triangle.

Figure 3.35b

The Pythagorean theorem applied to the vector addition diagram gives $v_{M/E}^2 = v_{M/W}^2 + v_{W/E}^2$.

EXECUTE: $v_{M/E} = \sqrt{v_{M/W}^2 + v_{W/E}^2} = \sqrt{(4.2 \text{ m/s})^2 + (2.0 \text{ m/s})^2} = 4.7 \text{ m/s}$; $\tan\theta = \dfrac{v_{M/W}}{v_{W/E}} = \dfrac{4.2 \text{ m/s}}{2.0 \text{ m/s}} = 2.10$;

$\theta = 65°$; or $\phi = 90° - \theta = 25°$. The velocity of the man relative to the earth has magnitude 4.7 m/s and direction 25° S of E.

(b) This requires careful thought. To cross the river the man must travel 800 m due east relative to the earth. The man's velocity relative to the earth is $\vec{v}_{M/E}$. But, from the vector addition diagram the eastward component of $v_{M/E}$ equals $v_{M/W} = 4.2$ m/s.

Thus $t = \dfrac{x - x_0}{v_x} = \dfrac{800 \text{ m}}{4.2 \text{ m/s}} = 190$ s.

(c) The southward component of $\vec{v}_{M/E}$ equals $v_{W/E} = 2.0$ m/s. Therefore, in the 190 s it takes him to cross the river, the distance south the man travels relative to the earth is

$$y - y_0 = v_y t = (2.0 \text{ m/s})(190 \text{ s}) = 380 \text{ m.}$$

EVALUATE: If there were no current he would cross in the same time, $(800 \text{ m})/(4.2 \text{ m/s}) = 190$ s. The current carries him downstream but doesn't affect his motion in the perpendicular direction, from bank to bank.

3.37. **IDENTIFY:** Relative velocity problem in two dimensions.

(a) SET UP: $\vec{v}_{P/A}$ is the velocity of the plane relative to the air. The problem states that $\vec{v}_{P/A}$ has magnitude 35 m/s and direction south.

$\vec{v}_{A/E}$ is the velocity of the air relative to the earth. The problem states that $\vec{v}_{A/E}$ is to the southwest ($45°$ S of W) and has magnitude 10 m/s.

The relative velocity equation is $\vec{v}_{P/E} = \vec{v}_{P/A} + \vec{v}_{A/E}$.

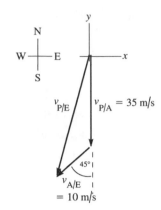

Figure 3.37a

EXECUTE: (b) $(v_{P/A})_x = 0,$ $(v_{P/A})_y = -35$ m/s

$(v_{A/E})_x = -(10 \text{ m/s})\cos 45° = -7.07$ m/s,

$(v_{A/E})_y = -(10 \text{ m/s})\sin 45° = -7.07$ m/s

$(v_{P/E})_x = (v_{P/A})_x + (v_{A/E})_x = 0 - 7.07 \text{ m/s} = -7.1$ m/s

$(v_{P/E})_y = (v_{P/A})_y + (v_{A/E})_y = -35 \text{ m/s} - 7.07 \text{ m/s} = -42$ m/s

(c)

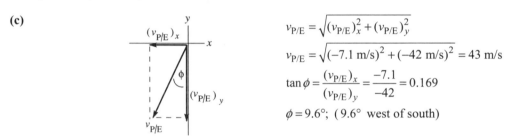

$v_{P/E} = \sqrt{(v_{P/E})_x^2 + (v_{P/E})_y^2}$

$v_{P/E} = \sqrt{(-7.1 \text{ m/s})^2 + (-42 \text{ m/s})^2} = 43$ m/s

$\tan\phi = \dfrac{(v_{P/E})_x}{(v_{P/E})_y} = \dfrac{-7.1}{-42} = 0.169$

$\phi = 9.6°;$ $(9.6°$ west of south$)$

Figure 3.37b

EVALUATE: The relative velocity addition diagram does not form a right triangle so the vector addition must be done using components. The wind adds both southward and westward components to the velocity of the plane relative to the ground.

3.39.   **IDENTIFY:**   The resultant velocity, relative to the ground, is directly southward. This velocity is the sum of the velocity of the bird relative to the air and the velocity of the air relative to the ground.

SET UP: $v_{B/A} = 100$ km/h. $\vec{v}_{A/G} = 40$ km/h, east. $\vec{v}_{B/G} = \vec{v}_{B/A} + \vec{v}_{A/G}$.

EXECUTE: We want $\vec{v}_{B/G}$ to be due south. The relative velocity addition diagram is shown in Figure 3.39.

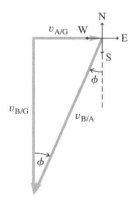

Figure 3.39

(a) $\sin\phi = \dfrac{v_{A/G}}{v_{B/A}} = \dfrac{40 \text{ km/h}}{100 \text{ km/h}}$, $\phi = 24°$, west of south.

(b) $v_{B/G} = \sqrt{v_{B/A}^2 - v_{A/G}^2} = 91.7 \text{ km/h}$. $t = \dfrac{d}{v_{B/G}} = \dfrac{500 \text{ km}}{91.7 \text{ km/h}} = 5.5 \text{ h}$.

EVALUATE: The speed of the bird relative to the ground is less than its speed relative to the air. Part of its velocity relative to the air is directed to oppose the effect of the wind.

3.43. **IDENTIFY:** $\vec{v} = d\vec{r}/dt$. This vector will make a $45°$ angle with both axes when its x- and y-components are equal.

SET UP: $\dfrac{d(t^n)}{dt} = nt^{n-1}$.

EXECUTE: $\vec{v} = 2bt\hat{i} + 3ct^2\hat{j}$. $v_x = v_y$ gives $t = 2b/3c$.

EVALUATE: Both components of $\vec{v}$ change with t.

3.45. **IDENTIFY:** Given the velocity components of a plane, when will its velocity be perpendicular to its acceleration?

SET UP: By definition, $a_x = dv_x/dt$, and $a_y = dv_y/dt$. When two vectors are perpendicular, their scalar product is zero.

EXECUTE: Taking the time derivative of the velocity vector gives $\vec{a}(t) = (1.20 \text{ m/s}^2)\hat{i} + (-2.00 \text{ m/s}^2)\hat{j}$. When the velocity and acceleration are perpendicular to each other,

$\vec{v} \cdot \vec{a} = (1.20 \text{ m/s}^2)^2 t + (12.0 \text{ m/s} - (2.00 \text{ m/s}^2)t)(-2.00 \text{ m/s}^2) = 0$. Solving for t gives

$(5.44 \text{ m}^2/\text{s}^4)t = 24.0 \text{ m}^2/\text{s}^3$, so $t = 4.41 \text{ s}$.

EVALUATE: There is only one instant at which the velocity and acceleration are perpendicular, so it is not a general rule.

3.47. **IDENTIFY:** Once the rocket leaves the incline it moves in projectile motion. The acceleration along the incline determines the initial velocity and initial position for the projectile motion.

SET UP: For motion along the incline let $+x$ be directed up the incline. $v_x^2 = v_{0x}^2 + 2a_x(x - x_0)$ gives

$v_x = \sqrt{2(1.25 \text{ m/s}^2)(200 \text{ m})} = 22.36 \text{ m/s}$. When the projectile motion begins the rocket has $v_0 = 22.36 \text{ m/s}$ at $35.0°$ above the horizontal and is at a vertical height of $(200.0 \text{ m})\sin 35.0° = 114.7 \text{ m}$. For the projectile motion let $+x$ be horizontal to the right and let $+y$ be upward. Let $y = 0$ at the ground. Then $y_0 = 114.7 \text{ m}$, $v_{0x} = v_0\cos 35.0° = 18.32 \text{ m/s}$, $v_{0y} = v_0\sin 35.0° = 12.83 \text{ m/s}$, $a_x = 0$, $a_y = -9.80 \text{ m/s}^2$. Let $x = 0$ at point A, so $x_0 = (200.0 \text{ m})\cos 35.0° = 163.8 \text{ m}$.

EXECUTE: **(a)** At the maximum height $v_y = 0$. $v_y^2 = v_{0y}^2 + 2a_y(y - y_0)$ gives

$y - y_0 = \dfrac{v_y^2 - v_{0y}^2}{2a_y} = \dfrac{0 - (12.83 \text{ m/s})^2}{2(-9.80 \text{ m/s}^2)} = 8.40 \text{ m}$ and $y = 114.7 \text{ m} + 8.40 \text{ m} = 123 \text{ m}$. The maximum height above ground is 123 m.

(b) The time in the air can be calculated from the vertical component of the projectile motion:

$y - y_0 = -114.7 \text{ m}$, $v_{0y} = 12.83 \text{ m/s}$, $a_y = -9.80 \text{ m/s}^2$. $y - y_0 = v_{0y}t + \frac{1}{2}a_y t^2$ gives

$(4.90 \text{ m/s}^2)t^2 - (12.83 \text{ m/s})t - 114.7 \text{ m}$. The quadratic formula gives

$t = \dfrac{1}{9.80}\left(12.83 \pm \sqrt{(12.83)^2 + 4(4.90)(114.7)}\right)$ s. The positive root is $t = 6.32 \text{ s}$. Then

$x - x_0 = v_{0x}t + \frac{1}{2}a_x t^2 = (18.32 \text{ m/s})(6.32 \text{ s}) = 115.8 \text{ m}$ and $x = 163.8 \text{ m} + 115.8 \text{ m} = 280 \text{ m}$. The horizontal range of the rocket is 280 m.

EVALUATE: The expressions for h and R derived in Example 3.8 do not apply here. They are only for a projectile fired on level ground.

3.49.   **IDENTIFY:** The range for a projectile that lands at the same height from which it was launched is

$R = \dfrac{v_0^2 \sin 2\alpha}{g}$.

SET UP: The maximum range is for $\alpha = 45°$.

EXECUTE: Assuming $\alpha = 45°$, and $R = 50$ m, $v_0 = \sqrt{gR} = 22$ m/s.

EVALUATE: We have assumed that debris was launched at all angles, including the angle of $45°$ that gives maximum range.

3.51.   **IDENTIFY:** Take $+y$ to be downward. Both objects have the same vertical motion, with v_{0y} and $a_y = +g$. Use constant acceleration equations for the x and y components of the motion.

SET UP: Use the vertical motion to find the time in the air:

$v_{0y} = 0$, $a_y = 9.80$ m/s^2, $y - y_0 = 25$ m, $t = ?$.

EXECUTE: $y - y_0 = v_{0y}t + \frac{1}{2}a_y t^2$ gives $t = 2.259$ s.

During this time the dart must travel 90 m, so the horizontal component of its velocity must be

$v_{0x} = \dfrac{x - x_0}{t} = \dfrac{70 \text{ m}}{2.25 \text{ s}} = 31$ m/s.

EVALUATE: Both objects hit the ground at the same time. The dart hits the monkey for any muzzle velocity greater than 31 m/s.

3.53.   **IDENTIFY:** The cannister moves in projectile motion. Its initial velocity is horizontal. Apply constant acceleration equations for the x and y components of motion.

SET UP:

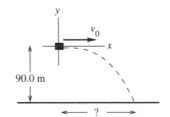

Take the origin of coordinates at the point where the canister is released. Take $+y$ to be upward. The initial velocity of the canister is the velocity of the plane, 64.0 m/s in the $+x$-direction.

Figure 3.53

Use the vertical motion to find the time of fall:

$t = ?$, $v_{0y} = 0$, $a_y = -9.80$ m/s^2, $y - y_0 = -90.0$ m (When the canister reaches the ground it is 90.0 m below the origin.)

$$y - y_0 = v_{0y}t + \frac{1}{2}a_y t^2$$

EXECUTE: Since $v_{0y} = 0$, $t = \sqrt{\dfrac{2(y - y_0)}{a_y}} = \sqrt{\dfrac{2(-90.0 \text{ m})}{-9.80 \text{ m/s}^2}} = 4.286$ s.

SET UP: Then use the horizontal component of the motion to calculate how far the canister falls in this time:

$x - x_0 = ?$, $a_x - 0$, $v_{0x} = 64.0$ m/s

EXECUTE: $x - x_0 = v_0 t + \frac{1}{2}at^2 = (64.0 \text{ m/s})(4.286 \text{ s}) + 0 = 274$ m.

EVALUATE: The time it takes the cannister to fall 90.0 m, starting from rest, is the time it travels horizontally at constant speed.

3.55.   **IDENTIFY:** The suitcase moves in projectile motion. The initial velocity of the suitcase equals the velocity of the airplane.

SET UP: Take $+y$ to be upward. $a_x = 0$, $a_y = -g$.

EXECUTE: Use the vertical motion to find the time it takes the suitcase to reach the ground:

$v_{0y} = v_0 \sin 23°$, $a_y = -9.80 \text{ m/s}^2$, $y - y_0 = -114 \text{ m}$, $t = ?$ $y - y_0 = v_{0y}t + \frac{1}{2}a_yt^2$ gives $t = 9.60 \text{ s}$.

The distance the suitcase travels horizontally is $x - x_0 = v_{0x} = (v_0 \cos 23.0°)t = 795 \text{ m}$.

EVALUATE: An object released from rest at a height of 114 m strikes the ground at

$t = \sqrt{\dfrac{2(y - y_0)}{-g}} = 4.82 \text{ s}$. The suitcase is in the air much longer than this since it initially has an upward

component of velocity.

3.59.   **IDENTIFY:** Projectile motion problem.

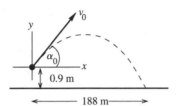

Take the origin of coordinates at the point where the ball leaves the bat, and take $+y$ to be upward.

$v_{0x} = v_0 \cos\alpha_0$

$v_{0y} = v_0 \sin\alpha_0$,

but we don't know v_0.

Figure 3.59

Write down the equation for the horizontal displacement when the ball hits the ground and the corresponding equation for the vertical displacement. The time t is the same for both components, so this will give us two equations in two unknowns (v_0 and t).

(a) SET UP: y-component:

$a_y = -9.80 \text{ m/s}^2$, $y - y_0 = -0.9 \text{ m}$, $v_{0y} = v_0 \sin 45°$

$y - y_0 = v_{0y}t + \frac{1}{2}a_yt^2$

EXECUTE: $-0.9 \text{ m} = (v_0 \sin 45°)t + \frac{1}{2}(-9.80 \text{ m/s}^2)t^2$

SET UP: x-component:

$a_x = 0$, $x - x_0 = 188 \text{ m}$, $v_{0x} = v_0 \cos 45°$

$x - x_0 = v_{0x}t + \frac{1}{2}a_xt^2$

EXECUTE: $t = \dfrac{x - x_0}{v_{0x}} = \dfrac{188 \text{ m}}{v_0 \cos 45°}$

Put the expression for t from the x-component motion into the y-component equation and solve for v_0. (Note that $\sin 45° = \cos 45°$.)

$$-0.9 \text{ m} = (v_0 \sin 45°)\left(\frac{188 \text{ m}}{v_0 \cos 45°}\right) - (4.90 \text{ m/s}^2)\left(\frac{188 \text{ m}}{v_0 \cos 45°}\right)^2$$

$$4.90 \text{ m/s}^2\left(\frac{188 \text{ m}}{v_0 \cos 45°}\right)^2 = 188 \text{ m} + 0.9 \text{ m} = 188.9 \text{ m}$$

$$\left(\frac{v_0 \cos 45°}{188 \text{ m}}\right)^2 = \frac{4.90 \text{ m/s}^2}{188.9 \text{ m}}, \quad v_0 = \left(\frac{188 \text{ m}}{\cos 45°}\right)\sqrt{\frac{4.90 \text{ m/s}^2}{188.9 \text{ m}}} = 42.8 \text{ m/s}$$

(b) Use the horizontal motion to find the time it takes the ball to reach the fence:

SET UP: x-component:

$x - x_0 = 116 \text{ m}$, $a_x = 0$, $v_{0x} = v_0 \cos 45° = (42.8 \text{ m/s}) \cos 45° = 30.3 \text{ m/s}$, $t = ?$

$x - x_0 = v_{0x}t + \frac{1}{2}a_xt^2$

EXECUTE: $t = \dfrac{x - x_0}{v_{0x}} = \dfrac{116 \text{ m}}{30.3 \text{ m/s}} = 3.83 \text{ s}$

SET UP: Find the vertical displacement of the ball at this t:

y-component:

$y - y_0 = ?, \quad a_y = -9.80 \text{ m/s}^2, \quad v_{0y} = v_0 \sin 45° = 30.3 \text{ m/s}, \quad t = 3.83 \text{ s}$

$y - y_0 = v_{0y} t + \frac{1}{2} a_y t^2$

EXECUTE: $y - y_0 = (30.3 \text{ s})(3.83 \text{ s}) + \frac{1}{2}(-9.80 \text{ m/s}^2)(3.83 \text{ s})^2$

$y - y_0 = 116.0 \text{ m} - 71.9 \text{ m} = +44.1 \text{ m},$ above the point where the ball was hit. The height of the ball above the ground is $44.1 \text{ m} + 0.90 \text{ m} = 45.0 \text{ m}.$ It's height then above the top of the fence is $45.0 \text{ m} - 3.0 \text{ m} = 42.0 \text{ m}.$

EVALUATE: With $v_0 = 42.8 \text{ m/s}, \quad v_{0y} = 30.3 \text{ m/s}$ and it takes the ball 6.18 s to return to the height where it was hit and only slightly longer to reach a point 0.9 m below this height. $t = (188 \text{ m})/(v_0 \cos 45°)$ gives $t = 6.21 \text{ s},$ which agrees with this estimate. The ball reaches its maximum height approximately $(188 \text{ m})/2 = 94 \text{ m}$ from home plate, so at the fence the ball is not far past its maximum height of 47.6 m, so a height of 45.0 m at the fence is reasonable.

3.61.   **IDENTIFY:** The equations for *h* and *R* from Example 3.8 can be used.

SET UP: $h = \dfrac{v_0^2 \sin^2 \alpha_0}{2g}$ and $R = \dfrac{v_0^2 \sin 2\alpha_0}{g}.$ If the projectile is launched straight up, $\alpha_0 = 90°.$

EXECUTE: (a) $h = \dfrac{v_0^2}{2g}$ and $v_0 = \sqrt{2gh}.$

(b) Calculate α_0 that gives a maximum height of *h* when $v_0 = 2\sqrt{2gh}.$ $h = \dfrac{8gh \sin^2 \alpha_0}{2g} = 4h \sin^2 \alpha_0.$

$\sin \alpha_0 = \frac{1}{2}$ and $\alpha_0 = 30.0°.$

(c) $R = \dfrac{(2\sqrt{2gh})^2 \sin 60.0°}{g} = 6.93h.$

EVALUATE: $\dfrac{v_0^2}{g} = \dfrac{2h}{\sin^2 \alpha_0}$ so $R = \dfrac{2h \sin(2\alpha_0)}{\sin^2 \alpha_0}.$ For a given $\alpha_0,$ *R* increases when *h* increases. For $\alpha_0 = 90°, \quad R = 0$ and for $\alpha_0 = 0°, \quad h = 0$ and $R = 0.$ For $\alpha_0 = 45°, \quad R = 4h.$

3.63.   **IDENTIFY:** From the figure in the text, we can read off the maximum height and maximum horizontal distance reached by the grasshopper. Knowing its acceleration is *g* downward, we can find its initial speed and the height of the cliff (the target variables).

SET UP: Use coordinates with the origin at the ground and $+y$ upward. $a_x = 0, \quad a_y = -9.80 \text{ m/s}^2.$ The constant-acceleration kinematics formulas $v_y^2 = v_{0y}^2 + 2a_y(y - y_0)$ and $x - x_0 = v_{0x} t + \frac{1}{2} a_x t^2$ apply.

EXECUTE: (a) $v_y = 0$ when $y - y_0 = 0.0674 \text{ m}.$ $v_y^2 = v_{0y}^2 + 2a_y(y - y_0)$ gives

$v_{0y} = \sqrt{-2a_y(y - y_0)} = \sqrt{-2(-9.80 \text{ m/s}^2)(0.0674 \text{ m})} = 1.15 \text{ m/s}.$ $v_{0y} = v_0 \sin \alpha_0$ so

$v_0 = \dfrac{v_{0y}}{\sin \alpha_0} = \dfrac{1.15 \text{ m/s}}{\sin 50.0°} = 1.50 \text{ m/s}.$

(b) Use the horizontal motion to find the time in the air. The grasshopper travels horizontally $x - x_0 = 1.06 \text{ m}.$ $x - x_0 = v_{0x} t + \frac{1}{2} a_x t^2$ gives $t = \dfrac{x - x_0}{v_{0x}} = \dfrac{x - x_0}{v_0 \cos 50.0°} = 1.10 \text{ s}.$ Find the vertical displacement of the grasshopper at $t = 1.10 \text{ s}:$

$y - y_0 = v_{0y} t + \frac{1}{2} a_y t^2 = (1.15 \text{ m/s})(1.10 \text{ s}) + \frac{1}{2}(-9.80 \text{ m/s}^2)(1.10 \text{ s})^2 = -4.66 \text{ m}.$ The height of the cliff is 4.66 m.

EVALUATE: The grasshopper's maximum height (6.74 cm) is physically reasonable, so its takeoff speed of 1.50 m/s must also be reasonable. Note that the equation $R = \dfrac{v_0^2 \sin 2\alpha_0}{g}$ does *not* apply here since the launch point is not at the same level as the landing point.

3.65.   **IDENTIFY:**   The snowball moves in projectile motion. In part (a) the vertical motion determines the time in the air. In part (c), find the height of the snowball above the ground after it has traveled horizontally 4.0 m.

SET UP: Let $+y$ be downward. $a_x = 0$, $a_y = +9.80 \text{ m/s}^2$. $v_{0x} = v_0 \cos\theta_0 = 5.36$ m/s, $v_{0y} = v_0 \sin\theta_0 = 4.50$ m/s.

EXECUTE: **(a)** Use the vertical motion to find the time in the air: $y - y_0 = v_{0y}t + \frac{1}{2}a_y t^2$ with $y - y_0 = 14.0$ m gives $14.0 \text{ m} = (4.50 \text{ m/s})\,t + (4.9 \text{ m/s}^2)\,t^2$. The quadratic formula gives $t = \dfrac{1}{2(4.9)}\left(-4.50 \pm \sqrt{(4.50)^2 - 4(4.9)(-14.0)}\right)$ s. The positive root is $t = 1.29$ s. Then

$x - x_0 = v_{0x}t + \frac{1}{2}a_x t^2 = (5.36 \text{ m/s})(1.29 \text{ s}) = 6.91$ m.

(b) The x-t, y-t, v_x-t and v_y-t graphs are sketched in Figure 3.65.

(c) $x - x_0 = v_{0x}t + \frac{1}{2}a_x t^2$ gives $t = \dfrac{x - x_0}{v_{0x}} = \dfrac{4.0 \text{ m}}{5.36 \text{ m/s}} = 0.746$ s. In this time the snowball travels downward a distance $y - y_0 = v_{0y}t + \frac{1}{2}a_y t^2 = 6.08$ m and is therefore $14.0 \text{ m} - 6.08 \text{ m} = 7.9$ m above the ground. The snowball passes well above the man and doesn't hit him.

EVALUATE: If the snowball had been released from rest at a height of 14.0 m it would have reached the ground in $t = \sqrt{\dfrac{2(14.0 \text{ m})}{9.80 \text{ m/s}^2}} = 1.69$ s. The snowball reaches the ground in a shorter time than this because of its initial downward component of velocity.

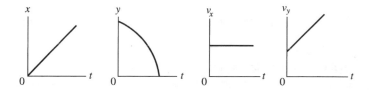

Figure 3.65

3.67.   **(a) IDENTIFY:**   Projectile motion.

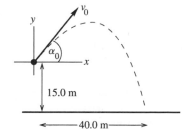

Take the origin of coordinates at the top of the ramp and take $+y$ to be upward.

The problem specifies that the object is displaced 40.0 m to the right when it is 15.0 m below the origin.

Figure 3.67

We don't know t, the time in the air, and we don't know v_0. Write down the equations for the horizontal and vertical displacements. Combine these two equations to eliminate one unknown.

SET UP: y-component:

$y - y_0 = -15.0$ m, $a_y = -9.80 \text{ m/s}^2$, $v_{0y} = v_0 \sin 53.0°$

$y - y_0 = v_{0y}t + \frac{1}{2}a_y t^2$

EXECUTE: $-15.0 \text{ m} = (v_0 \sin 53.0°) t - (4.90 \text{ m/s}^2) t^2$

SET UP: x-component:

$x - x_0 = 40.0 \text{ m}, \quad a_x = 0, \quad v_{0x} = v_0 \cos 53.0°$

$x - x_0 = v_{0x} t + \frac{1}{2} a_x t^2$

EXECUTE: $40.0 \text{ m} = (v_0 t) \cos 53.0°$

The second equation says $v_0 t = \dfrac{40.0 \text{ m}}{\cos 53.0°} = 66.47 \text{ m}.$

Use this to replace $v_0 t$ in the first equation:

$-15.0 \text{ m} = (66.47 \text{ m}) \sin 53° - (4.90 \text{ m/s}^2) t^2$

$t = \sqrt{\dfrac{(66.46 \text{ m}) \sin 53° + 15.0 \text{ m}}{4.90 \text{ m/s}^2}} = \sqrt{\dfrac{68.08 \text{ m}}{4.90 \text{ m/s}^2}} = 3.727 \text{ s}.$

Now that we have t we can use the x-component equation to solve for v_0:

$v_0 = \dfrac{40.0 \text{ m}}{t \cos 53.0°} = \dfrac{40.0 \text{ m}}{(3.727 \text{ s}) \cos 53.0°} = 17.8 \text{ m/s}.$

EVALUATE: Using these values of v_0 and t in the $y = y_0 = v_{0y} + \frac{1}{2} a_y t^2$ equation verifies that

$y - y_0 = -15.0 \text{ m}.$

(b) IDENTIFY: $v_0 = (17.8 \text{ m/s})/2 = 8.9 \text{ m/s}$

This is less than the speed required to make it to the other side, so he lands in the river.
Use the vertical motion to find the time it takes him to reach the water:

SET UP: $y - y_0 = -100 \text{ m}; \quad v_{0y} = +v_0 \sin 53.0° = 7.11 \text{ m/s}; \quad a_y = -9.80 \text{ m/s}^2$

$y - y_0 = v_{0y} t + \frac{1}{2} a_y t^2$ gives $-100 = 7.11 t - 4.90 t^2$

EXECUTE: $4.90 t^2 - 7.11 t - 100 = 0$ and $t = \frac{1}{9.80}\left(7.11 \pm \sqrt{(7.11)^2 - 4(4.90)(-100)}\right)$

$t = 0.726 \text{ s} \pm 4.57 \text{ s}$ so $t = 5.30 \text{ s}.$
The horizontal distance he travels in this time is

$x - x_0 = v_{0x} t = (v_0 \cos 53.0°) t = (5.36 \text{ m/s})(5.30 \text{ s}) = 28.4 \text{ m}.$

He lands in the river a horizontal distance of 28.4 m from his launch point.

EVALUATE: He has half the minimum speed and makes it only about halfway across.

3.69.   **IDENTIFY and SET UP:**   Take $+y$ to be upward. The rocket moves with projectile motion, with

$v_{0y} = +40.0 \text{ m/s}$ and $v_{0x} = 30.0 \text{ m/s}$ relative to the ground. The vertical motion of the rocket is unaffected
by its horizontal velocity.

EXECUTE: **(a)** $v_y = 0$ (at maximum height), $v_{0y} = +40.0 \text{ m/s}$, $a_y = -9.80 \text{ m/s}^2$, $y - y_0 = ?$

$v_y^2 = v_{0y}^2 + 2a_y (y - y_0)$ gives $y - y_0 = 81.6 \text{ m}$

(b) Both the cart and the rocket have the same constant horizontal velocity, so both travel the same
horizontal distance while the rocket is in the air and the rocket lands in the cart.

(c) Use the vertical motion of the rocket to find the time it is in the air.

$v_{0y} = 40 \text{ m/s}$, $a_y = -9.80 \text{ m/s}^2$, $v_y = -40 \text{ m/s}$, $t = ?$

$v_y = v_{0y} + a_y t$ gives $t = 8.164 \text{ s}$

Then $x - x_0 = v_{0x} t = (30.0 \text{ m/s})(8.164 \text{ s}) = 245 \text{ m}.$

(d) Relative to the ground the rocket has initial velocity components $v_{0x} = 30.0 \text{ m/s}$ and $v_{0y} = 40.0 \text{ m/s}$,
so it is traveling at $53.1°$ above the horizontal.

(e) (i)

Figure 3.69a

Relative to the cart, the rocket travels straight up and then straight down.

(ii)

Figure 3.69b

Relative to the ground the rocket travels in a parabola.

EVALUATE: Both the cart and rocket have the same constant horizontal velocity. The rocket lands in the cart.

3.71. **IDENTIFY:** The boulder moves in projectile motion.

SET UP: Take $+y$ downward. $v_{0x} = v_0$, $a_x = 0$, $a_x = 0$, $a_y = +9.80$ m/s^2.

EXECUTE: **(a)** Use the vertical motion to find the time for the boulder to reach the level of the lake:

$$y - y_0 = v_{0y}t + \frac{1}{2}a_y t^2 \text{ with } y - y_0 = +20 \text{ m gives } t = \sqrt{\frac{2(y - y_0)}{a_y}} = \sqrt{\frac{2(20 \text{ m})}{9.80 \text{ m/s}^2}} = 2.02 \text{ s. The rock must}$$

travel horizontally 100 m during this time. $x - x_0 = v_{0x}t + \frac{1}{2}a_x t^2$ gives $v_0 = v_{0x} = \dfrac{x - x_0}{t} = \dfrac{100 \text{ m}}{2.02 \text{ s}} = 49.5$ m/s

(b) In going from the edge of the cliff to the plain, the boulder travels downward a distance of

$$y - y_0 = 45 \text{ m. } t = \sqrt{\frac{2(y - y_0)}{a_y}} = \sqrt{\frac{2(45 \text{ m})}{9.80 \text{ m/s}^2}} = 3.03 \text{ s and } x - x_0 = v_{0x}t = (49.5 \text{ m/s})(3.03 \text{ s}) = 150 \text{ m.}$$

The rock lands 150 m $-$ 100 m $=$ 50 m beyond the foot of the dam.

EVALUATE: The boulder passes over the dam 2.02 s after it leaves the cliff and then travels an additional 1.01 s before landing on the plain. If the boulder has an initial speed that is less than 49 m/s, then it lands in the lake.

3.73. **IDENTIFY:** The shell moves in projectile motion. To find the horizontal distance between the tanks we must find the horizontal velocity of one tank relative to the other. Take $+y$ to be upward.

(a) SET UP: The vertical motion of the shell is unaffected by the horizontal motion of the tank. Use the vertical motion of the shell to find the time the shell is in the air:

$v_{0y} = v_0 \sin \alpha = 43.4$ m/s, $a_y = -9.80$ m/s^2, $y - y_0 = 0$ (returns to initial height), $t = ?$

EXECUTE: $y - y_0 = v_{0y}t + \frac{1}{2}a_y t^2$ gives $t = 8.86$ s

SET UP: Consider the motion of one tank relative to the other.

EXECUTE: Relative to tank #1 the shell has a constant horizontal velocity $v_0 \cos \alpha = 246.2$ m/s. Relative to the ground the horizontal velocity component is 246.2 m/s $+$ 15.0 m/s $=$ 261.2 m/s. Relative to tank #2 the shell has horizontal velocity component 261.2 m/s $-$ 35.0 m/s $=$ 226.2 m/s. The distance between the tanks when the shell was fired is the (226.2 m/s)(8.86 s) $=$ 2000 m that the shell travels relative to tank #2 during the 8.86 s that the shell is in the air.

(b) The tanks are initially 2000 m apart. In 8.86 s tank #1 travels 133 m and tank #2 travels 310 m, in the same direction. Therefore, their separation increases by 310 m $-$ 133 m $=$ 177 m. So, the separation becomes 2180 m (rounding to 3 significant figures).

EVALUATE: The retreating tank has greater speed than the approaching tank, so they move farther apart while the shell is in the air. We can also calculate the separation in part (b) as the relative speed of the tanks times the time the shell is in the air: (35.0 m/s $-$ 15.0 m/s)(8.86 s) $=$ 177 m.

3.75. **IDENTIFY:** The original firecracker moves as a projectile. At its maximum height its velocity is horizontal. The velocity $\vec{v}_{A/G}$ of fragment A relative to the ground is related to the velocity $\vec{v}_{F/G}$ of the original firecracker relative to the ground and the velocity $\vec{v}_{A/F}$ of the fragment relative to the original firecracker by $\vec{v}_{A/G} = \vec{v}_{A/F} + \vec{v}_{F/G}$. Fragment B obeys a similar equation.

SET UP: Let $+x$ be along the direction of the horizontal motion of the firecracker before it explodes and let $+y$ be upward. Fragment A moves at $53.0°$ above the $+x$ direction and fragment B moves at $53.0°$ below the $+x$ direction. Before it explodes the firecracker has $a_x = 0$ and $a_y = -9.80 \text{ m/s}^2$.

EXECUTE: The horizontal component of the firecracker's velocity relative to the ground is constant (since $a_x = 0$), so $v_{F/G-x} = (25.0 \text{ m/s})\cos 30.0° = 21.65 \text{ m/s}$. At the time of the explosion, $v_{F/G-y} = 0$. For fragment A, $v_{A/F-x} = (20.0 \text{ m/s})\cos 53.0° = 12.0 \text{ m/s}$ and $v_{A/F-y} = (20.0 \text{ m/s})\sin 53.0° = 16.0 \text{ m/s}$.

$v_{A/G-x} = v_{A/F-x} + v_{F/G-x} = 12.0 \text{ m/s} + 21.65 \text{ m/s} = 33.7 \text{ m/s}$. $v_{A/G-y} = v_{A/F-y} + v_{F/G-y} = 16.0 \text{ m/s}$.

$\tan \alpha_0 = \dfrac{v_{A/G-y}}{v_{A/G-x}} = \dfrac{16.0 \text{ m/s}}{33.7 \text{ m/s}}$ and $\alpha_0 = 25.4°$. The calculation for fragment B is the same, except

$v_{A/F-y} = -16.0 \text{ m/s}$. The fragments move at $25.4°$ above and $25.4°$ below the horizontal.

EVALUATE: As the initial velocity of the firecracker increases the angle with the horizontal for the fragments, as measured from the ground, decreases.

3.81. **IDENTIFY:** Relative velocity problem. The plane's motion relative to the earth is determined by its velocity relative to the earth.

SET UP: Select a coordinate system where $+y$ is north and $+x$ is east.

The velocity vectors in the problem are:

$\vec{v}_{P/E}$, the velocity of the plane relative to the earth.

$\vec{v}_{P/A}$, the velocity of the plane relative to the air (the magnitude $v_{P/A}$ is the airspeed of the plane and the direction of $\vec{v}_{P/A}$ is the compass course set by the pilot).

$\vec{v}_{A/E}$, the velocity of the air relative to the earth (the wind velocity).

The rule for combining relative velocities gives $\vec{v}_{P/E} = \vec{v}_{P/A} + \vec{v}_{A/E}$.

(a) We are given the following information about the relative velocities:

$\vec{v}_{P/A}$ has magnitude 220 km/h and its direction is west. In our coordinates it has components

$(v_{P/A})_x = -220 \text{ km/h}$ and $(v_{P/A})_y = 0$.

From the displacement of the plane relative to the earth after 0.500 h, we find that $\vec{v}_{P/E}$ has components in our coordinate system of

$$(v_{P/E})_x = -\frac{120 \text{ km}}{0.500 \text{ h}} = -240 \text{ km/h} \text{ (west)}$$

$$(v_{P/E})_y = -\frac{20 \text{ km}}{0.500 \text{ h}} = -40 \text{ km/h} \text{ (south)}$$

With this information the diagram corresponding to the velocity addition equation is shown in Figure 3.81a.

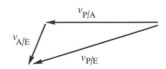

Figure 3.81a

We are asked to find $\vec{v}_{A/E}$, so solve for this vector:

$\vec{v}_{P/E} = \vec{v}_{P/A} + \vec{v}_{A/E}$ gives $\vec{v}_{A/E} = \vec{v}_{P/E} - \vec{v}_{P/A}$.

EXECUTE: The x-component of this equation gives

$(v_{A/E})_x = (v_{P/E})_x - (v_{P/A})_x = -240 \text{ km/h} - (-220 \text{ km/h}) = -20 \text{ km/h}$.

The y-component of this equation gives

$(v_{A/E})_y = (v_{P/E})_y - (v_{P/A})_y = -40 \text{ km/h}$.

Now that we have the components of $\vec{v}_{A/E}$ we can find its magnitude and direction.

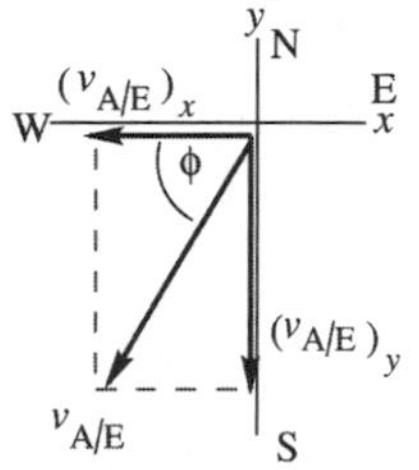

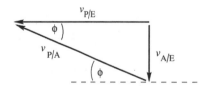

$$v_{A/E} = \sqrt{(v_{A/E})_x^2 + (v_{A/E})_y^2}$$

$$v_{A/E} = \sqrt{(-20 \text{ km/h})^2 + (-40 \text{ km/h})^2} = 44.7 \text{ km/h}$$

$$\tan\phi = \frac{40 \text{ km/h}}{20 \text{ km/h}} = 2.00; \quad \phi = 63.4°$$

The direction of the wind velocity is $63.4°$ S of W, or $26.6°$ W of S.

Figure 3.81b

EVALUATE: The plane heads west. It goes farther west than it would without wind and also travels south, so the wind velocity has components west and south.

(b) SET UP: The rule for combining the relative velocities is still $\vec{v}_{P/E} = \vec{v}_{P/A} + \vec{v}_{A/E}$, but some of these velocities have different values than in part **(a)**.

$\vec{v}_{P/A}$ has magnitude 220 km/h but its direction is to be found.

$\vec{v}_{A/E}$ has magnitude 40 km/h and its direction is due south.

The direction of $\vec{v}_{P/E}$ is west; its magnitude is not given.

The vector diagram for $\vec{v}_{P/E} = \vec{v}_{P/A} + \vec{v}_{A/E}$ and the specified directions for the vectors is shown in Figure 3.81c.

Figure 3.81c

The vector addition diagram forms a right triangle.

EXECUTE: $\sin\phi = \dfrac{v_{A/E}}{v_{P/A}} = \dfrac{40 \text{ km/h}}{220 \text{ km/h}} = 0.1818; \quad \phi = 10.5°$.

The pilot should set her course $10.5°$ north of west.

EVALUATE: The velocity of the plane relative to the air must have a northward component to counteract the wind and a westward component in order to travel west.

NEWTON'S LAWS OF MOTION

4

4.3. **IDENTIFY:** We know the resultant of two vectors of equal magnitude and want to find their magnitudes. They make the same angle with the vertical.

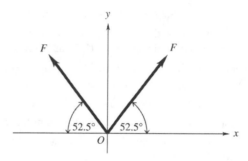

Figure 4.3

SET UP: Take $+y$ to be upward, so $\sum F_y = 5.00$ N. The strap on each side of the jaw exerts a force F directed at an angle of 52.5° above the horizontal, as shown in Figure 4.3.

EXECUTE: $\sum F_y = 2F\sin 52.5° = 5.00$ N, so $F = 3.15$ N.

EVALUATE: The resultant force has magnitude 5.00 N which is *not* the same as the sum of the magnitudes of the two vectors, which would be 6.30 N.

4.7. **IDENTIFY:** Friction is the only horizontal force acting on the skater, so it must be the one causing the acceleration. Newton's second law applies.

SET UP: Take $+x$ to be the direction in which the skater is moving initially. The final velocity is $v_x = 0$, since the skater comes to rest. First use the kinematics formula $v_x = v_{0x} + a_x t$ to find the acceleration, then apply $\sum F_x = 5.00$ N to the skater.

EXECUTE: $v_x = v_{0x} + a_x t$ so $a_x = \dfrac{v_x - v_{0x}}{t} = \dfrac{0 - 2.40 \text{ m/s}}{3.52 \text{ s}} = -0.682 \text{ m/s}^2$. The only horizontal force on the skater is the friction force, so $f_x = ma_x = (68.5 \text{ kg})(-0.682 \text{ m/s}^2) = -46.7$ N. The force is 46.7 N, directed opposite to the motion of the skater.

EVALUATE: Although other forces are acting on the skater (gravity and the upward force of the ice), they are vertical and therefore do not affect the horizontal motion.

4.11. **IDENTIFY and SET UP:** Use Newton's second law in component form (Eq. 4.8) to calculate the acceleration produced by the force. Use constant acceleration equations to calculate the effect of the acceleration on the motion.

EXECUTE: **(a)** During this time interval the acceleration is constant and equal to

$$a_x = \frac{F_x}{m} = \frac{0.250 \text{ N}}{0.160 \text{ } kg} = 1.562 \text{ m/s}^2$$

We can use the constant acceleration kinematic equations from Chapter 2.

$$x - x_0 = v_{0x}t + \tfrac{1}{2}a_x t^2 = 0 + \tfrac{1}{2}(1.562 \text{ m/s}^2)(2.00 \text{ s})^2,$$

so the puck is at $x = 3.12$ m.

$$v_x = v_{0x} + a_x t = 0 + (1.562 \text{ m/s}^2)(2.00 \text{ s}) = 3.12 \text{ m/s}.$$

(b) In the time interval from $t = 2.00$ s to 5.00 s the force has been removed so the acceleration is zero. The speed stays constant at $v_x = 3.12$ m/s. The distance the puck travels is

$$x - x_0 = v_{0x}t = (3.12 \text{ m/s})(5.00 \text{ s} - 2.00 \text{ s}) = 9.36 \text{ m}.$$ At the end of the interval it is at

$$x = x_0 + 9.36 \text{ m} = 12.5 \text{ m}.$$

In the time interval from $t = 5.00$ s to 7.00 s the acceleration is again $a_x = 1.562 \text{ m/s}^2$. At the start of this interval $v_{0x} = 3.12$ m/s and $x_0 = 12.5$ m.

$$x - x_0 = v_{0x}t + \tfrac{1}{2}a_x t^2 = (3.12 \text{ m/s})(2.00 \text{ s}) + \tfrac{1}{2}(1.562 \text{ m/s}^2)(2.00 \text{ s})^2.$$

$$x - x_0 = 6.24 \text{ m} + 3.12 \text{ m} = 9.36 \text{ m}.$$

Therefore, at $t = 7.00$ s the puck is at $x = x_0 + 9.36 \text{ m} = 12.5 \text{ m} + 9.36 \text{ m} = 21.9 \text{ m}.$

$$v_x = v_{0x} + a_x t \approx 3.12 \text{ m/s} + (1.562 \text{ m/s}^2)(2.00 \text{ s}) = 6.24 \text{ m/s}$$

EVALUATE: The acceleration says the puck gains 1.56 m/s of velocity for every second the force acts. The force acts a total of 4.00 s so the final velocity is $(1.56 \text{ m/s})(4.0 \text{ s}) = 6.24 \text{ m/s}.$

4.13. **IDENTIFY:** The force and acceleration are related by Newton's second law.
SET UP: $\sum F_x = ma_x$, where $\sum F_x$ is the net force. $m = 4.50$ kg.
EXECUTE: **(a)** The maximum net force occurs when the acceleration has its maximum value.
$\sum F_x = ma_x = (4.50 \text{ kg})(10.0 \text{ m/s}^2) = 45.0$ N. This maximum force occurs between 2.0 s and 4.0 s.
(b) The net force is constant when the acceleration is constant. This is between 2.0 s and 4.0 s.
(c) The net force is zero when the acceleration is zero. This is the case at $t = 0$ and $t = 6.0$ s.
EVALUATE: A graph of $\sum F_x$ versus t would have the same shape as the graph of a_x versus t.

4.23. **IDENTIFY:** The system is accelerating so we use Newton's second law.
SET UP: The acceleration of the entire system is due to the 100-N force, but the acceleration of box B is due to the force that box A exerts on it. $\sum F = ma$ applies to the two-box system and to each box individually.
EXECUTE: For the two-box system: $a_x = \dfrac{100 \text{ N}}{25 \text{ kg}} = 4.0 \text{ m/s}^2.$ Then for box B, where F_A is the force exerted on B by A, $F_A = m_B a = (5.0 \text{ kg})(4.0 \text{ m/s}^2) = 20$ N.
EVALUATE: The force on B is less than the force on A.

4.25. **IDENTIFY:** Apply Newton's second law to the earth.
SET UP: The force of gravity that the earth exerts on her is her weight,
$w = mg = (45 \text{ kg})(9.8 \text{ m/s}^2) = 441$ N. By Newton's third law, she exerts an equal and opposite force on the earth.
Apply $\sum \vec{F} = m\vec{a}$ to the earth, with $\left| \sum \vec{F} \right| = w = 441$ N, but must use the mass of the earth for m.
EXECUTE: $a = \dfrac{w}{m} = \dfrac{441 \text{ N}}{6.0 \times 10^{24} \text{ kg}} = 7.4 \times 10^{-23} \text{ m/s}^2.$

EVALUATE: This is *much* smaller than her acceleration of 9.8 m/s^2. The force she exerts on the earth equals in magnitude the force the earth exerts on her, but the acceleration the force produces depends on the mass of the object and her mass is *much* less than the mass of the earth.

4.27. **IDENTIFY:** Identify the forces on each object.
SET UP: In each case the forces are the noncontact force of gravity (the weight) and the forces applied by objects that are in contact with each crate. Each crate touches the floor and the other crate, and some object applies $\vec{F}$ to crate A.

EXECUTE: (a) The free-body diagrams for each crate are given in Figure 4.27.

F_{AB} (the force on m_A due to m_B) and F_{BA} (the force on m_B due to m_A) form an action-reaction pair.

(b) Since there is no horizontal force opposing F, any value of F, no matter how small, will cause the crates to accelerate to the right. The weight of the two crates acts at a right angle to the horizontal, and is in any case balanced by the upward force of the surface on them.

EVALUATE: Crate B is accelerated by F_{BA} and crate A is accelerated by the net force $F - F_{AB}$. The greater the total weight of the two crates, the greater their total mass and the smaller will be their acceleration.

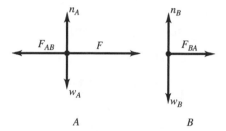

Figure 4.27

4.31.   IDENTIFY: Identify the forces on the chair. The floor exerts a normal force and a friction force.
SET UP: Let $+y$ be upward and let $+x$ be in the direction of the motion of the chair.

EXECUTE: (a) The free-body diagram for the chair is given in Figure 4.31.

(b) For the chair, $a_y = 0$ so $\sum F_y = ma_y$ gives $n - mg - F\sin 37° = 0$ and $n = 142$ N.

EVALUATE: n is larger than the weight because $\vec{F}$ has a downward component.

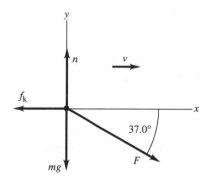

Figure 4.31

4.33.   IDENTIFY: Apply Newton's second law to the bucket and constant-acceleration kinematics.
SET UP: The minimum time to raise the bucket will be when the tension in the cord is a maximum since this will produce the greatest acceleration of the bucket.

EXECUTE: Apply Newton's second law to the bucket: $T - mg = ma$. For the maximum acceleration, the tension is greatest, so $a = \dfrac{T - mg}{m} = \dfrac{75.0\ \text{N} - (4.80\ \text{kg})(9.8\ \text{m/s}^2)}{4.80\ \text{kg}} = 5.825\ \text{m/s}^2$.

The kinematics equation for $y(t)$ gives $t = \sqrt{\dfrac{2(y - y_0)}{a_y}} = \sqrt{\dfrac{2(12.0\ \text{m})}{5.825\ \text{m/s}^2}} = 2.03\ \text{s}$.

EVALUATE: A shorter time would require a greater acceleration and hence a stronger pull, which would break the cord.

4.35.   IDENTIFY: Vector addition problem. Write the vector addition equation in component form. We know one vector and its resultant and are asked to solve for the other vector.

SET UP: Use coordinates with the $+x$-axis along $\vec{F_1}$ and the $+y$-axis along $\vec{R}$, as shown in Figure 4.35a.

$F_{1x} = +1300 \text{ N}, \quad F_{1y} = 0$

$R_x = 0, \quad R_y = +1300 \text{ N}$

Figure 4.35a

$\vec{F}_1 + \vec{F}_2 = \vec{R}$, so $\vec{F}_2 = \vec{R} - \vec{F}_1$

EXECUTE: $F_{2x} = R_x - F_{1x} = 0 - 1300 \text{ N} = -1300 \text{ N}$

$F_{2y} = R_y - F_{1y} = +1300 \text{ N} - 0 = +1300 \text{ N}$

The components of $\vec{F}_2$ are sketched in Figure 4.35b.

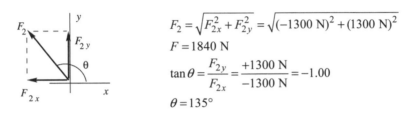

$F_2 = \sqrt{F_{2x}^2 + F_{2y}^2} = \sqrt{(-1300 \text{ N})^2 + (1300 \text{ N})^2}$

$F = 1840 \text{ N}$

$\tan\theta = \dfrac{F_{2y}}{F_{2x}} = \dfrac{+1300 \text{ N}}{-1300 \text{ N}} = -1.00$

$\theta = 135°$

Figure 4.35b

The magnitude of $\vec{F}_2$ is 1840 N and its direction is $135°$ counterclockwise from the direction of $\vec{F}_1$.

EVALUATE: $\vec{F}_2$ has a negative x-component to cancel $\vec{F}_1$ and a y-component to equal $\vec{R}$.

4.39.   **IDENTIFY:** We can apply constant acceleration equations to relate the kinematic variables and we can use Newton's second law to relate the forces and acceleration.
(a) **SET UP:** First use the information given about the height of the jump to calculate the speed he has at the instant his feet leave the ground. Use a coordinate system with the $+y$-axis upward and the origin at the position when his feet leave the ground.

$v_y = 0$ (at the maximum height), $v_{0y} = ?$, $a_y = -9.80 \text{ m/s}^2$, $y - y_0 = +1.2 \text{ m}$

$v_y^2 = v_{0y}^2 + 2a_y(y - y_0)$

EXECUTE: $v_{0y} = \sqrt{-2a_y(y - y_0)} = \sqrt{-2(-9.80 \text{ m/s}^2)(1.2 \text{ m})} = 4.85 \text{ m/s}$

(b) **SET UP:** Now consider the acceleration phase, from when he starts to jump until when his feet leave the ground. Use a coordinate system where the $+y$-axis is upward and the origin is at his position when he starts his jump.
EXECUTE: Calculate the average acceleration:

$$(a_{av})_y = \frac{v_y - v_{0y}}{t} = \frac{4.85 \text{ m/s} - 0}{0.300 \text{ s}} = 16.2 \text{ m/s}^2$$

(c) **SET UP:** Finally, find the average upward force that the ground must exert on him to produce this average upward acceleration. (Don't forget about the downward force of gravity.) The forces are sketched in Figure 4.39.

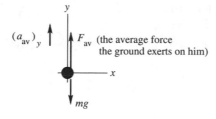

EXECUTE:

$$m = w/g = \frac{890 \text{ N}}{9.80 \text{ m/s}^2} = 90.8 \text{ kg}$$

$$\sum F_y = ma_y$$

$$F_{av} - mg = m(a_{av})_y$$

$$F_{av} = m(g + (a_{av})_y)$$

$$F_{av} = 90.8 \text{ kg}(9.80 \text{ m/s}^2 + 16.2 \text{ m/s}^2)$$

$$F_{av} = 2360 \text{ N}$$

Figure 4.39

This is the average force exerted on him by the ground. But by Newton's third law, the average force he exerts on the ground is equal and opposite, so is 2360 N, downward. The net force on him is equal to ma, so $F_{net} = ma = (90.8 \text{ kg})(16.2 \text{ m/s}^2) = 1470 \text{ N}$ upward.

EVALUATE: In order for him to accelerate upward, the ground must exert an upward force greater than his weight.

4.41.   **IDENTIFY:** Using constant-acceleration kinematics, we can find the acceleration of the ball. Then we can apply Newton's second law to find the force causing that acceleration.

SET UP: Use coordinates where $+x$ is in the direction the ball is thrown. $v_x^2 = v_{0x}^2 + 2a_x(x - x_0)$ and $\sum F_x = ma_x$.

EXECUTE: **(a)** Solve for a_x: $x - x_0 = 1.0$ m, $v_{0x} = 0$, $v_x = 46$ m/s. $v_x^2 = v_{0x}^2 + 2a_x(x - x_0)$ gives

$$a_x = \frac{v_x^2 - v_{0x}^2}{2(x - x)} = \frac{(46 \text{ m/s})^2 - 0}{2(1.0 \text{ m})} = 1058 \text{ m/s}^2.$$

The free-body diagram for the ball during the pitch is shown in Figure 4.41a. The force $\vec{F}$ is applied to the ball by the pitcher's hand. $\sum F_x = ma_x$ gives $F = (0.145 \text{ kg})(1058 \text{ m/s}^2) = 153$ N.

(b) The free-body diagram after the ball leaves the hand is given in Figure 4.41b. The only force on the ball is the downward force of gravity.

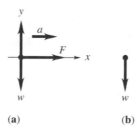

(a)   (b)

Figure 4.41

EVALUATE: The force is much greater than the weight of the ball because it gives it an acceleration much greater than g.

4.43.   **IDENTIFY:** Use Newton's second law to relate the acceleration and forces for each crate.

(a) SET UP: Since the crates are connected by a rope, they both have the same acceleration, 2.50 m/s^2.

(b) The forces on the 4.00 kg crate are shown in Figure 4.43a.

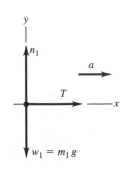

EXECUTE:
$$\sum F_x = ma_x$$
$$T = m_1 a = (4.00 \text{ kg})(2.50 \text{ m/s}^2) = 10.0 \text{ N}.$$

Figure 4.43a

(c) SET UP: Forces on the 6.00 kg crate are shown in Figure 4.43b.

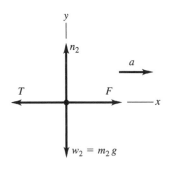

The crate accelerates to the right, so the net force is to the right. F must be larger than T.

Figure 4.43b

(d) EXECUTE: $\sum F_x = ma_x$ gives $F - T = m_2 a$

$$F = T + m_2 a = 10.0 \text{ N} + (6.00 \text{ kg})(2.50 \text{ m/s}^2) = 10.0 \text{ N} + 15.0 \text{ N} = 25.0 \text{ N}$$

EVALUATE: We can also consider the two crates and the rope connecting them as a single object of mass $m = m_1 + m_2 = 10.0$ kg. The free-body diagram is sketched in Figure 4.43c.

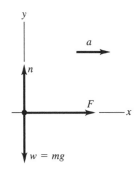

$$\sum F_x = ma_x$$
$$F = ma = (10.0 \text{ kg})(2.50 \text{ m/s}^2) = 25.0 \text{ N}$$
This agrees with our answer in part (d).

Figure 4.43c

4.47.   **IDENTIFY:**   The ship and instrument have the same acceleration. The forces and acceleration are related by Newton's second law. We can use a constant acceleration equation to calculate the acceleration from the information given about the motion.

 SET UP: Let $+y$ be upward. The forces on the instrument are the upward tension $\vec{T}$ exerted by the wire and the downward force $\vec{w}$ of gravity. $w = mg = (6.50 \text{ kg})(9.80 \text{ m/s}^2) = 63.7 \text{ N}$

 EXECUTE: **(a)** The free-body diagram is sketched in Figure 4.47. The acceleration is upward, so $T > w$.

(b) $y - y_0 = 276$ m, $t = 15.0$ s, $v_{0y} = 0$. $y - y_0 = v_{0y}t + \frac{1}{2}a_y t^2$ gives $a_y = \dfrac{2(y - y_0)}{t^2} = \dfrac{2(276 \text{ m})}{(15.0 \text{ s})^2} = 2.45 \text{ m/s}^2$.

$\Sigma F_y = ma_y$ gives $T - w = ma$ and $T = w + ma = 63.7 \text{ N} + (6.50 \text{ kg})(2.45 \text{ m/s}^2) = 79.6$ N.

EVALUATE: There must be a net force in the direction of the acceleration.

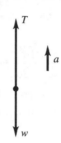

Figure 4.47

4.49. **IDENTIFY:** Using kinematics we can find the acceleration of the froghopper and then apply Newton's second law to find the force on it from the ground.

SET UP: Take $+y$ to be upward. $\Sigma F_y = ma_y$ and for constant acceleration, $v_y = v_{0y} + a_y t$.

EXECUTE: **(a)** The free-body diagram for the froghopper while it is still pushing against the ground is given in Figure 4.49.

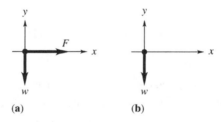

Figure 4.49

(b) $v_{0y} = 0$, $v_y = 4.0$ m/s, $t = 1.0 \times 10^{-3}$ s. $v_y = v_{0y} + a_y t$ gives

$a_y = \dfrac{v_y - v_{0y}}{t} = \dfrac{4.0 \text{ m/s} - 0}{1.0 \times 10^{-3} \text{ s}} = 4.0 \times 10^3 \text{ m/s}^2$. $\Sigma F_y = ma_y$ gives $n - w = ma$, so

$n = w + ma = m(g + a) = (12.3 \times 10^{-6} \text{ kg})(9.8 \text{ m/s}^2 + 4.0 \times 10^3 \text{ m/s}^2) = 0.049$ N.

(c) $\dfrac{F}{w} = \dfrac{0.049 \text{ N}}{(12.3 \times 10^{-6} \text{ kg})(9.8 \text{ m/s}^2)} = 410$; $F = 410w$.

EVALUATE: Because the force from the ground is huge compared to the weight of the froghopper, it produces an acceleration of around 400g!

4.51. **IDENTIFY:** He is in free-fall until he contacts the ground. Use the constant acceleration equations and apply $\Sigma \vec{F} = m\vec{a}$.

SET UP: Take $+y$ downward. While he is in the air, before he touches the ground, his acceleration is $a_y = 9.80 \text{ m/s}^2$.

EXECUTE: **(a)** $v_{0y} = 0$, $y - y_0 = 3.10$ m, and $a_y = 9.80 \text{ m/s}^2$. $v_y^2 = v_{0y}^2 + 2a_y(y - y_0)$ gives

$v_y = \sqrt{2a_y(y - y_0)} = \sqrt{2(9.80 \text{ m/s}^2)(3.10 \text{ m})} = 7.79$ m/s

(b) $v_{0y} = 7.79$ m/s, $v_y = 0$, $y - y_0 = 0.60$ m. $v_y^2 = v_{0y}^2 + 2a_y(y - y_0)$ gives

$a_y = \dfrac{v_y^2 - v_{0y}^2}{2(y - y_0)} = \dfrac{0 - (7.79 \text{ m/s})^2}{2(0.60 \text{ m})} = -50.6$ m/s^2. The acceleration is upward.

(c) The free-body diagram is given in Fig. 4.51. $\vec{F}$ is the force the ground exerts on him.

$\Sigma F_y = ma_y$ gives $mg - F = -ma$. $F = m(g + a) = (75.0 \text{ kg})(9.80 \text{ m/s}^2 + 50.6 \text{ m/s}^2) = 4.53 \times 10^3$ N, upward.

$$\frac{F}{w} = \frac{4.53 \times 10^3 \text{ N}}{(75.0 \text{ kg})(9.80 \text{ m/s}^2)} \quad \text{so,} \quad F = 6.16w = 6.16mg.$$

By Newton's third law, the force his feet exert on the ground is $-\vec{F}$.

EVALUATE: The force the ground exerts on him is about six times his weight.

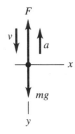

Figure 4.51

4.55.   **IDENTIFY:** Apply $\Sigma \vec{F} = m\vec{a}$ to the barbell and to the athlete. Use the motion of the barbell to calculate its acceleration.

SET UP: Let $+y$ be upward.

EXECUTE: (a) The free-body diagrams for the baseball and for the athlete are sketched in Figure 4.55.

(b) The athlete's weight is $mg = (90.0 \text{ kg})(9.80 \text{ m/s}^2) = 882$ N. The upward acceleration of the barbell is

found from $y - y_0 = v_{0y}t + \frac{1}{2}a_yt^2$. $a_y = \dfrac{2(y - y_0)}{t^2} = \dfrac{2(0.600 \text{ m})}{(1.6 \text{ s})^2} = 0.469$ m/s^2. The force needed to lift the

barbell is given by $F_{\text{lift}} - w_{\text{barbell}} = ma_y$. The barbell's mass is $(490 \text{ N})/(9.80 \text{ m/s}^2) = 50.0$ kg, so

$F_{\text{lift}} = w_{\text{barbell}} + ma = 490 \text{ N} + (50.0 \text{ kg})(0.469 \text{ m/s}^2) = 490 \text{ N} + 23 \text{ N} = 513$ N.

The athlete is not accelerating, so $F_{\text{floor}} - F_{\text{lift}} - w_{\text{athlete}} = 0$. $F_{\text{floor}} = F_{\text{lift}} + w_{\text{athlete}} = 513 \text{ N} + 882 \text{ N} = 1395$ N.

EVALUATE: Since the athlete pushes upward on the barbell with a force greater than its weight, the barbell pushes down on him and the normal force on the athlete is greater than the total weight, 1372 N, of the athlete plus barbell.

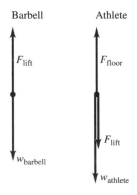

Figure 4.55

4.57.   **IDENTIFY:**   The system is accelerating, so we apply Newton's second law to each box and can use the constant acceleration kinematics for formulas to find the acceleration.

SET UP: First use the constant acceleration kinematics for formulas to find the acceleration of the system. Then apply $\sum F = ma$ to each box.

EXECUTE: **(a)** The kinematics formula for $y(t)$ gives

$$a_y = \frac{2(y - y_0)}{t^2} = \frac{2(12.0 \text{ m})}{(4.0 \text{ s})^2} = 1.5 \text{ m/s}^2. \text{ For box } B, \; mg - T = ma \text{ and}$$

$$m = \frac{T}{g - a} = \frac{36.0 \text{ N}}{9.8 \text{ m/s}^2 - 1.5 \text{ m/s}^2} = 4.34 \text{ kg}.$$

(b) For box A, $T + mg - F = ma$ and $m = \dfrac{F - T}{g - a} = \dfrac{80.0 \text{ N} - 36.0 \text{ N}}{9.8 \text{ m/s}^2 - 1.5 \text{ m/s}^2} = 5.30 \text{ kg}.$

EVALUATE: The boxes have the same acceleration but experience different forces because they have different masses.

4.59.   **IDENTIFY:** $F_x = ma_x$ and $a_x = \dfrac{d^2 x}{dt^2}$.

SET UP: $\dfrac{d}{dt}(t^n) = nt^{n-1}$

EXECUTE: The velocity as a function of time is $v_x(t) = A - 3Bt^2$ and the acceleration as a function of time is $a_x(t) = -6Bt$, and so the force as a function of time is $F_x(t) = ma(t) = -6mBt$.

EVALUATE: Since the acceleration is along the x-axis, the force is along the x-axis.

4.61.   **IDENTIFY:**   The rocket accelerates due to a variable force, so we apply Newton's second law. But the acceleration will not be constant because the force is not constant.

SET UP: We can use $a_x = F_x/m$ to find the acceleration, but must integrate to find the velocity and then the distance the rocket travels.

EXECUTE: Using $a_x = F_x/m$ gives $a_x(t) = \dfrac{(16.8 \text{ N/s})t}{45.0 \text{ kg}} = (0.3733 \text{ m/s}^3)t.$ Now integrate the acceleration

to get the velocity, and then integrate the velocity to get the distance moved.

$v_x(t) = v_0 + \displaystyle\int_0^t a_x(t')\,dt' = (0.1867 \text{ m/s}^3)t^2$ and $x - x_0 = \displaystyle\int_0^t v(t')\,dt' = (0.06222 \text{ m/s}^3)t^3.$ At $t = 5.00$ s,

$x - x_0 = 7.78$ m.

EVALUATE: The distance moved during the next 5.0 s would be considerably greater because the acceleration is increase with time.

APPLYING NEWTON'S LAWS

5

5.5. **IDENTIFY:** Apply $\Sigma \vec{F} = m\vec{a}$ to the frame.

SET UP: Let w be the weight of the frame. Since the two wires make the same angle with the vertical, the tension is the same in each wire. $T = 0.75w$.

EXECUTE: The vertical component of the force due to the tension in each wire must be half of the weight, and this in turn is the tension multiplied by the cosine of the angle each wire makes with the vertical.

$$\frac{w}{2} = \frac{3w}{4}\cos\theta \text{ and } \theta = \arccos\frac{2}{3} = 48°.$$

EVALUATE: If $\theta = 0°$, $T = w/2$ and $T \rightarrow \infty$ as $\theta \rightarrow 90°$. Therefore, there must be an angle where $T = 3w/4$.

5.7. **IDENTIFY:** Apply $\Sigma \vec{F} = m\vec{a}$ to the object and to the knot where the cords are joined.

SET UP: Let $+y$ be upward and $+x$ be to the right.

EXECUTE: **(a)** $T_C = w$, $T_A \sin 30° + T_B \sin 45° = T_C = w$, and $T_A \cos 30° - T_B \cos 45° = 0$. Since $\sin 45° = \cos 45°$, adding the last two equations gives $T_A(\cos 30° + \sin 30°) = w$, and so

$$T_A = \frac{w}{1.366} = 0.732w. \text{ Then, } T_B = T_A\frac{\cos 30°}{\cos 45°} = 0.897w.$$

(b) Similar to part (a), $T_C = w$, $-T_A \cos 60° + T_B \sin 45° = w$, and $T_A \sin 60° - T_B \cos 45° = 0$.

Adding these two equations, $T_A = \dfrac{w}{(\sin 60° - \cos 60°)} = 2.73w$, and $T_B = T_A\dfrac{\sin 60°}{\cos 45°} = 3.35w$.

EVALUATE: In part (a), $T_A + T_B > w$ since only the vertical components of T_A and T_B hold the object against gravity. In part (b), since T_A has a downward component T_B is greater than w.

5.9. **IDENTIFY:** Since the velocity is constant, apply Newton's first law to the piano. The push applied by the man must oppose the component of gravity down the incline.

SET UP: The free-body diagrams for the two cases are shown in Figures 5.9a and b. $\vec{F}$ is the force applied by the man. Use the coordinates shown in the figure.

EXECUTE: **(a)** $\Sigma F_x = 0$ gives $F - w\sin 11.0° = 0$ and $F = (180 \text{ kg})(9.80 \text{ m/s}^2)\sin 11.0° = 337$ N.

(b) $\Sigma F_y = 0$ gives $n\cos 11.0° - w = 0$ and $n = \dfrac{w}{\cos 11.0°}$. $\Sigma F_x = 0$ gives $F - n\sin 11.0° = 0$ and

$$F = \left(\frac{w}{\cos 11.0°}\right)\sin 11.0° = w\tan 11.0° = 343 \text{ N}.$$

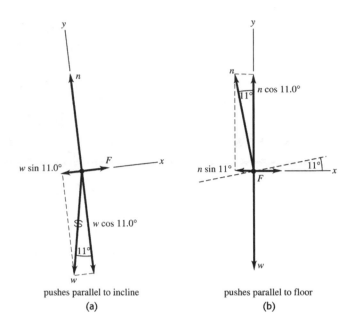

pushes parallel to incline   pushes parallel to floor
(a)   (b)

Figure 5.9a, b

5.11.   **IDENTIFY:** We apply Newton's second law to the rocket and the astronaut in the rocket. A constant force means we have constant acceleration, so we can use the standard kinematics equations.

   SET UP: The free-body diagrams for the rocket (weight w_r) and astronaut (weight w) are given in Figures 5.11a and 5.11b. F_T is the thrust and n is the normal force the rocket exerts on the astronaut. The speed of sound is 331 m/s. We use $\Sigma F_y = ma_y$ and $v = v_0 + at$.

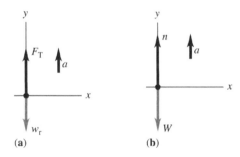

(a)   (b)

Figure 5.11

   EXECUTE: **(a)** Apply $\Sigma F_y = ma_y$ to the rocket: $F_T - w_r = ma$. $a = 4g$ and $w_r = mg$, so

$$F = m(5g) = (2.25 \times 10^6 \text{ kg})(5)(9.80 \text{ m/s}^2) = 1.10 \times 10^8 \text{ N}.$$

   (b) Apply $\Sigma F_y = ma_y$ to the astronaut: $n - w = ma$. $a = 4g$ and $m = \dfrac{w}{g}$, so $n = w + \left(\dfrac{w}{g}\right)(4g) = 5w$.

   (c) $v_0 = 0$, $v = 331$ m/s and $a = 4g = 39.2$ m/s^2. $v = v_0 + at$ gives $t = \dfrac{v - v_0}{a} = \dfrac{331 \text{ m/s}}{39.2 \text{ m/s}^2} = 8.4$ s.

   EVALUATE: The 8.4 s is probably an unrealistically short time to reach the speed of sound because you would not want your astronauts at the brink of blackout during a launch.

5.13.   **IDENTIFY:** Use the kinematic information to find the acceleration of the capsule and the stopping time. Use Newton's second law to find the force F that the ground exerted on the capsule during the crash.

   SET UP: Let $+y$ be upward. 311 km/h = 86.4 m/s. The free-body diagram for the capsule is given in Figure 5.13.

   EXECUTE: $y - y_0 = -0.810$ m, $v_{0y} = -86.4$ m/s, $v_y = 0$. $v_y^2 = v_{0y}^2 + 2a_y(y - y_0)$ gives

$$a_y = \frac{v_y^2 - v_{0y}^2}{2(y - y_0)} = \frac{0 - (-86.4 \text{ m/s})^2}{2(-0.810) \text{ m}} = 4610 \text{ m/s}^2 = 470g.$$

(b) $\Sigma F_y = ma_y$ applied to the capsule gives $F - mg = ma$ and

$$F = m(g + a) = (210 \text{ kg})(9.80 \text{ m/s}^2 + 4610 \text{ m/s}^2) = 9.70 \times 10^5 \text{ N} = 471w.$$

(c) $y - y_0 = \left(\frac{v_{0y} + v_y}{2} \right) t$ gives $t = \frac{2(y - y_0)}{v_{0y} + v_y} = \frac{2(-0.810 \text{ m})}{-86.4 \text{ m/s} + 0} = 0.0187 \text{ s}$

EVALUATE: The upward force exerted by the ground is much larger than the weight of the capsule and stops the capsule in a short amount of time. After the capsule has come to rest, the ground still exerts a force mg on the capsule, but the large 9.70×10^5 N force is exerted only for 0.0187 s.

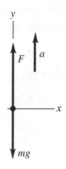

Figure 5.13

5.15. **IDENTIFY:** Apply $\Sigma \vec{F} = m\vec{a}$ to the load of bricks and to the counterweight. The tension is the same at each end of the rope. The rope pulls up with the same force (T) on the bricks and on the counterweight. The counterweight accelerates downward and the bricks accelerate upward; these accelerations have the same magnitude.

(a) SET UP: The free-body diagrams for the bricks and counterweight are given in Figure 5.15.

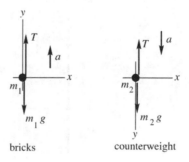

bricks   counterweight

Figure 5.15

(b) EXECUTE: Apply $\Sigma F_y = ma_y$ to each object. The acceleration magnitude is the same for the two objects. For the bricks take $+y$ to be upward since $\vec{a}$ for the bricks is upward. For the counterweight take $+y$ to be downward since $\vec{a}$ is downward.

bricks: $\Sigma F_y = ma_y$

$T - m_1 g = m_1 a$

counterweight: $\Sigma F_y = ma_y$

$m_2 g - T = m_2 a$

Add these two equations to eliminate T:

$$(m_2 - m_1)g = (m_1 + m_2)a$$

$$a = \left(\frac{m_2 - m_1}{m_1 + m_2}\right)g = \left(\frac{28.0\ \text{kg} - 15.0\ \text{kg}}{15.0\ \text{kg} + 28.0\ \text{kg}}\right)(9.80\ \text{m/s}^2) = 2.96\ \text{m/s}^2$$

(c) $T - m_1 g = m_1 a$ gives $T = m_1(a + g) = (15.0\ \text{kg})(2.96\ \text{m/s}^2 + 9.80\ \text{m/s}^2) = 191\ \text{N}$

As a check, calculate T using the other equation.

$m_2 g - T = m_2 a$ gives $T = m_2(g - a) = 28.0\ \text{kg}(9.80\ \text{m/s}^2 - 2.96\ \text{m/s}^2) = 191\ \text{N}$, which checks.

EVALUATE: The tension is 1.30 times the weight of the bricks; this causes the bricks to accelerate upward. The tension is 0.696 times the weight of the counterweight; this causes the counterweight to accelerate downward. If $m_1 = m_2$, $a = 0$ and $T = m_1 g = m_2 g$. In this special case the objects don't move. If $m_1 = 0$, $a = g$ and $T = 0$; in this special case the counterweight is in free fall. Our general result is correct in these two special cases.

5.19.   IDENTIFY: The maximum tension in the chain is at the top of the chain. Apply $\Sigma \vec{F} = m\vec{a}$ to the composite object of chain and boulder. Use the constant acceleration kinematic equations to relate the acceleration to the time.

SET UP: Let $+y$ be upward. The free-body diagram for the composite object is given in Figure 5.19.

$T = 2.50 w_{\text{chain}}$. $m_{\text{tot}} = m_{\text{chain}} + m_{\text{boulder}} = 1325\ \text{kg}$.

EXECUTE: (a) $\Sigma F_y = ma_y$ gives $T - m_{\text{tot}}g = m_{\text{tot}}a$.

$$a = \frac{T - m_{\text{tot}}g}{m_{\text{tot}}} = \frac{2.50 m_{\text{chain}}g - m_{\text{tot}}g}{m_{\text{tot}}} = \left(\frac{2.50 m_{\text{chain}}}{m_{\text{tot}}} - 1\right)g$$

$$a = \left(\frac{2.50[575\ \text{kg}]}{1325\ \text{kg}} - 1\right)(9.80\ \text{m/s}^2) = 0.832\ \text{m/s}^2$$

(b) Assume the acceleration has its maximum value: $a_y = 0.832\ \text{m/s}^2$, $y - y_0 = 125\ \text{m}$ and $v_{0y} = 0$.

$$y - y_0 = v_{0y}t + \tfrac{1}{2}a_y t^2 \text{ gives } t = \sqrt{\frac{2(y - y_0)}{a_y}} = \sqrt{\frac{2(125\ \text{m})}{0.832\ \text{m/s}^2}} = 17.3\ \text{s}$$

EVALUATE: The tension in the chain is $T = 1.41 \times 10^4\ \text{N}$ and the total weight is $1.30 \times 10^4\ \text{N}$. The upward force exceeds the downward force and the acceleration is upward.

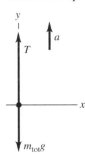

Figure 5.19

5.21.   IDENTIFY: While the person is in contact with the ground, he is accelerating upward and experiences two forces: gravity downward and the upward force of the ground. Once he is in the air, only gravity acts on him so he accelerates downward. Newton's second law applies during the jump (and at all other times).

SET UP: Take $+y$ to be upward. After he leaves the ground the person travels upward 60 cm and his acceleration is $g = 9.80\ \text{m/s}^2$, downward. His weight is w so his mass is w/g. $\Sigma F_y = ma_y$ and $v_y^2 = v_{0y}^2 + 2a_y(y - y_0)$ apply to the jumper.

EXECUTE: (a) $v_y = 0$ (at the maximum height), $y - y_0 = 0.60\ \text{m}$, $a_y = -9.80\ \text{m/s}^2$.

$v_y^2 = v_{0y}^2 + 2a_y(y - y_0)$ gives $v_{0y} = \sqrt{-2a_y(y - y_0)} = \sqrt{-2(-9.80\ \text{m/s}^2)(0.60\ \text{m})} = 3.4\ \text{m/s}$.

(b) The free-body diagram for the person while he is pushing up against the ground is given in Figure 5.21.

(c) For the jump, $v_{0y} = 0$, $v_y = 3.4$ m/s (from part (a)), and $y - y_0 = 0.50$ m.

$$v_y^2 = v_{0y}^2 + 2a_y(y - y_0) \text{ gives } a_y = \frac{v_y^2 - v_{0y}^2}{2(y - y_0)} = \frac{(3.4 \text{ m/s})^2 - 0}{2(0.50 \text{ m})} = 11.6 \text{ m/s}^2. \quad \Sigma F_y = ma_y \text{ gives } n - w = ma.$$

$$n = w + ma = w\left(1 + \frac{a}{g}\right) = 2.2w.$$

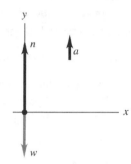

Figure 5.21

EVALUATE: To accelerate the person upward during the jump, the upward force from the ground must exceed the downward pull of gravity. The ground pushes up on him because he pushes down on the ground.

5.23. **IDENTIFY:** We know the external forces on the box and want to find the distance it moves and its speed. The force is not constant, so the acceleration will not be constant, so we cannot use the standard constant-acceleration kinematics formulas. But Newton's second law will apply.

SET UP: First use Newton's second law to find the acceleration as a function of time: $a_x(t) = \dfrac{F_x}{m}$. Then integrate the acceleration to find the velocity as a function of time, and next integrate the velocity to find the position as a function of time.

EXECUTE: Let $+x$ be to the right. $a_x(t) = \dfrac{F_x}{m} = \dfrac{(-6.00 \text{ N/s}^2)t^2}{2.00 \text{ kg}} = -(3.00 \text{ m/s}^4)t^2$. Integrate the acceleration to find the velocity as a function of time: $v_x(t) = -(1.00 \text{ m/s}^4)t^3 + 9.00 \text{ m/s}$. Next integrate the velocity to find the position as a function of time: $x(t) = -(0.250 \text{ m/s}^4)t^4 + (9.00 \text{ m/s})t$. Now use the given values of time.

(a) $v_x = 0$ when $(1.00 \text{ m/s}^4)t^3 = 9.00 \text{ m/s}$. This gives $t = 2.08$ s. At $t = 2.08$ s,

$x = (9.00 \text{ m/s})(2.08 \text{ s}) - (0.250 \text{ m/s}^4)(2.08 \text{ s})^4 = 18.72 \text{ m} - 4.68 \text{ m} = 14.0 \text{ m}$.

(b) At $t = 3.00$ s, $v_x(t) = -(1.00 \text{ m/s}^4)(3.00 \text{ s})^3 + 9.00 \text{ m/s} = -18.0 \text{ m/s}$, so the speed is 18.0 m/s.

EVALUATE: The box starts out moving to the right. But because the acceleration is to the left, it reverses direction and v_x is negative in part (b).

5.29. **IDENTIFY:** Apply $\Sigma \vec{F} = m\vec{a}$ to the crate. $f_s \leq \mu_s n$ and $f_k = \mu_k n$.

SET UP: Let $+y$ be upward and let $+x$ be in the direction of the push. Since the floor is horizontal and the push is horizontal, the normal force equals the weight of the crate: $n = mg = 441$ N. The force it takes to start the crate moving equals max f_s and the force required to keep it moving equals f_k.

EXECUTE: **(a)** max $f_s = 313$ N, so $\mu_s = \dfrac{313 \text{ N}}{441 \text{ N}} = 0.710$. $f_k = 208$ N, so $\mu_k = \dfrac{208 \text{ N}}{441 \text{ N}} = 0.472$.

(b) The friction is kinetic. $\Sigma F_x = ma_x$ gives $F - f_k = ma$ and

$F = f_k + ma = 208 \text{ N} + (45.0 \text{ kg})(1.10 \text{ m/s}^2) = 258$ N.

(c) (i) The normal force now is $mg = 72.9$ N. To cause it to move,

$F = \max f_s = \mu_s n = (0.710)(72.9 \text{ N}) = 51.8$ N.

(ii) $F = f_k + ma$ and $a = \dfrac{F - f_k}{m} = \dfrac{258 \text{ N} - (0.472)(72.9 \text{ N})}{45.0 \text{ kg}} = 4.97 \text{ m/s}^2$

EVALUATE: The kinetic friction force is independent of the speed of the object. On the moon, the mass of the crate is the same as on earth, but the weight and normal force are less.

5.31. **IDENTIFY:** Apply $\Sigma\vec{F} = m\vec{a}$ to the composite object consisting of the two boxes and to the top box. The friction the ramp exerts on the lower box is kinetic friction. The upper box doesn't slip relative to the lower box, so the friction between the two boxes is static. Since the speed is constant the acceleration is zero.

SET UP: Let $+x$ be up the incline. The free-body diagrams for the composite object and for the upper box are given in Figures 5.31a and b. The slope angle ϕ of the ramp is given by $\tan\phi = \dfrac{2.50 \text{ m}}{4.75 \text{ m}}$, so $\phi = 27.76°$. Since the boxes move down the ramp, the kinetic friction force exerted on the lower box by the ramp is directed up the incline. To prevent slipping relative to the lower box the static friction force on the upper box is directed up the incline. $m_{tot} = 32.0 \text{ kg} + 48.0 \text{ kg} = 80.0 \text{ kg}$.

EXECUTE: (a) $\Sigma F_y = ma_y$ applied to the composite object gives $n_{tot} = m_{tot}g\cos\phi$ and $f_k = \mu_k m_{tot}g\cos\phi$. $\Sigma F_x = ma_x$ gives $f_k + T - m_{tot}g\sin\phi = 0$ and

$T = (\sin\phi - \mu_k\cos\phi)m_{tot}g = (\sin 27.76° - [0.444]\cos 27.76°)(80.0 \text{ kg})(9.80 \text{ m/s}^2) = 57.1 \text{ N}.$

The person must apply a force of 57.1 N, directed up the ramp.

(b) $\Sigma F_x = ma_x$ applied to the upper box gives $f_s = mg\sin\phi = (32.0 \text{ kg})(9.80 \text{ m/s}^2)\sin 27.76° = 146 \text{ N}$, directed up the ramp.

EVALUATE: For each object the net force is zero.

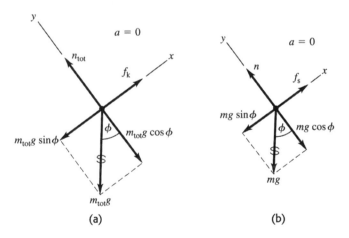

Figure 5.31

5.33. **IDENTIFY:** Use $\Sigma\vec{F} = m\vec{a}$ to find the acceleration that can be given to the car by the kinetic friction force. Then use a constant acceleration equation.

SET UP: Take $+x$ in the direction the car is moving.

EXECUTE: (a) The free-body diagram for the car is shown in Figure 5.33. $\Sigma F_y = ma_y$ gives $n = mg$.

$\Sigma F_x = ma_x$ gives $-\mu_k n = ma_x$. $-\mu_k mg = ma_x$ and $a_x = -\mu_k g$. Then $v_x = 0$ and $v_x^2 = v_{0x}^2 + 2a_x(x - x_0)$

gives $(x - x_0) = -\dfrac{v_{0x}^2}{2a_x} = +\dfrac{v_{0x}^2}{2\mu_k g} = \dfrac{(28.7 \text{ m/s})^2}{2(0.80)(9.80 \text{ m/s}^2)} = 52.5 \text{ m}.$

(b) $v_{0x} = \sqrt{2\mu_k g(x - x_0)} = \sqrt{2(0.25)(9.80 \text{ m/s}^2)52.5 \text{ m}} = 16.0 \text{ m/s}$

EVALUATE: For constant stopping distance $\dfrac{v_{0x}^2}{\mu_k}$ is constant and v_{0x} is proportional to $\sqrt{\mu_k}$. The answer to part (b) can be calculated as $(28.7 \text{ m/s})\sqrt{0.25/0.80} = 16.0 \text{ m/s}$.

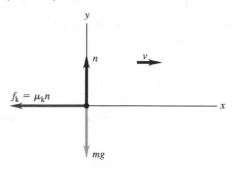

Figure 5.33

5.37. **IDENTIFY:** Apply $\Sigma \vec{F} = m\vec{a}$ to each block. The target variables are the tension T in the cord and the acceleration a of the blocks. Then a can be used in a constant acceleration equation to find the speed of each block. The magnitude of the acceleration is the same for both blocks.
SET UP: The system is sketched in Figure 5.37a.

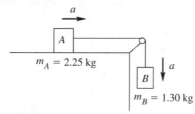

For each block take a positive coordinate direction to be the direction of the block's acceleration.

Figure 5.37a

block on the table: The free-body is sketched in Figure 5.37b.

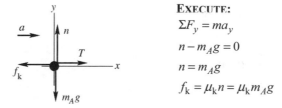

EXECUTE:
$\Sigma F_y = ma_y$
$n - m_A g = 0$
$n = m_A g$
$f_k = \mu_k n = \mu_k m_A g$

Figure 5.37b

$\Sigma F_x = ma_x$
$T - f_k = m_A a$
$T - \mu_k m_A g = m_A a$

SET UP: hanging block: The free-body is sketched in Figure 5.37c.

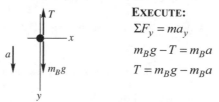

EXECUTE:
$\Sigma F_y = ma_y$
$m_B g - T = m_B a$
$T = m_B g - m_B a$

Figure 5.37c

(a) Use the second equation in the first

$$m_B g - m_B a - \mu_k m_A g = m_A a$$

$$(m_A + m_B)a = (m_B - \mu_k m_A)g$$

$$a = \frac{(m_B - \mu_k m_A)g}{m_A + m_B} = \frac{(1.30 \text{ kg} - (0.45)(2.25 \text{ kg}))(9.80 \text{ m/s}^2)}{2.25 \text{ kg} + 1.30 \text{ kg}} = 0.7937 \text{ m/s}^2$$

SET UP: Now use the constant acceleration equations to find the final speed. Note that the blocks have the same speeds. $x - x_0 = 0.0300$ m, $a_x = 0.7937$ m/s^2, $v_{0x} = 0$, $v_x = ?$

$$v_x^2 = v_{0x}^2 + 2a_x(x - x_0)$$

EXECUTE: $v_x = \sqrt{2a_x(x - x_0)} = \sqrt{2(0.7937 \text{ m/s}^2)(0.0300 \text{ m})} = 0.218 \text{ m/s} = 21.8 \text{ cm/s}.$

(b) $T = m_B g - m_B a = m_B(g - a) = 1.30 \text{ kg}(9.80 \text{ m/s}^2 - 0.7937 \text{ m/s}^2) = 11.7 \text{ N}$

Or, to check, $T - \mu_k m_A g = m_A a.$

$T = m_A(a + \mu_k g) = 2.25 \text{ kg}(0.7937 \text{ m/s}^2 + (0.45)(9.80 \text{ m/s}^2)) = 11.7 \text{ N},$ which checks.

EVALUATE: The force T exerted by the cord has the same value for each block. $T < m_B g$ since the hanging block accelerates downward. Also, $f_k = \mu_k m_A g = 9.92 \text{ N}.$ $T > f_k$ and the block on the table accelerates in the direction of T.

5.39. **(a) IDENTIFY:** Apply $\Sigma \vec{F} = m\vec{a}$ to the crate. Constant v implies $a = 0$. Crate moving says that the friction is kinetic friction. The target variable is the magnitude of the force applied by the woman.
SET UP: The free-body diagram for the crate is sketched in Figure 5.39.

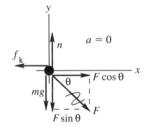

EXECUTE:
$$\Sigma F_y = ma_y$$
$$n - mg - F\sin\theta = 0$$
$$n = mg + F\sin\theta$$
$$f_k = \mu_k n = \mu_k mg + \mu_k F\sin\theta$$

Figure 5.39

$$\Sigma F_x = ma_x$$
$$F\cos\theta - f_k = 0$$
$$F\cos\theta - \mu_k mg - \mu_k F\sin\theta = 0$$
$$F(\cos\theta - \mu_k\sin\theta) = \mu_k mg$$
$$F = \frac{\mu_k mg}{\cos\theta - \mu_k\sin\theta}$$

(b) IDENTIFY and **SET UP:** "start the crate moving" means the same force diagram as in part (a), except that μ_k is replaced by μ_s. Thus $F = \dfrac{\mu_s mg}{\cos\theta - \mu_s\sin\theta}$.

EXECUTE: $F \to \infty$ if $\cos\theta - \mu_s\sin\theta = 0$. This gives $\mu_s = \dfrac{\cos\theta}{\sin\theta} = \dfrac{1}{\tan\theta}$.

EVALUATE: $\vec{F}$ has a downward component so $n > mg$. If $\theta = 0$ (woman pushes horizontally), $n = mg$ and $F = f_k = \mu_k mg$.

5.43. **IDENTIFY:** Apply $\Sigma\vec{F} = m\vec{a}$ to one of the masses. The mass moves in a circular path, so has acceleration $a_{rad} = \dfrac{v^2}{R}$, directed toward the center of the path.

SET UP: In each case, $R = 0.200$ m. In part (a), let $+x$ be toward the center of the circle, so $a_x = a_{rad}$. In part (b) let $+y$ be toward the center of the circle, so $a_y = a_{rad}$. $+y$ is downward when the mass is at the top of the circle and $+y$ is upward when the mass is at the bottom of the circle. Since a_{rad} has its greatest possible value, $\vec{F}$ is in the direction of $\vec{a}_{rad}$ at both positions.

EXECUTE: **(a)** $\Sigma F_x = ma_x$ gives $F = ma_{rad} = m\dfrac{v^2}{R}$. $F = 75.0$ N and

$$v = \sqrt{\dfrac{FR}{m}} = \sqrt{\dfrac{(75.0 \text{ N})(0.200 \text{ m})}{1.15 \text{ kg}}} = 3.61 \text{ m/s}.$$

(b) The free-body diagrams for a mass at the top of the path and at the bottom of the path are given in Figure 5.43. At the top, $\Sigma F_y = ma_y$ gives $F = ma_{rad} - mg$ and at the bottom it gives $F = mg + ma_{rad}$. For a given rotation rate and hence value of a_{rad}, the value of F required is larger at the bottom of the path.

(c) $F = mg + ma_{rad}$ so $\dfrac{v^2}{R} = \dfrac{F}{m} - g$ and

$$v = \sqrt{R\left(\dfrac{F}{m} - g\right)} = \sqrt{(0.200 \text{ m})\left(\dfrac{75.0 \text{ N}}{1.15 \text{ kg}} - 9.80 \text{ m/s}^2\right)} = 3.33 \text{ m/s}$$

EVALUATE: The maximum speed is less for the vertical circle. At the bottom of the vertical path $\vec{F}$ and the weight are in opposite directions so F must exceed ma_{rad} by an amount equal to mg. At the top of the vertical path F and mg are in the same direction and together provide the required net force, so F must be larger at the bottom.

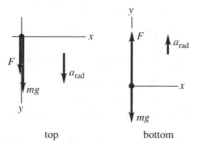

top   bottom

Figure 5.43

5.47.   **IDENTIFY:**   Apply $\Sigma\vec{F} = m\vec{a}$ to the composite object of the person plus seat. This object moves in a horizontal circle and has acceleration a_{rad}, directed toward the center of the circle.

SET UP: The free-body diagram for the composite object is given in Figure 5.47. Let $+x$ be to the right, in the direction of $\vec{a}_{rad}$. Let $+y$ be upward. The radius of the circular path is $R = 7.50$ m. The total mass is $(255 \text{ N} + 825 \text{ N})/(9.80 \text{ m/s}^2) = 110.2$ kg. Since the rotation rate is 32.0 rev/min $= 0.5333$ rev/s, the period T is $\dfrac{1}{0.5333 \text{ rev/s}} = 1.875$ s.

EXECUTE: $\Sigma F_y = ma_y$ gives $T_A \cos 40.0° - mg = 0$ and $T_A = \dfrac{mg}{\cos 40.0°} = \dfrac{255 \text{ N} + 825 \text{ N}}{\cos 40.0°} = 1410$ N.

$\Sigma F_x = ma_x$ gives $T_A \sin 40.0° + T_B = ma_{rad}$ and

$$T_B = m\dfrac{4\pi^2 R}{T^2} - T_A \sin 40.0° = (110.2 \text{ kg})\dfrac{4\pi^2 (7.50 \text{ m})}{(1.875 \text{ s})^2} - (1410 \text{ N})\sin 40.0° = 8370 \text{ N}$$

The tension in the horizontal cable is 8370 N and the tension in the other cable is 1410 N.

EVALUATE: The weight of the composite object is 1080 N. The tension in cable A is larger than this since its vertical component must equal the weight. $ma_{rad} = 9280$ N. The tension in cable B is less than this because part of the required inward force comes from a component of the tension in cable A.

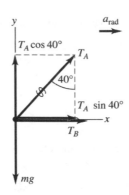

Figure 5.47

5.49.   **IDENTIFY:** The acceleration due to circular motion is $a_{rad} = \dfrac{4\pi^2 R}{T^2}$.

SET UP: $R = 400$ m. $1/T$ is the number of revolutions per second.

EXECUTE: (a) Setting $a_{rad} = g$ and solving for the period T gives

$$T = 2\pi\sqrt{\frac{R}{g}} = 2\pi\sqrt{\frac{400 \text{ m}}{9.80 \text{ m/s}^2}} = 40.1 \text{ s},$$

so the number of revolutions per minute is $(60 \text{ s/min})/(40.1 \text{ s}) = 1.5$ rev/min.

(b) The lower acceleration corresponds to a longer period, and hence a lower rotation rate, by a factor of the square root of the ratio of the accelerations, $T' = (1.5 \text{ rev/min}) \times \sqrt{3.70/9.8} = 0.92$ rev/min.

EVALUATE: In part (a) the tangential speed of a point at the rim is given by $a_{rad} = \dfrac{v^2}{R}$, so

$v = \sqrt{Ra_{rad}} = \sqrt{Rg} = 62.6$ m/s; the space station is rotating rapidly.

5.55.   **IDENTIFY:** Since the arm is swinging in a circle, objects in it are accelerated toward the center of the circle, and Newton's second law applies to them.

SET UP: $R = 0.700$ m. A 45° angle is $\frac{1}{8}$ of a full rotation, so in $\frac{1}{2}$ s a hand travels through a distance of $\frac{1}{8}(2\pi R)$. In (c) use coordinates where $+y$ is upward, in the direction of $\vec{a}_{rad}$ at the bottom of the swing.

The acceleration is $a_{rad} = \dfrac{v^2}{R}$.

EXECUTE: (a) $v = \dfrac{1}{8}\left(\dfrac{2\pi R}{0.50 \text{ s}}\right) = 1.10$ m/s and $a_{rad} = \dfrac{v^2}{R} = \dfrac{(1.10 \text{ m/s})^2}{0.700 \text{ m}} = 1.73$ m/s^2.

(b) The free-body diagram is shown in Figure 5.55. F is the force exerted by the blood vessel.

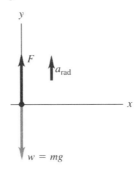

Figure 5.55

(c) $\Sigma F_y = ma_y$ gives $F - w = ma_{rad}$ and

$F = m(g + a_{rad}) = (1.00 \times 10^{-3} \text{ kg})(9.80 \text{ m/s}^2 + 1.73 \text{ m/s}^2) = 1.15 \times 10^{-2}$ N, upward.

(d) When the arm hangs vertically and is at rest, $a_{rad} = 0$ so $F = w = mg = 9.8 \times 10^{-3}$ N.

EVALUATE: The acceleration of the hand is only about 20% of g, so the increase in the force on the blood drop when the arm swings is about 20%.

5.57.   **IDENTIFY:**   Apply $\Sigma \vec{F} = m\vec{a}$ to the knot.

SET UP: $a = 0$. Use coordinates with axes that are horizontal and vertical.

EXECUTE: (a) The free-body diagram for the knot is sketched in Figure 5.57.

T_1 is more vertical so supports more of the weight and is larger. You can also see this from $\Sigma F_x = ma_x$:

$T_2 \cos 40° - T_1 \cos 60° = 0$. $T_2 \cos 40° - T_1 \cos 60° = 0$.

(b) T_1 is larger so set $T_1 = 5000$ N. Then $T_2 = T_1/1.532 = 3263.5$ N. $\Sigma F_y = ma_y$ gives

$T_1 \sin 60° + T_2 \sin 40° = w$ and $w = 6400$ N.

EVALUATE: The sum of the vertical components of the two tensions equals the weight of the suspended object. The sum of the tensions is greater than the weight.

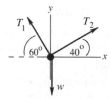

Figure 5.57

5.59.   **IDENTIFY:**   Apply Newton's first law to the ball. The force of the wall on the ball and the force of the ball on the wall are related by Newton's third law.

SET UP: The forces on the ball are its weight, the tension in the wire, and the normal force applied by the wall.

To calculate the angle ϕ that the wire makes with the wall, use Figure 5.59a. $\sin \phi = \dfrac{16.0 \text{ cm}}{46.0 \text{ cm}}$ and $\phi = 20.35°$

EXECUTE: (a) The free-body diagram is shown in Figure 5.59b. Use the x and y coordinates shown in the

figure. $\Sigma F_y = 0$ gives $T \cos \phi - w = 0$ and $T = \dfrac{w}{\cos \phi} = \dfrac{(45.0 \text{ kg})(9.80 \text{ m/s}^2)}{\cos 20.35°} = 470$ N

(b) $\Sigma F_x = 0$ gives $T \sin \phi - n = 0$. $n = (470 \text{ N}) \sin 20.35° = 163$ N. By Newton's third law, the force the ball exerts on the wall is 163 N, directed to the right.

EVALUATE: $n = \left(\dfrac{w}{\cos \phi} \right) \sin \phi = w \tan \phi$. As the angle ϕ decreases (by increasing the length of the wire),

T decreases and n decreases.

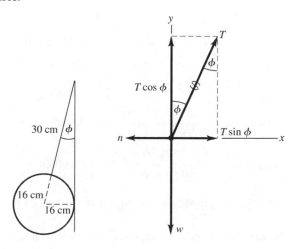

Figure 5.59a, b

5.63. **IDENTIFY:** We know the forces on the box and want to find information about its position and velocity. Newton's second law will give us the box's acceleration.

SET UP: $a_y(t) = \dfrac{\Sigma F_y}{m}$. We can integrate the acceleration to find the velocity and the velocity to find the position. At an altitude of several hundred meters, the acceleration due to gravity is essentially the same as it is at the earth's surface.

EXECUTE: Let $+y$ be upward. Newton's second law gives $T - mg = ma_y$, so $a_y(t) = (12.0 \text{ m/s}^3)t - 9.8 \text{ m/s}^2$. Integrating the acceleration gives $v_y(t) = (6.00 \text{ m/s}^3)t^2 - (9.8 \text{ m/s}^2)t$.

(a) (i) At $t = 1.00$ s, $v_y = -3.80$ m/s. (ii) At $t = 3.00$ s, $v_y = 24.6$ m/s.

(b) Integrating the velocity gives $y - y_0 = (2.00 \text{ m/s}^3)t^3 - (4.9 \text{ m/s}^2)t^2$. $v_y = 0$ at $t = 1.63$ s. At $t = 1.63$ s, $y - y_0 = 8.71 \text{ m} - 13.07 \text{ m} = -4.36$ m.

(c) Setting $y - y_0 = 0$ and solving for t gives $t = 2.45$ s.

EVALUATE: The box accelerates and initially moves downward until the tension exceeds the weight of the box. Once the tension exceeds the weight, the box will begin to accelerate upward and will eventually move upward, as we saw in part (b).

5.65. **IDENTIFY:** The system of boxes is accelerating, so we apply Newton's second law to each box. The friction is kinetic friction. We can use the known acceleration to find the tension and the mass of the second box.

SET UP: The force of friction is $f_k = \mu_k n$, $\Sigma F_x = ma_x$ applies to each box, and the forces perpendicular to the surface balance.

EXECUTE: **(a)** Call the $+x$ axis along the surface. For the 5 kg block, the vertical forces balance, so $n + F\sin 53.1° - mg = 0$, which gives $n = 49.0 \text{ N} - 31.99 \text{ N} = 17.01$ N. The force of kinetic friction is $f_k = \mu_k n = 5.104$ N. Applying Newton's second law along the surface gives $F\cos 53.1° - T - f_k = ma$. Solving for T gives $T = F\cos 53.1° - f_k - ma = 24.02 \text{ N} - 5.10 \text{ N} - 7.50 \text{ N} = 11.4$ N.

(b) For the second box, $T - f_k = ma$. $T - \mu_k mg = ma$. Solving for m gives

$$m = \frac{T}{\mu_k g + a} = \frac{11.42 \text{ N}}{(0.3)(9.8 \text{ m/s}^2) + 1.5 \text{ m/s}^2} = 2.57 \text{ kg}.$$

EVALUATE: The normal force for box B is less than its weight due to the upward pull, but the normal force for box A is equal to its weight because the rope pulls horizontally on A.

5.69. **IDENTIFY:** $f = \mu_r n$. Apply $\Sigma \vec{F} = m\vec{a}$ to the tire.

SET UP: $n = mg$ and $f = ma$.

EXECUTE: $a_x = \dfrac{v^2 - v_0^2}{L}$, where L is the distance covered before the wheel's speed is reduced to half its original speed and $v = v_0/2$. $\mu_r = \dfrac{a}{g} = \dfrac{v_0^2 - v^2}{2Lg} = \dfrac{v_0^2 - \frac{1}{4}v_0^2}{2Lg} = \dfrac{3}{8}\dfrac{v_0^2}{Lg}$.

Low pressure, $L = 18.1$ m and $\dfrac{3}{8}\dfrac{(3.50 \text{ m/s})^2}{(18.1 \text{ m})(9.80 \text{ m/s}^2)} = 0.0259$.

High pressure, $L = 92.9$ m and $\dfrac{3}{8}\dfrac{(3.50 \text{ m/s})^2}{(92.9 \text{ m})(9.80 \text{ m/s}^2)} = 0.00505$.

EVALUATE: μ_r is inversely proportional to the distance L, so $\dfrac{\mu_{r1}}{\mu_{r2}} = \dfrac{L_2}{L_1}$.

5.73. **IDENTIFY:** Apply $\Sigma \vec{F} = m\vec{a}$ to each block. Use Newton's third law to relate forces on A and on B.

SET UP: Constant speed means $a = 0$.

EXECUTE: (a) Treat A and B as a single object of weight $w = w_A + w_B = 6.00$ N. The free-body diagram for this combined object is given in Figure 5.73a. $\Sigma F_y = ma_y$ gives $n = w = 6.00$ N. $f_k = \mu_k n = 1.80$ N. $\Sigma F_x = ma_x$ gives $F = f_k = 1.80$ N.

(b) The free-body force diagrams for blocks A and B are given in Figure 5.73b. n and f_k are the normal and friction forces applied to block B by the tabletop and are the same as in part (a). f_{kB} is the friction force that A applies to B. It is to the right because the force from A opposes the motion of B. n_B is the downward force that A exerts on B. f_{kA} is the friction force that B applies to A. It is to the left because block B wants A to move with it. n_A is the normal force that block B exerts on A. By Newton's third law, $f_{kB} = f_{kA}$ and these forces are in opposite directions. Also, $n_A = n_B$ and these forces are in opposite directions.

$\Sigma F_y = ma_y$ for block A gives $n_A = w_A = 2.40$ N, so $n_B = 2.40$ N.

$f_{kA} = \mu_k n_A = (0.300)(2.40 \text{ N}) = 0.720$ N, and $f_{kB} = 0.720$ N.

$\Sigma F_x = ma_x$ for block A gives $T = f_{kA} = 0.720$ N.

$\Sigma F_x = ma_x$ for block B gives $F = f_{kB} + f_k = 0.720 \text{ N} + 1.80 \text{ N} = 2.52$ N.

EVALUATE: In part (a) block A is at rest with respect to B and it has zero acceleration. There is no horizontal force on A besides friction, and the friction force on A is zero. A larger force F is needed in part (b), because of the friction force between the two blocks.

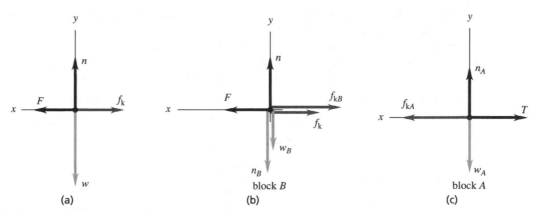

Figure 5.73a–c

5.75. **IDENTIFY:** The net force at any time is $F_{net} = ma$.

SET UP: At $t = 0$, $a = 62g$. The maximum acceleration is $140g$ at $t = 1.2$ ms.

EXECUTE: (a) $F_{net} = ma = 62mg = 62(210 \times 10^{-9} \text{ kg})(9.80 \text{ m/s}^2) = 1.3 \times 10^{-4}$ N. This force is 62 times the flea's weight.

(b) $F_{net} = 140mg = 2.9 \times 10^{-4}$ N, at $t = 1.2$ ms.

(c) Since the initial speed is zero, the maximum speed is the area under the $a_x - t$ graph. This gives 1.2 m/s.

EVALUATE: a is much larger than g and the net external force is much larger than the flea's weight.

5.77. **IDENTIFY:** $a = dv/dt$. Apply $\Sigma \vec{F} = m\vec{a}$ to yourself.

SET UP: The reading of the scale is equal to the normal force the scale applies to you.

EXECUTE: The elevator's acceleration is $a = \dfrac{dv(t)}{dt} = 3.0 \text{ m/s}^2 + 2(0.20 \text{ m/s}^3)t = 3.0 \text{ m/s}^2 + (0.40 \text{ m/s}^3)t$.

At $t = 4.0$ s, $a = 3.0 \text{ m/s}^2 + (0.40 \text{ m/s}^3)(4.0 \text{ s}) = 4.6 \text{ m/s}^2$. From Newton's second law, the net force on you is $F_{net} = F_{scale} - w = ma$ and $F_{scale} = w + ma = (64 \text{ kg})(9.8 \text{ m/s}^2) + (64 \text{ kg})(4.6 \text{ m/s}^2) = 920$ N.

EVALUATE: a increases with time, so the scale reading is increasing.

5.79.   **IDENTIFY:** Apply $\Sigma \vec{F} = m\vec{a}$ to the package. Calculate a and then use a constant acceleration equation to describe the motion.

SET UP: Let $+x$ be directed up the ramp.

EXECUTE: (a) $F_{\text{net}} = -mg\sin 37° - f_k = -mg\sin 37° - \mu_k mg\cos 37° = ma$ and

$$a = -(9.8 \text{ m/s}^2)(0.602 + (0.30)(0.799)) = -8.25 \text{ m/s}^2$$

Since we know the length of the slope, we can use $v_x^2 = v_{0x}^2 + 2a_x(x - x_0)$ with $x_0 = 0$ and $v_x = 0$ at the

top. $v_0^2 = -2ax = -2(-8.25 \text{ m/s}^2)(8.0 \text{ m}) = 132 \text{ m}^2/\text{s}^2$ and $v_0 = \sqrt{132 \text{ m}^2/\text{s}^2} = 11.5 \text{ m/s}$

(b) For the trip back down the slope, gravity and the friction force operate in opposite directions to each other. $F_{\text{net}} = -mg\sin 37° + \mu_k mg\cos 37° = ma$ and

$$a = g(-\sin 37° + 0.30\cos 37°) = (9.8 \text{ m/s}^2)((-0.602) + (0.30)(0.799)) = -3.55 \text{ m/s}^2.$$

Now we have $v_0 = 0, x_0 = -8.0$ m, $x = 0$ and

$$v^2 = v_0^2 + 2a(x - x_0) = 0 + 2(-3.55 \text{ m/s}^2)(-8.0 \text{ m}) = 56.8 \text{ m}^2/\text{s}^2, \text{ so } v = \sqrt{56.8 \text{ m}^2/\text{s}^2} = 7.54 \text{ m/s}.$$

EVALUATE: In both cases, moving up the incline and moving down the incline, the acceleration is directed down the incline. The magnitude of a is greater when the package is going up the incline, because $mg\sin 37°$ and f_k are in the same direction whereas when the package is going down these two forces are in opposite directions.

5.81.   **IDENTIFY:** Apply $\Sigma \vec{F} = m\vec{a}$ to the washer and to the crate. Since the washer is at rest relative to the crate, these two objects have the same acceleration.

SET UP: The free-body diagram for the washer is given in Figure 5.81.

EXECUTE: It's interesting to look at the string's angle measured from the perpendicular to the top of the crate. This angle is $\theta_{\text{string}} = 90° -$ angle measured from the top of the crate. The free-body diagram for the washer then leads to the following equations, using Newton's second law and taking the upslope direction as positive:

$$-m_w g\sin\theta_{\text{slope}} + T\sin\theta_{\text{string}} = m_w a \text{ and } T\sin\theta_{\text{string}} = m_w(a + g\sin\theta_{\text{slope}})$$

$$-m_w g\cos\theta_{\text{slope}} + T\cos\theta_{\text{string}} = 0 \text{ and } T\cos\theta_{\text{string}} = m_w g\cos\theta_{\text{slope}}$$

Dividing the two equations: $\tan\theta_{\text{string}} = \dfrac{a + g\sin\theta_{\text{slope}}}{g\cos\theta_{\text{slope}}}$

For the crate, the component of the weight along the slope is $-m_c g\sin\theta_{\text{slope}}$ and the normal force is $m_c g\cos\theta_{\text{slope}}$. Using Newton's second law again: $-m_c g\sin\theta_{\text{slope}} + \mu_k m_c g\cos\theta_{\text{slope}} = m_c a.$

$\mu_k = \dfrac{a + g\sin\theta_{\text{slope}}}{g\cos\theta_{\text{slope}}}$. This leads to the interesting observation that the string will hang at an angle whose

tangent is equal to the coefficient of kinetic friction:

$$\mu_k = \tan\theta_{\text{string}} = \tan(90° - 68°) = \tan 22° = 0.40.$$

EVALUATE: In the limit that $\mu_k \to 0$, $\theta_{\text{string}} \to 0$ and the string is perpendicular to the top of the crate. As μ_k increases, θ_{string} increases.

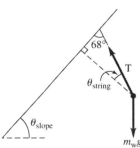

Figure 5.81

5.87.   **IDENTIFY:** Apply $\Sigma \vec{F} = m\vec{a}$ to the point where the three wires join and also to one of the balls. By symmetry the tension in each of the 35.0 cm wires is the same.

SET UP: The geometry of the situation is sketched in Figure 5.87a. The angle ϕ that each wire makes with the vertical is given by $\sin\phi = \dfrac{12.5 \text{ cm}}{47.5 \text{ cm}}$ and $\phi = 15.26°$. Let T_A be the tension in the vertical wire and let T_B be the tension in each of the other two wires. Neglect the weight of the wires. The free-body diagram for the left-hand ball is given in Figure 5.87b and for the point where the wires join in Figure 5.87c. n is the force one ball exerts on the other.

EXECUTE: **(a)** $\Sigma F_y = ma_y$ applied to the ball gives $T_B \cos\phi - mg = 0$.

$T_B = \dfrac{mg}{\cos\phi} = \dfrac{(15.0 \text{ kg})(9.80 \text{ m/s}^2)}{\cos 15.26°} = 152$ N. Then $\Sigma F_y = ma_y$ applied in Figure 5.87c gives

$T_A - 2T_B \cos\phi = 0$ and $T_A = 2(152 \text{ N})\cos\phi = 294$ N.

(b) $\Sigma F_x = ma_x$ applied to the ball gives $n - T_B \sin\phi = 0$ and $n = (152 \text{ N})\sin 15.26° = 40.0$ N.

EVALUATE: T_A equals the total weight of the two balls.

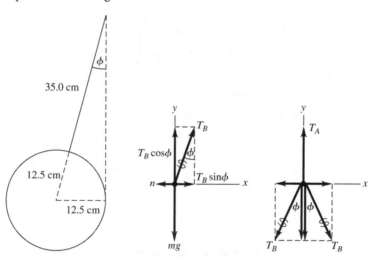

Figure 5.87a–c

5.89.   **IDENTIFY:** Apply $\Sigma \vec{F} = m\vec{a}$ to each block. Forces between the blocks are related by Newton's third law. The target variable is the force F. Block B is pulled to the left at constant speed, so block A moves to the right at constant speed and $a = 0$ for each block.

SET UP: The free-body diagram for block A is given in Figure 5.89a. n_{BA} is the normal force that B exerts on A. $f_{BA} = \mu_k n_{BA}$ is the kinetic friction force that B exerts on A. Block A moves to the right relative to B, and f_{BA} opposes this motion, so f_{BA} is to the left. Note also that F acts just on B, not on A.

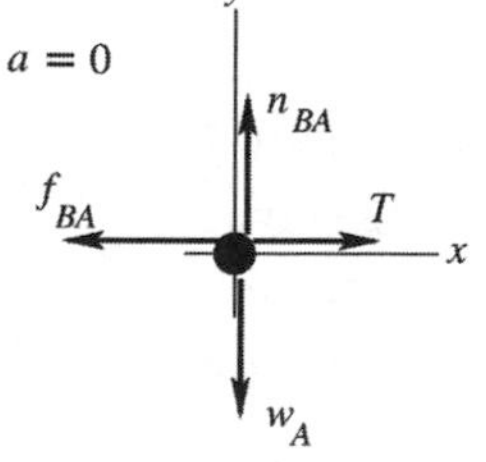

EXECUTE:

$\Sigma F_y = ma_y$

$n_{BA} - w_A = 0$

$n_{BA} = 1.90$ N

$f_{BA} = \mu_k n_{BA} = (0.30)(1.90 \text{ N}) = 0.57$ N

Figure 5.89a

$\Sigma F_x = ma_x$. $T - f_{BA} = 0$. $T = f_{BA} = 0.57$ N.

SET UP: The free-body diagram for block B is given in Figure 5.89b.

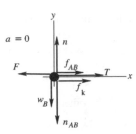

Figure 5.89b

EXECUTE: n_{AB} is the normal force that block A exerts on block B. By Newton's third law n_{AB} and n_{BA} are equal in magnitude and opposite in direction, so $n_{AB} = 1.90$ N. f_{AB} is the kinetic friction force that A exerts on B. Block B moves to the left relative to A and f_{AB} opposes this motion, so f_{AB} is to the right. $f_{AB} = \mu_k n_{AB} = (0.30)(1.90$ N$) = 0.42$ N. n and f_k are the normal and friction force exerted by the floor on block B; $f_k = \mu_k n$. Note that block B moves to the left relative to the floor and f_k opposes this motion, so f_k is to the right.

$\Sigma F_y = ma_y$: $n - w_B - n_{AB} = 0$. $n = w_B + n_{AB} = 4.20$ N $+ 1.90$ N $= 6.10$ N. Then

$f_k = \mu_k n = (0.30)(6.10$ N$) = 1.83$ N. $\Sigma F_x = ma_x$: $f_{AB} + T + f_k - F = 0$.

$F = T + f_{AB} + f_k = 0.57$ N $+ 0.57$ N $+ 1.83$ N $= 3.0$ N.

EVALUATE: Note that f_{AB} and f_{BA} are a third law action-reaction pair, so they must be equal in magnitude and opposite in direction and this is indeed what our calculation gives.

5.91.   **IDENTIFY:** Apply $\Sigma \vec{F} = m\vec{a}$ to each block. Parts (a) and (b) will be done together.

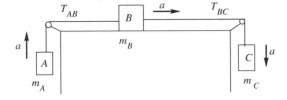

Figure 5.91a

Note that each block has the same magnitude of acceleration, but in different directions. For each block let the direction of $\vec{a}$ be a positive coordinate direction.

SET UP: The free-body diagram for block A is given in Figure 5.91b.

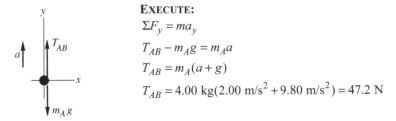

EXECUTE:

$\Sigma F_y = ma_y$

$T_{AB} - m_A g = m_A a$

$T_{AB} = m_A(a + g)$

$T_{AB} = 4.00$ kg$(2.00$ m/s$^2 + 9.80$ m/s$^2) = 47.2$ N

Figure 5.91b

SET UP: The free-body diagram for block B is given in Figure 5.91c.

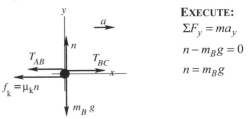

EXECUTE:

$\Sigma F_y = ma_y$

$n - m_B g = 0$

$n = m_B g$

Figure 5.91c

$$f_k = \mu_k n = \mu_k m_B g = (0.25)(12.0 \text{ kg})(9.80 \text{ m/s}^2) = 29.4 \text{ N}$$

$$\Sigma F_x = ma_x$$

$$T_{BC} - T_{AB} - f_k = m_B a$$

$$T_{BC} = T_{AB} + f_k + m_B a = 47.2 \text{ N} + 29.4 \text{ N} + (12.0 \text{ kg})(2.00 \text{ m/s}^2)$$

$$T_{BC} = 100.6 \text{ N}$$

SET UP: The free-body diagram for block C is sketched in Figure 5.91d.

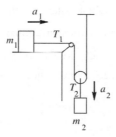

EXECUTE:

$$\Sigma F_y = ma_y$$

$$m_C g - T_{BC} = m_C a$$

$$m_C(g - a) = T_{BC}$$

$$m_C = \frac{T_{BC}}{g - a} = \frac{100.6 \text{ N}}{9.80 \text{ m/s}^2 - 2.00 \text{ m/s}^2} = 12.9 \text{ kg}$$

Figure 5.91d

EVALUATE: If all three blocks are considered together as a single object and $\Sigma \vec{F} = m\vec{a}$ is applied to this combined object, $m_C g - m_A g - \mu_k m_B g = (m_A + m_B + m_C)a$. Using the values for μ_k, m_A and m_B given in the problem and the mass m_C we calculated, this equation gives $a = 2.00 \text{ m/s}^2$, which checks.

5.93.   **IDENTIFY:** Let the tensions in the ropes be T_1 and T_2.

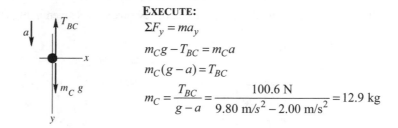

Figure 5.93a

Consider the forces on each block. In each case take a positive coordinate direction in the direction of the acceleration of that block.

SET UP: The free-body diagram for m_1 is given in Figure 5.93b.

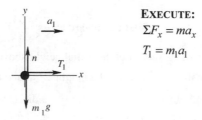

EXECUTE:

$$\Sigma F_x = ma_x$$

$$T_1 = m_1 a_1$$

Figure 5.93b

SET UP: The free-body diagram for m_2 is given in Figure 5.93c.

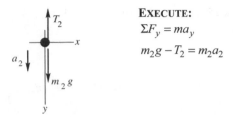

EXECUTE:

$\Sigma F_y = ma_y$

$m_2 g - T_2 = m_2 a_2$

Figure 5.93c

This gives us two equations, but there are four unknowns (T_1, T_2, a_1 and a_2) so two more equations are required.

SET UP: The free-body diagram for the moveable pulley (mass m) is given in Figure 5.93d.

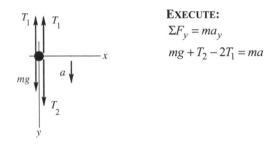

EXECUTE:

$\Sigma F_y = ma_y$

$mg + T_2 - 2T_1 = ma$

Figure 5.93d

But our pulleys have negligible mass, so $mg = ma = 0$ and $T_2 = 2T_1$. Combine these three equations to eliminate T_1 and T_2: $m_2 g - T_2 = m_2 a_2$ gives $m_2 g - 2T_1 = m_2 a_2$. And then with $T_1 = m_1 a_1$ we have $m_2 g - 2m_1 a_1 = m_2 a_2$.

SET UP: There are still two unknowns, a_1 and a_2. But the accelerations a_1 and a_2 are related. In any time interval, if m_1 moves to the right a distance d, then in the same time m_2 moves downward a distance $d/2$. One of the constant acceleration kinematic equations says $x - x_0 = v_{0x}t + \frac{1}{2}a_x t^2$, so if m_2 moves half the distance it must have half the acceleration of m_1: $a_2 = a_1/2$, or $a_1 = 2a_2$.

EXECUTE: This is the additional equation we need. Use it in the previous equation and get $m_2 g - 2m_1 (2a_2) = m_2 a_2$.

$a_2(4m_1 + m_2) = m_2 g$

$a_2 = \dfrac{m_2 g}{4m_1 + m_2}$ and $a_1 = 2a_2 = \dfrac{2m_2 g}{4m_1 + m_2}$.

EVALUATE: If $m_2 \to 0$ or $m_1 \to \infty$, $a_1 = a_2 = 0$. If $m_2 \gg m_1$, $a_2 = g$ and $a_1 = 2g$.

5.95. **IDENTIFY:** Apply the method of Exercise 5.15 to calculate the acceleration of each object. Then apply constant acceleration equations to the motion of the 2.00 kg object.

SET UP: After the 5.00 kg object reaches the floor, the 2.00 kg object is in free fall, with downward acceleration g.

EXECUTE: The 2.00-kg object will accelerate upward at $g\dfrac{5.00 \text{ kg} - 2.00 \text{ kg}}{5.00 \text{ kg} + 2.00 \text{ kg}} = 3g/7$, and the 5.00-kg object will accelerate downward at $3g/7$. Let the initial height above the ground be h_0. When the large object hits the ground, the small object will be at a height $2h_0$, and moving upward with a speed given by $v_0^2 = 2ah_0 = 6gh_0/7$. The small object will continue to rise a distance $v_0^2/2g = 3h_0/7$, and so the maximum height reached will be $2h_0 + 3h_0/7 = 17h_0/7 = 1.46$ m above the floor, which is 0.860 m above its initial height.

EVALUATE: The small object is 1.20 m above the floor when the large object strikes the floor, and it rises an additional 0.26 m after that.

5.97. **IDENTIFY:** Apply $\Sigma \vec{F} = m\vec{a}$ to the block. The cart and the block have the same acceleration. The normal force exerted by the cart on the block is perpendicular to the front of the cart, so is horizontal and to the right. The friction force on the block is directed so as to hold the block up against the downward pull of gravity. We want to calculate the minimum a required, so take static friction to have its maximum value, $f_s = \mu_s n$.

SET UP: The free-body diagram for the block is given in Figure 5.97.

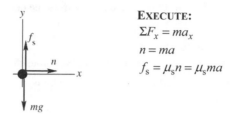

EXECUTE:
$$\Sigma F_x = ma_x$$
$$n = ma$$
$$f_s = \mu_s n = \mu_s ma$$

Figure 5.97

$$\Sigma F_y = ma_y$$

$$f_s - mg = 0$$

$$\mu_s ma = mg$$

$$a = g/\mu_s$$

EVALUATE: An observer on the cart sees the block pinned there, with no reason for a horizontal force on it because the block is at rest relative to the cart. Therefore, such an observer concludes that $n = 0$ and thus $f_s = 0$, and he doesn't understand what holds the block up against the downward force of gravity. The reason for this difficulty is that $\Sigma \vec{F} = m\vec{a}$ does not apply in a coordinate frame attached to the cart. This reference frame is accelerated, and hence not inertial. The smaller μ_s is, the larger a must be to keep the block pinned against the front of the cart.

5.99. **IDENTIFY:** Apply $\Sigma \vec{F} = m\vec{a}$ to the block and to the plank.

SET UP: Both objects have $a = 0$.

EXECUTE: Let n_B be the normal force between the plank and the block and n_A be the normal force between the block and the incline. Then, $n_B = w\cos\theta$ and $n_A = n_B + 3w\cos\theta = 4w\cos\theta$. The net frictional force on the block is $\mu_k(n_A + n_B) = \mu_k 5w\cos\theta$. To move at constant speed, this must balance the component of the block's weight along the incline, so $3w\sin\theta = \mu_k 5w\cos\theta$, and

$$\mu_k = \tfrac{3}{5}\tan\theta = \tfrac{3}{5}\tan 37° = 0.452.$$

EVALUATE: In the absence of the plank the block slides down at constant speed when the slope angle and coefficient of friction are related by $\tan\theta = \mu_k$. For $\theta = 36.9°$, $\mu_k = 0.75$. A smaller μ_k is needed when the plank is present because the plank provides an additional friction force.

5.101. **IDENTIFY:** Apply $\Sigma \vec{F} = m\vec{a}$ to the automobile.

SET UP: The "correct" banking angle is for zero friction and is given by $\tan\beta = \dfrac{v_0^2}{gR}$, as derived in Example 5.22. Use coordinates that are vertical and horizontal, since the acceleration is horizontal.

EXECUTE: For speeds larger than v_0, a frictional force is needed to keep the car from skidding. In this case, the inward force will consist of a part due to the normal force n and the friction force f; $n\sin\beta + f\cos\beta = ma_{\text{rad}}$. The normal and friction forces both have vertical components; since there is no vertical acceleration, $n\cos\beta - f\sin\beta = mg$. Using $f = \mu_s n$ and $a_{\text{rad}} = \dfrac{v^2}{R} = \dfrac{(1.5v_0)^2}{R} = 2.25\,g\tan\beta$, these two relations become $n \sin\beta + \mu_s n \cos\beta = 2.25\, mg \tan\beta$ and $n \cos\beta - \mu_s n \sin\beta = mg$. Dividing to cancel n gives $\dfrac{\sin\beta + \mu_s \cos\beta}{\cos\beta - \mu_s \sin\beta} = 2.25 \tan\beta$. Solving for μ_s and simplifying yields $\mu_s = \dfrac{1.25 \sin\beta \cos\beta}{1 + 1.25 \sin^2\beta}$.

Using $\beta = \arctan\left(\dfrac{(20 \text{ m/s})^2}{(9.80 \text{ m/s}^2)(120 \text{ m})}\right) = 18.79°$ gives $\mu_s = 0.34$.

EVALUATE: If μ_s is insufficient, the car skids away from the center of curvature of the roadway, so the friction is inward.

5.107. **IDENTIFY:** Apply $\Sigma \vec{F} = m\vec{a}$ to the rock.

SET UP: Equations 5.9 through 5.13 apply, but with a_0 rather than g as the initial acceleration.

EXECUTE: **(a)** The rock is released from rest, and so there is initially no resistive force and $a_0 = (18.0 \text{ N})/(3.00 \text{ kg}) = 6.00 \text{ m/s}^2$.

(b) $(18.0 \text{ N} - (2.20 \text{ N} \cdot \text{s/m})(3.00 \text{ m/s}))/(3.00 \text{ kg}) = 3.80 \text{ m/s}^2$.

(c) The net force must be 1.80 N, so $kv = 16.2$ N and $v = (16.2 \text{ N})/(2.20 \text{ N} \cdot \text{s/m}) = 7.36 \text{ m/s}$.

(d) When the net force is equal to zero, and hence the acceleration is zero, $kv_t = 18.0$ N and $v_t = (18.0 \text{ N})/(2.20 \text{ N} \cdot \text{s/m}) = 8.18 \text{ m/s}$.

(e) From Eq. (5.12),

$$y = (8.18 \text{ m/s})\left[(2.00 \text{ s}) - \frac{3.00 \text{ kg}}{2.20 \text{ N} \cdot \text{s/m}}\left(1 - e^{-((2.20 \text{ N·s/m})/(3.00 \text{ kg}))(2.00 \text{ s})}\right)\right] = +7.78 \text{ m}.$$

From Eq. (5.10), $v = (8.18 \text{ m/s})\left[1 - e^{-((2.20 \text{ N·s/m})/(3.00 \text{ kg}))(2.00 \text{ s})}\right] = 6.29 \text{ m/s}$.

From Eq. (5.11), but with a_0 instead of g, $a = (6.00 \text{ m/s}^2)e^{-((2.20 \text{ N·s/m})/(3.00 \text{ kg}))(2.00 \text{ s})} = 1.38 \text{ m/s}^2$.

(f) $1 - \dfrac{v}{v_t} = 0.1 = e^{-(k/m)t}$ and $t = \dfrac{m}{k} \ln(10) = 3.14 \text{ s}$.

EVALUATE: The acceleration decreases with time until it becomes zero when $v = v_t$. The speed increases with time and approaches v_t as $t \to \infty$.

5.115. **IDENTIFY:** Apply $\Sigma \vec{F} = m\vec{a}$ to the person. The person moves in a horizontal circle so his acceleration is $a_{\text{rad}} = v^2/R$, directed toward the center of the circle. The target variable is the coefficient of static friction between the person and the surface of the cylinder.

$$v = (0.60 \text{ rev/s})\left(\frac{2\pi R}{1 \text{ rev}}\right) = (0.60 \text{ rev/s})\left(\frac{2\pi(2.5 \text{ m})}{1 \text{ rev}}\right) = 9.425 \text{ m/s}$$

(a) SET UP: The problem situation is sketched in Figure 5.115a.

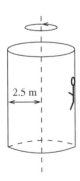

2.5 m

Figure 5.115a

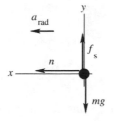

The free-body diagram for the person is sketched in Figure 5.115b.
The person is held up against gravity by the static friction force exerted on him by the wall.
The acceleration of the person is a_{rad}, directed in toward the axis of rotation.

Figure 5.115b

(b) EXECUTE: To calculate the minimum μ_s required, take f_s to have its maximum value, $f_s = \mu_s n$.

$\Sigma F_y = ma_y$

$f_s - mg = 0$

$\mu_s n = mg$

$\Sigma F_x = ma_x$

$n = mv^2/R$

Combine these two equations to eliminate n:

$\mu_s mv^2/R = mg$

$\mu_s = \dfrac{Rg}{v^2} = \dfrac{(2.5\ \text{m})(9.80\ \text{m/s}^2)}{(9.425\ \text{m/s})^2} = 0.28$

(c) EVALUATE: No, the mass of the person divided out of the equation for μ_s. Also, the smaller μ_s is, the larger v must be to keep the person from sliding down. For smaller μ_s the cylinder must rotate faster to make n large enough.

5.119. **IDENTIFY:** Apply $\Sigma \vec{F} = m\vec{a}$ to the circular motion of the bead. Also use Eq. (5.16) to relate a_{rad} to the period of rotation T.
SET UP: The bead and hoop are sketched in Figure 5.119a.

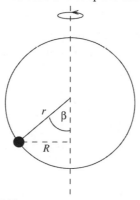

The bead moves in a circle of radius $R = r\sin\beta$.
The normal force exerted on the bead by the hoop is radially inward.

Figure 5.119a

The free-body diagram for the bead is sketched in Figure 5.119b.

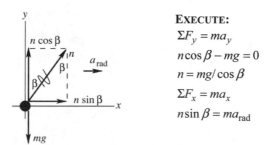

EXECUTE:
$\Sigma F_y = ma_y$
$n\cos\beta - mg = 0$
$n = mg/\cos\beta$
$\Sigma F_x = ma_x$
$n\sin\beta = ma_{\text{rad}}$

Figure 5.119b

Combine these two equations to eliminate n:

$$\left(\frac{mg}{\cos\beta}\right)\sin\beta = ma_{\text{rad}}$$

$$\frac{\sin\beta}{\cos\beta} = \frac{a_{\text{rad}}}{g}$$

$a_{\text{rad}} = v^2/R$ and $v = 2\pi R/T$, so $a_{\text{rad}} = 4\pi^2 R/T^2$, where T is the time for one revolution.

$R = r\sin\beta$, so $a_{\text{rad}} = \dfrac{4\pi^2 r\sin\beta}{T^2}$

Use this in the above equation: $\dfrac{\sin\beta}{\cos\beta} = \dfrac{4\pi^2 r\sin\beta}{T^2 g}$

This equation is satisfied by $\sin\beta = 0$, so $\beta = 0$, or by

$\dfrac{1}{\cos\beta} = \dfrac{4\pi^2 r}{T^2 g}$, which gives $\cos\beta = \dfrac{T^2 g}{4\pi^2 r}$.

(a) 4.00 rev/s implies $T = (1/4.00)$ s $= 0.250$ s

Then $\cos\beta = \dfrac{(0.250 \text{ s})^2(9.80 \text{ m/s}^2)}{4\pi^2(0.100 \text{ m})}$ and $\beta = 81.1°$.

(b) This would mean $\beta = 90°$. But $\cos 90° = 0$, so this requires $T \to 0$. So β approaches $90°$ as the hoop rotates very fast, but $\beta = 90°$ is not possible.

(c) 1.00 rev/s implies $T = 1.00$ s

The $\cos\beta = \dfrac{T^2 g}{4\pi^2 r}$ equation then says $\cos\beta = \dfrac{(1.00 \text{ s})^2(9.80 \text{ m/s}^2)}{4\pi^2(0.100 \text{ m})} = 2.48$, which is not possible. The only

way to have the $\Sigma\vec{F} = m\vec{a}$ equations satisfied is for $\sin\beta = 0$. This means $\beta = 0$; the bead sits at the bottom of the hoop.

EVALUATE: $\beta \to 90°$ as $T \to 0$ (hoop moves faster). The largest value T can have is given by

$T^2 g/(4\pi^2 r) = 1$ so $T = 2\pi\sqrt{r/g} = 0.635$ s. This corresponds to a rotation rate of

$(1/0.635)$ rev/s $= 1.58$ rev/s. For a rotation rate less than 1.58 rev/s, $\beta = 0$ is the only solution and the bead sits at the bottom of the hoop. Part (c) is an example of this.

WORK AND KINETIC ENERGY

6.5. **IDENTIFY:** The gravity force is constant and the displacement is along a straight line, so $W = Fs\cos\phi$.

SET UP: The displacement is upward along the ladder and the gravity force is downward, so $\phi = 180.0° - 30.0° = 150.0°$. $w = mg = 735$ N.

EXECUTE: **(a)** $W = (735 \text{ N})(2.75 \text{ m})\cos150.0° = -1750$ J.

(b) No, the gravity force is independent of the motion of the painter.

EVALUATE: Gravity is downward and the vertical component of the displacement is upward, so the gravity force does negative work.

6.9. **IDENTIFY:** Apply Eq. (6.2) or (6.3).

SET UP: The gravity force is in the $-y$-direction, so $\vec{F}_{mg} \cdot \vec{s} = -mg(y_2 - y_1)$

EXECUTE: **(a)** (i) Tension force is always perpendicular to the displacement and does no work.

(ii) Work done by gravity is $-mg(y_2 - y_1)$. When $y_1 = y_2$, $W_{mg} = 0$.

(b) (i) Tension does no work. (ii) Let l be the length of the string. $W_{mg} = -mg(y_2 - y_1) = -mg(2l) = -25.1$ J

EVALUATE: In part (b) the displacement is upward and the gravity force is downward, so the gravity force does negative work.

6.11. **IDENTIFY:** Since the speed is constant, the acceleration and the net force on the monitor are zero.

SET UP: Use the fact that the net force on the monitor is zero to develop expressions for the friction force, f_k, and the normal force, n. Then use $W = F_p s = (F\cos\phi)s$ to calculate W.

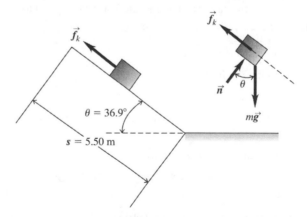

Figure 6.11

EXECUTE: **(a)** Summing forces along the incline, $\Sigma F = ma = 0 = f_k - mg\sin\theta$, giving $f_k = mg\cos\theta$, directed up the incline. Substituting gives $W_f = (f_k\cos\phi)s = [(mg\sin\theta)\cos\phi]s$.

$W_f = [(10.0 \text{ kg})(9.80 \text{ m/s}^2)(\sin36.9°)](\cos0°)(5.50 \text{ m}) = +324$ J.

(b) The gravity force is downward and the displacement is directed up the incline so $\phi = 126.9°$.

$W_{\text{grav}} = (10.0 \text{ kg})(9.80 \text{ m/s}^2)(\cos 126.9°)(5.50 \text{ m}) = -324$ J.

(c) The normal force, n, is perpendicular to the displacement and thus does zero work.

EVALUATE: Friction does positive work and gravity does negative work. The net work done is zero.

6.19.   **IDENTIFY:** $W_{\text{tot}} = K_2 - K_1$. In each case calculate W_{tot} from what we know about the force and the displacement.

SET UP: The gravity force is mg, downward. The friction force is $f_k = \mu_k n = \mu_k mg$ and is directed opposite to the displacement. The mass of the object isn't given, so we expect that it will divide out in the calculation.

EXECUTE: **(a)** $K_1 = 0$. $W_{\text{tot}} = W_{\text{grav}} = mgs$. $mgs = \frac{1}{2}mv_2^2$ and

$v_2 = \sqrt{2gs} = \sqrt{2(9.80 \text{ m/s}^2)(95.0 \text{ m})} = 43.2$ m/s.

(b) $K_2 = 0$ (at the maximum height). $W_{\text{tot}} = W_{\text{grav}} = -mgs$. $-mgs = -\frac{1}{2}mv_1^2$ and

$v_1 = \sqrt{2gs} = \sqrt{2(9.80 \text{ m/s}^2)(525 \text{ m})} = 101$ m/s.

(c) $K_1 = \frac{1}{2}mv_1^2$. $K_2 = 0$. $W_{\text{tot}} = W_f = -\mu_k mgs$. $-\mu_k mgs = -\frac{1}{2}mv_1^2$.

$s = \dfrac{v_1^2}{2\mu_k g} = \dfrac{(5.00 \text{ m/s})^2}{2(0.220)(9.80 \text{ m/s}^2)} = 5.80$ m.

(d) $K_1 = \frac{1}{2}mv_1^2$. $K_2 = \frac{1}{2}mv_2^2$. $W_{\text{tot}} = W_f = -\mu_k mgs$. $K_2 = W_{\text{tot}} + K_1$. $\frac{1}{2}mv_2^2 = -\mu_k mgs + \frac{1}{2}mv_1^2$.

$v_2 = \sqrt{v_1^2 - 2\mu_k gs} = \sqrt{(5.00 \text{ m/s})^2 - 2(0.220)(9.80 \text{ m/s}^2)(2.90 \text{ m})} = 3.53$ m/s.

(e) $K_1 = \frac{1}{2}mv_1^2$. $K_2 = 0$. $W_{\text{grav}} = -mgy_2$, where y_2 is the vertical height. $-mgy_2 = -\frac{1}{2}mv_1^2$ and

$y_2 = \dfrac{v_1^2}{2g} = \dfrac{(12.0 \text{ m/s})^2}{2(9.80 \text{ m/s}^2)} = 7.35$ m.

EVALUATE: In parts (c) and (d), friction does negative work and the kinetic energy is reduced. In part (a), gravity does positive work and the speed increases. In parts (b) and (e), gravity does negative work and the speed decreases. The vertical height in part (e) is independent of the slope angle of the hill.

6.21.   **IDENTIFY and SET UP:** Apply Eq. (6.6) to the box. Let point 1 be at the bottom of the incline and let point 2 be at the skier. Work is done by gravity and by friction. Solve for K_1 and from that obtain the required initial speed.

EXECUTE: $W_{\text{tot}} = K_2 - K_1$

$K_1 = \frac{1}{2}mv_0^2$, $K_2 = 0$

Work is done by gravity and friction, so $W_{\text{tot}} = W_{mg} + W_f$.

$W_{mg} = -mg(y_2 - y_1) = -mgh$

$W_f = -fs$. The normal force is $n = mg\cos\alpha$ and $s = h/\sin\alpha$, where s is the distance the box travels along the incline.

$W_f = -(\mu_k mg\cos\alpha)(h/\sin\alpha) = -\mu_k mgh/\tan\alpha$

Substituting these expressions into the work-energy theorem gives

$-mgh - \mu_k mgh/\tan\alpha = -\frac{1}{2}mv_0^2$.

Solving for v_0 then gives $v_0 = \sqrt{2gh(1 + \mu_k/\tan\alpha)}$.

EVALUATE: The result is independent of the mass of the box. As $\alpha \rightarrow 90°$, $h = s$ and $v_0 = \sqrt{2gh}$, the same as throwing the box straight up into the air. For $\alpha = 90°$ the normal force is zero so there is no friction.

6.29.   **IDENTIFY:** $W_{\text{tot}} = K_2 - K_1$. Only friction does work.

SET UP: $W_{\text{tot}} = W_{f_k} = -\mu_k mgs$. $K_2 = 0$ (car stops). $K_1 = \frac{1}{2}mv_0^2$.

EXECUTE: (a) $W_{\text{tot}} = K_2 - K_1$ gives $-\mu_k mgs = -\frac{1}{2}mv_0^2$. $s = \dfrac{v_0^2}{2\mu_k g}$.

(b) (i) $\mu_{kb} = 2\mu_{ka}$. $s\mu_k = \dfrac{v_0^2}{2g}$ = constant so $s_a\mu_{ka} = s_b\mu_{kb}$. $s_b = \left(\dfrac{\mu_{ka}}{\mu_{kb}}\right)s_a = s_a/2$. The minimum stopping

distance would be halved. (ii) $v_{0b} = 2v_{0a}$. $\dfrac{s}{v_0^2} = \dfrac{1}{2\mu_k g}$ = constant, so $\dfrac{s_a}{v_{0a}^2} = \dfrac{s_b}{v_{0b}^2}$. $s_b = s_a\left(\dfrac{v_{0b}}{v_{0a}}\right)^2 = 4s_a$.

The stopping distance would become 4 times as great. (iii) $v_{0b} = 2v_{0a}$, $\mu_{kb} = 2\mu_{ka}$. $\dfrac{s\mu_k}{v_0^2} = \dfrac{1}{2g}$ = constant,

so $\dfrac{s_a\mu_{ka}}{v_{0a}^2} = \dfrac{s_b\mu_{kb}}{v_{0b}^2}$. $s_b = s_a\left(\dfrac{\mu_{ka}}{\mu_{kb}}\right)\left(\dfrac{v_{0b}}{v_{0a}}\right)^2 = s_a\left(\dfrac{1}{2}\right)(2)^2 = 2s_a$. The stopping distance would double.

EVALUATE: The stopping distance is directly proportional to the square of the initial speed and indirectly proportional to the coefficient of kinetic friction.

6.35.   IDENTIFY: Use the work-energy theorem and the results of Problem 6.30.
SET UP: For $x = 0$ to $x = 8.0$ m, $W_{\text{tot}} = 40$ J. For $x = 0$ to $x = 12.0$ m, $W_{\text{tot}} = 60$ J.

EXECUTE: (a) $v = \sqrt{\dfrac{(2)(40\text{ J})}{10\text{ kg}}} = 2.83$ m/s

(b) $v = \sqrt{\dfrac{(2)(60\text{ J})}{10\text{ kg}}} = 3.46$ m/s.

EVALUATE: $\vec{F}$ is always in the $+x$-direction. For this motion $\vec{F}$ does positive work and the speed continually increases during the motion.

6.39.   IDENTIFY: Apply $\Sigma\vec{F} = m\vec{a}$ to calculate the μ_s required for the static friction force to equal the spring force.
SET UP: (a) The free-body diagram for the glider is given in Figure 6.39.

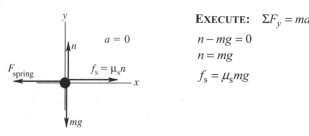

EXECUTE: $\Sigma F_y = ma_y$
$n - mg = 0$
$n = mg$
$f_s = \mu_s mg$

Figure 6.39

$\Sigma F_x = ma_x$
$f_s - F_{\text{spring}} = 0$
$\mu_s mg - kd = 0$
$\mu_s = \dfrac{kd}{mg} = \dfrac{(20.0\text{ N/m})(0.086\text{ m})}{(0.100\text{ kg})(9.80\text{ m/s}^2)} = 1.76$

(b) IDENTIFY and SET UP: Apply $\Sigma\vec{F} = m\vec{a}$ to find the maximum amount the spring can be compressed and still have the spring force balanced by friction. Then use $W_{\text{tot}} = K_2 - K_1$ to find the initial speed that results in this compression of the spring when the glider stops.
EXECUTE: $\mu_s mg = kd$

$d = \dfrac{\mu_s mg}{k} = \dfrac{(0.60)(0.100\text{ kg})(9.80\text{ m/s}^2)}{20.0\text{ N/m}} = 0.0294$ m

Now apply the work-energy theorem to the motion of the glider:
$W_{\text{tot}} = K_2 - K_1$

$K_1 = \frac{1}{2}mv_1^2$, $K_2 = 0$ (instantaneously stops)

$W_{tot} = W_{spring} + W_{fric} = -\frac{1}{2}kd^2 - \mu_k mgd$ (as in Example 6.8)

$W_{tot} = -\frac{1}{2}(20.0 \text{ N/m})(0.0294 \text{ m})^2 - 0.47(0.100 \text{ kg})(9.80 \text{ m/s}^2)(0.0294 \text{ m}) = -0.02218 \text{ J}$

Then $W_{tot} = K_2 - K_1$ gives $-0.02218 \text{ J} = -\frac{1}{2}mv_1^2$.

$v_1 = \sqrt{\dfrac{2(0.02218 \text{ J})}{0.100 \text{ kg}}} = 0.67 \text{ m/s}$

EVALUATE: In Example 6.8 an initial speed of 1.50 m/s compresses the spring 0.086 m and in part (a) of this problem we found that the glider doesn't stay at rest. In part (b) we found that a smaller displacement of 0.0294 m when the glider stops is required if it is to stay at rest. And we calculate a smaller initial speed (0.67 m/s) to produce this smaller displacement.

6.41. **IDENTIFY** and **SET UP:** The magnitude of the work done by F_x equals the area under the F_x versus x curve. The work is positive when F_x and the displacement are in the same direction; it is negative when they are in opposite directions.

EXECUTE: **(a)** F_x is positive and the displacement Δx is positive, so $W > 0$.

$W = \frac{1}{2}(2.0 \text{ N})(2.0 \text{ m}) + (2.0 \text{ N})(1.0 \text{ m}) = +4.0 \text{ J}$

(b) During this displacement $F_x = 0$, so $W = 0$.

(c) F_x is negative, Δx is positive, so $W < 0$. $W = -\frac{1}{2}(1.0 \text{ N})(2.0 \text{ m}) = -1.0 \text{ J}$

(d) The work is the sum of the answers to parts (a), (b), and (c), so $W = 4.0 \text{ J} + 0 - 1.0 \text{ J} = +3.0 \text{ J}$.

(e) The work done for $x = 7.0$ m to $x = 3.0$ m is $+1.0$ J. This work is positive since the displacement and the force are both in the $-x$-direction. The magnitude of the work done for $x = 3.0$ m to $x = 2.0$ m is 2.0 J, the area under F_x versus x. This work is negative since the displacement is in the $-x$-direction and the force is in the $+x$-direction. Thus $W = +1.0 \text{ J} - 2.0 \text{ J} = -1.0 \text{ J}$.

EVALUATE: The work done when the car moves from $x = 2.0$ m to $x = 0$ is $-\frac{1}{2}(2.0 \text{ N})(2.0 \text{ m}) = -2.0 \text{ J}$.

Adding this to the work for $x = 7.0$ m to $x = 2.0$ m gives a total of $W = -3.0$ J for $x = 7.0$ m to $x = 0$. The work for $x = 7.0$ m to $x = 0$ is the negative of the work for $x = 0$ to $x = 7.0$ m.

6.45. **IDENTIFY** and **SET UP:** Apply Eq. (6.6) to the glider. Work is done by the spring and by gravity. Take point 1 to be where the glider is released. In part (a) point 2 is where the glider has traveled 1.80 m and $K_2 = 0$. There are two points shown in Figure 6.45a. In part (b) point 2 is where the glider has traveled 0.80 m.

EXECUTE: **(a)** $W_{tot} = K_2 - K_1 = 0$. Solve for x_1, the amount the spring is initially compressed.

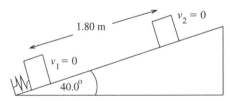

$W_{tot} = W_{spr} + W_w = 0$

So $W_{spr} = -W_w$

(The spring does positive work on the glider since the spring force is directed up the incline, the same as the direction of the displacement.)

Figure 6.45a

The directions of the displacement and of the gravity force are shown in Figure 6.45b.

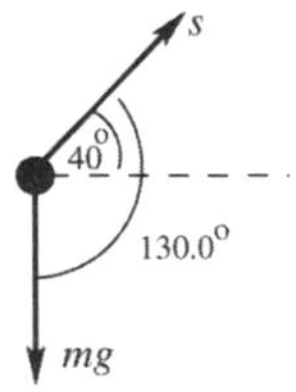

$W_w = (w\cos\phi)s = (mg\cos 130.0°)s$

$W_w = (0.0900 \text{ kg})(9.80 \text{ m/s}^2)(\cos 130.0°)(1.80 \text{ m}) = -1.020 \text{ J}$

(The component of w parallel to the incline is directed down the incline, opposite to the displacement, so gravity does negative work.)

Figure 6.45b

$W_{spr} = -W_w = +1.020$ J

$W_{spr} = \frac{1}{2}kx_1^2$ so $x_1 = \sqrt{\dfrac{2W_{spr}}{k}} = \sqrt{\dfrac{2(1.020 \text{ J})}{640 \text{ N/m}}} = 0.0565$ m

(b) The spring was compressed only 0.0565 m so at this point in the motion the glider is no longer in contact with the spring. Points 1 and 2 are shown in Figure 6.45c.

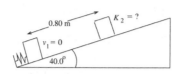

$W_{tot} = K_2 - K_1$

$K_2 = K_1 + W_{tot}$

$K_1 = 0$

Figure 6.45c

$W_{tot} = W_{spr} + W_w$

From part (a), $W_{spr} = 1.020$ J and

$W_w = (mg \cos 130.0°)s = (0.0900 \text{ kg})(9.80 \text{ m/s}^2)(\cos 130.0°)(0.80 \text{ m}) = -0.454$ J

Then $K_2 = W_{spr} + W_w = +1.020$ J $- 0.454$ J $= +0.57$ J.

EVALUATE: The kinetic energy in part (b) is positive, as it must be. In part (a), $x_2 = 0$ since the spring force is no longer applied past this point. In computing the work done by gravity we use the full 0.80 m the glider moves.

6.47.   **IDENTIFY:** The force does work on the box, which gives it kinetic energy, so the work-energy theorem applies. The force is variable so we must integrate to calculate the work it does on the box.

SET UP: $W_{tot} = \Delta K = K_f - K_i = \frac{1}{2}mv_f^2 - \frac{1}{2}mv_i^2$ and $W_{tot} = \int_{x_1}^{x_2} F(x)dx$.

EXECUTE: $W_{tot} = \int_{x_1}^{x_2} F(x)dx = \int_0^{14.0 \text{m}} [18.0 \text{ N} - (0.530 \text{ N/m})x]dx$

$W_{tot} = (18.0 \text{ N})(14.0 \text{ m}) - (0.265 \text{ N/m})(14.0 \text{ m})^2 = 252.0$ J $- 51.94$ J $= 200.1$ J. The initial kinetic energy is

zero, so $W_{tot} = \Delta K = K_f - K_i = \frac{1}{2}mv_f^2$. Solving for v_f gives $v_f = \sqrt{\dfrac{2W_{tot}}{m}} = \sqrt{\dfrac{2(200.1 \text{ J})}{6.00 \text{ kg}}} = 8.17$ m/s.

EVALUATE: We could not readily do this problem by integrating the acceleration over time because we know the force as a function of x, not of t. The work-energy theorem provides a much simpler method.

6.51.   **IDENTIFY:** $P_{av} = \dfrac{\Delta W}{\Delta t}$. ΔW is the energy released.

SET UP: ΔW is to be the same. 1 y $= 3.156 \times 10^7$ s.

EXECUTE: $P_{av}\Delta t = \Delta W = $ constant, so $P_{av\text{-}sun}\Delta t_{sun} = P_{av\text{-}m}\Delta t_m$.

$P_{av\text{-}m} = P_{av\text{-}sun}\left(\dfrac{\Delta t_{sun}}{\Delta t_m}\right) = \left(\dfrac{[2.5 \times 10^5 \text{ y}][3.156 \times 10^7 \text{ s/y}]}{0.20 \text{ s}}\right) = 3.9 \times 10^{13} P$.

EVALUATE: Since the power output of the magnetar is so much larger than that of our sun, the mechanism by which it radiates energy must be quite different.

6.53.   **IDENTIFY:** Use the relation $P = F_{\parallel}v$ to relate the given force and velocity to the total power developed.

SET UP: 1 hp $= 746$ W

EXECUTE: The total power is $P = F_{\parallel}v = (165 \text{ N})(9.00 \text{ m/s}) = 1.49 \times 10^3$ W. Each rider therefore

contributes $P_{\text{each rider}} = (1.49 \times 10^3 \text{ W})/2 = 745$ W ≈ 1 hp.

EVALUATE: The result of one horsepower is very large; a rider could not sustain this output for long periods of time.

6.55.   **IDENTIFY:** $P_{av} = \dfrac{\Delta W}{\Delta t}$. The work you do in lifting mass m a height h is mgh.

SET UP: 1 hp = 746 W

EXECUTE: **(a)** The number per minute would be the average power divided by the work (mgh) required to lift one box, $\dfrac{(0.50 \text{ hp})(746 \text{ W/hp})}{(30 \text{ kg})(9.80 \text{ m/s}^2)(0.90 \text{ m})} = 1.41/\text{s}$, or $84.6/\text{min}$.

(b) Similarly, $\dfrac{(100 \text{ W})}{(30 \text{ kg})(9.80 \text{ m/s}^2)(0.90 \text{ m})} = 0.378/\text{s}$, or $22.7/\text{min}$.

EVALUATE: A 30-kg crate weighs about 66 lbs. It is not possible for a person to perform work at this rate.

6.57.   **IDENTIFY:** To lift the skiers, the rope must do positive work to counteract the negative work developed by the component of the gravitational force acting on the total number of skiers, $F_{rope} = Nmg \sin \alpha$.

SET UP: $P = F_\| v = F_{rope} v$

EXECUTE: $P_{rope} = F_{rope} v = [+Nmg(\cos \phi)]v$.

$P_{rope} = [(50 \text{ riders})(70.0 \text{ kg})(9.80 \text{ m/s}^2)(\cos 75.0)]\left[(12.0 \text{ km/h})\left(\dfrac{1 \text{ m/s}}{3.60 \text{ km/h}} \right) \right]$.

$P_{rope} = 2.96 \times 10^4 \text{ W} = 29.6 \text{ kW}$.

EVALUATE: Some additional power would be needed to give the riders kinetic energy as they are accelerated from rest.

6.61.   **IDENTIFY:** For mass dm located a distance x from the axis and moving with speed v, the kinetic energy is $K = \frac{1}{2}(dm)v^2$. Follow the procedure specified in the hint.

SET UP: The bar and an infinitesimal mass element along the bar are sketched in Figure 6.61. Let M = total mass and T = time for one revolution. $v = \dfrac{2\pi x}{T}$.

EXECUTE: $K = \int \frac{1}{2}(dm)v^2$. $dm = \dfrac{M}{L}dx$, so

$$K = \int_0^L \frac{1}{2}\left(\frac{M}{L}dx \right)\left(\frac{2\pi x}{T} \right)^2 = \frac{1}{2}\left(\frac{M}{L} \right)\left(\frac{4\pi^2}{T^2} \right)\int_0^L x^2 dx = \frac{1}{2}\left(\frac{M}{L} \right)\left(\frac{4\pi^2}{T^2} \right)\left(\frac{L^3}{3} \right) = \frac{2}{3}\pi^2 M L^2 / T^2$$

There are 5 revolutions in 3 seconds, so $T = 3/5 \text{ s} = 0.60 \text{ s}$

$$K = \frac{2}{3}\pi^2 (12.0 \text{ kg})(2.00 \text{ m})^2/(0.60 \text{ s})^2 = 877 \text{ J}.$$

EVALUATE: If a point mass 12.0 kg is 2.00 m from the axis and rotates at the same rate as the bar, $v = \dfrac{2\pi(2.00 \text{ m})}{0.60 \text{ s}} = 20.9 \text{ m/s}$ and $K = \frac{1}{2}mv^2 = \frac{1}{2}(12 \text{ kg})(20.9 \text{ m/s})^2 = 2.62 \times 10^3 \text{ J}$. K for the bar is smaller by a factor of 0.33. The speed of a segment of the bar decreases toward the axis.

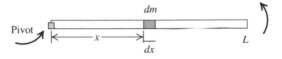

Figure 6.61

6.65.   **IDENTIFY:** Four forces act on the crate: the 290-N push, gravity, friction, and the normal force due to the surface of the ramp. The total work is the sum of the work due to all four of these forces. The acceleration is constant because the forces are constant.

SET UP: The work is $W = Fs \cos \phi$. We can use the standard kinematics formulas because the acceleration is constant. The work-energy theorem, $W_{tot} = \Delta K$, applies.

EXECUTE: **(a)** First calculate the work done by each of the four forces. The normal force does no work because it is perpendicular to the displacement. The other work is

$W_F = (F\cos 34.0^\circ)(15.0\text{ m}) = (290\text{ N})(\cos 34.0^\circ)(15.0\text{ m}) = 3606\text{ J}$,

$W_{mg} = -mg(15.0\text{ m})(\sin 34.0^\circ) = -(20.0\text{ kg})(9.8\text{ m/s}^2)(15.0\text{ m})(\sin 34.0^\circ) = -1644\text{ J}$ and

$W_f = -f(15.0\text{ m}) = -(65.0\text{ N})(15.0\text{ m}) = -975\text{ J}$. The total work is

$W_{\text{tot}} = 3606\text{ J} - 1644\text{ J} - 975\text{ J} = 987\text{ J}$.

(b) First find the final velocity: $W_{\text{tot}} = \Delta K$ so $v_f = \sqrt{\dfrac{2(987\text{ J})}{20.0\text{ kg}}} = 9.935\text{ m/s}$.

Constant acceleration gives $x - x_0 = \left(\dfrac{v_{0x} + v_x}{2}\right)t$ so

$$t = \frac{2(x - x_0)}{v_{0x} + v_x} = \frac{2(15.0\text{ m})}{0 + (9.935\text{ m/s})} = 3.02\text{ s}.$$

EVALUATE: Work is a scalar, so we can algebraically add the work done by each of the forces.

6.69. **IDENTIFY:** The initial kinetic energy of the head is absorbed by the neck bones during a sudden stop. Newton's second law applies to the passengers as well as to their heads.

SET UP: In part (a), the initial kinetic energy of the head is absorbed by the neck bones, so $\frac{1}{2}mv_{\text{max}}^2 = 8.0\text{ J}$. For part (b), assume constant acceleration and use $v_f = v_i + at$ with $v_i = 0$, to calculate a; then apply $F_{\text{net}} = ma$ to find the net accelerating force.

Solve: (a) $v_{\text{max}} = \sqrt{\dfrac{2(8.0\text{ J})}{5.0\text{ kg}}} = 1.8\text{ m/s} = 4.0\text{ mph}$.

(b) $a = \dfrac{v_f - v_i}{t} = \dfrac{1.8\text{ m/s} - 0}{10.0 \times 10^{-3}\text{ s}} = 180\text{ m/s}^2 \approx 18g$, and $F_{\text{net}} = ma = (5.0\text{ kg})(180\text{ m/s}^2) = 900\text{ N}$.

EVALUATE: The acceleration is very large, but if it lasts for only 10 ms it does not do much damage.

6.73. **IDENTIFY:** Calculate the work done by friction and apply $W_{\text{tot}} = K_2 - K_1$. Since the friction force is not constant, use Eq. (6.7) to calculate the work.

SET UP: Let x be the distance past P. Since μ_k increases linearly with x, $\mu_k = 0.100 + Ax$. When $x = 12.5\text{ m}$, $\mu_k = 0.600$, so $A = 0.500/(12.5\text{ m}) = 0.0400/\text{m}$.

EXECUTE: **(a)** $W_{\text{tot}} = \Delta K = K_2 - K_1$ gives $-\int \mu_k mg\,dx = 0 - \frac{1}{2}mv_1^2$. Using the above expression for μ_k,

$g\displaystyle\int_0^{x_2}(0.100 + Ax)\,dx = \frac{1}{2}v_1^2$ and $g\left[(0.100)x_2 + A\dfrac{x_2^2}{2}\right] = \frac{1}{2}v_1^2$.

$(9.80\text{ m/s}^2)\left[(0.100)x_2 + (0.0400/\text{m})\dfrac{x_2^2}{2}\right] = \frac{1}{2}(4.50\text{ m/s})^2$. Solving for x_2 gives $x_2 = 5.11\text{ m}$.

(b) $\mu_k = 0.100 + (0.0400/\text{m})(5.11\text{ m}) = 0.304$

(c) $W_{\text{tot}} = K_2 - K_1$ gives $-\mu_k mgx_2 = 0 - \frac{1}{2}mv_1^2$. $x_2 = \dfrac{v_1^2}{2\mu_k g} = \dfrac{(4.50\text{ m/s})^2}{2(0.100)(9.80\text{ m/s}^2)} = 10.3\text{ m}$.

EVALUATE: The box goes farther when the friction coefficient doesn't increase.

6.75. **IDENTIFY and SET UP:** Use $\Sigma \vec{F} = m\vec{a}$ to find the tension force T. The block moves in uniform circular motion and $\vec{a} = \vec{a}_{\text{rad}}$.

(a) The free-body diagram for the block is given in Figure 6.75.

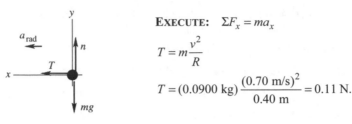

EXECUTE: $\Sigma F_x = ma_x$

$$T = m\frac{v^2}{R}$$

$$T = (0.0900 \text{ kg})\frac{(0.70 \text{ m/s})^2}{0.40 \text{ m}} = 0.11 \text{ N}.$$

Figure 6.75

(b) $T = m\dfrac{v^2}{R} = (0.0900 \text{ kg})\dfrac{(2.80 \text{ m/s})^2}{0.10 \text{ m}} = 7.1 \text{ N}.$

(c) SET UP: The tension changes as the distance of the block from the hole changes. We could use $W = \int_{x_1}^{x_2} F_x\, dx$ to calculate the work. But a much simpler approach is to use $W_{\text{tot}} = K_2 - K_1$.

EXECUTE: The only force doing work on the block is the tension in the cord, so $W_{\text{tot}} = W_T$.

$K_1 = \frac{1}{2}mv_1^2 = \frac{1}{2}(0.0900 \text{ kg})(0.70 \text{ m/s})^2 = 0.0221 \text{ J}, \quad K_2 = \frac{1}{2}mv_2^2 = \frac{1}{2}(0.0900 \text{ kg})(2.80 \text{ m/s})^2 = 0.353 \text{ J}$, so

$W_{\text{tot}} = K_2 - K_1 = 0.353 \text{ J} - 0.0221 \text{ J} = 0.33 \text{ J}.$ This is the amount of work done by the person who pulled the cord.

EVALUATE: The block moves inward, in the direction of the tension, so T does positive work and the kinetic energy increases.

6.77.   **IDENTIFY** and **SET UP:**   Use $v_x = dx/dt$ and $a_x = dv_x/dt$. Use $\Sigma \vec{F} = m\vec{a}$ to calculate $\vec{F}$ from $\vec{a}$.

EXECUTE: **(a)** $x(t) = \alpha t^2 + \beta t^3$, $v_x(t) = \dfrac{dx}{dt} = 2\alpha t + 3\beta t^2$. At $t = 4.00$ s:

$v_x = 2(0.200 \text{ m/s}^2)(4.00 \text{ s}) + 3(0.0200 \text{ m/s}^3)(4.00 \text{ s})^2 = 2.56 \text{ m/s}.$

(b) $a_x(t) = \dfrac{dv_x}{dt} = 2\alpha + 6\beta t$, so $F_x = ma_x = m(2\alpha + 6\beta t)$. At $t = 4.00$ s:

$$F_x = (4.00 \text{ kg})[2(0.200 \text{ m/s}^2) + 6(0.0200 \text{ m/s}^3)(4.00 \text{ s})] = 3.52 \text{ N}.$$

(c) IDENTIFY and **SET UP:**   Use Eq. (6.6) to calculate the work.

EXECUTE: $W_{\text{tot}} = K_2 - K_1$. At $t_1 = 0$, $v_1 = 0$ so $K_1 = 0$. $W_{\text{tot}} = W_F$.

$K_2 = \frac{1}{2}mv_2^2 = \frac{1}{2}(4.00 \text{ kg})(2.56 \text{ m/s})^2 = 13.1 \text{ J}.$ Then $W_{\text{tot}} = K_2 - K_1$ gives that $W_F = 13.1 \text{ J}.$

EVALUATE: Since v increases with t, the kinetic energy increases and the work done is positive. We can also calculate W_F directly from Eq. (6.7), by writing dx as $v_x\, dt$ and performing the integral.

6.79.   **IDENTIFY:**   The negative work done by the spring equals the change in kinetic energy of the car.

SET UP: The work done by a spring when it is compressed a distance x from equilibrium is $-\frac{1}{2}kx^2$.

$K_2 = 0.$

EXECUTE: $-\frac{1}{2}kx^2 = K_2 - K_1$ gives $\frac{1}{2}kx^2 = \frac{1}{2}mv_1^2$ and

$k = (mv_1^2)/x^2 = [(1200 \text{ kg})(0.65 \text{ m/s})^2]/(0.090 \text{ m})^2 = 6.3 \times 10^4 \text{ N/m}.$

EVALUATE: When the spring is compressed, the spring force is directed opposite to the displacement of the object and the work done by the spring is negative.

6.81.   **IDENTIFY** and **SET UP:**   Use Eq. (6.6). Work is done by the spring and by gravity. Let point 1 be where the textbook is released and point 2 be where it stops sliding. $x_2 = 0$ since at point 2 the spring is neither stretched nor compressed. The situation is sketched in Figure 6.81.

EXECUTE:

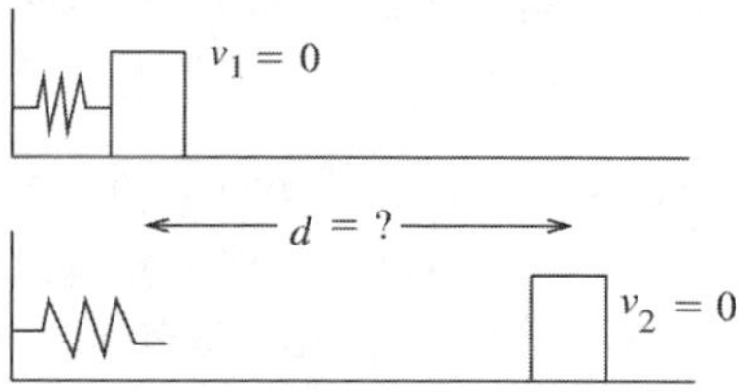

$$v_1 = 0$$

$$d = ?$$

$$v_2 = 0$$

Figure 6.81

$W_{\text{tot}} = K_2 - K_1$
$K_1 = 0, \quad K_2 = 0$
$W_{\text{tot}} = W_{\text{fric}} + W_{\text{spr}}$

$W_{\text{spr}} = \frac{1}{2}kx_1^2$, where $x_1 = 0.250$ m (Spring force is in direction of motion of block so it does positive work.)

$W_{\text{fric}} = -\mu_k mgd$

Then $W_{\text{tot}} = K_2 - K_1$ gives $\frac{1}{2}kx_1^2 - \mu_k mgd = 0$

$d = \dfrac{kx_1^2}{2\mu_k mg} = \dfrac{(250 \text{ N/m})(0.250 \text{ m})^2}{2(0.30)(2.50 \text{ kg})(9.80 \text{ m/s}^2)} = 1.1$ m, measured from the point where the block was released.

EVALUATE: The positive work done by the spring equals the magnitude of the negative work done by friction. The total work done during the motion between points 1 and 2 is zero and the textbook starts and ends with zero kinetic energy.

6.83. **IDENTIFY:** Apply $W_{\text{tot}} = K_2 - K_1$ to the vehicle.

SET UP: Call the bumper compression x and the initial speed v_0. The work done by the spring is $-\frac{1}{2}kx^2$ and $K_2 = 0$.

EXECUTE: **(a)** The necessary relations are $\frac{1}{2}kx^2 = \frac{1}{2}mv_0^2$, $kx < 5\,mg$. Combining to eliminate k and then

x, the two inequalities are $x > \dfrac{v^2}{5g}$ and $k < 25\dfrac{mg^2}{v^2}$. Using the given numerical values,

$x > \dfrac{(20.0 \text{ m/s})^2}{5(9.80 \text{ m/s}^2)} = 8.16$ m and $k < 25\dfrac{(1700 \text{ kg})(9.80 \text{ m/s}^2)^2}{(20.0 \text{ m/s})^2} = 1.02 \times 10^4$ N/m.

(b) A distance of 8 m is not commonly available as space in which to stop a car. Also, the car stops only momentarily and then returns to its original speed when the spring returns to its original length.

EVALUATE: If k were doubled, to 2.04×10^4 N/m, then $x = 5.77$ m. The stopping distance is reduced by a factor of $1/\sqrt{2}$, but the maximum acceleration would then be $kx/m = 69.2$ m/s^2, which is $7.07g$.

6.87. **IDENTIFY and SET UP:** Apply $W_{\text{tot}} = K_2 - K_1$ to the system consisting of both blocks. Since they are connected by the cord, both blocks have the same speed at every point in the motion. Also, when the 6.00-kg block has moved downward 1.50 m, the 8.00-kg block has moved 1.50 m to the right. The target variable, μ_k, will be a factor in the work done by friction. The forces on each block are shown in Figure 6.87.

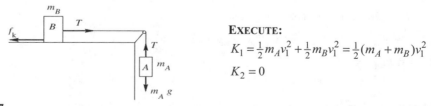

EXECUTE:
$K_1 = \frac{1}{2}m_A v_1^2 + \frac{1}{2}m_B v_1^2 = \frac{1}{2}(m_A + m_B)v_1^2$
$K_2 = 0$

Figure 6.87

The tension T in the rope does positive work on block B and the same magnitude of negative work on block A, so T does no net work on the system. Gravity docs work $W_{mg} = m_A gd$ on block A, where

$d = 2.00$ m. (Block B moves horizontally, so no work is done on it by gravity.) Friction does work $W_{\text{fric}} = -\mu_k m_B g d$ on block B. Thus $W_{\text{tot}} = W_{mg} + W_{\text{fric}} = m_A g d - \mu_k m_B g d$. Then $W_{\text{tot}} = K_2 - K_1$ gives

$m_A g d - \mu_k m_B g d = -\frac{1}{2}(m_A + m_B)v_1^2$ and

$$\mu_k = \frac{m_A}{m_B} + \frac{\frac{1}{2}(m_A + m_B)v_1^2}{m_B g d} = \frac{6.00 \text{ kg}}{8.00 \text{ kg}} + \frac{(6.00 \text{ kg} + 8.00 \text{ kg})(0.900 \text{ m/s})^2}{2(8.00 \text{ kg})(9.80 \text{ m/s}^2)(2.00 \text{ m})} = 0.786$$

EVALUATE: The weight of block A does positive work and the friction force on block B does negative work, so the net work is positive and the kinetic energy of the blocks increases as block A descends. Note that K_1 includes the kinetic energy of both blocks. We could have applied the work-energy theorem to block A alone, but then W_{tot} includes the work done on block A by the tension force.

6.89.   **IDENTIFY:** Apply Eq. (6.6) to the skater.

SET UP: Let point 1 be just before she reaches the rough patch and let point 2 be where she exits from the patch. Work is done by friction. We don't know the skater's mass so can't calculate either friction or the initial kinetic energy. Leave her mass m as a variable and expect that it will divide out of the final equation.

EXECUTE: $f_k = 0.25mg$ so $W_f = W_{\text{tot}} = -(0.25mg)s$, where s is the length of the rough patch.

$W_{\text{tot}} = K_2 - K_1$

$K_1 = \frac{1}{2}mv_0^2$, $K_2 = \frac{1}{2}mv_2^2 = \frac{1}{2}m(0.55v_0)^2 = 0.3025\left(\frac{1}{2}mv_0^2\right)$

The work-energy relation gives $-(0.25mg)s = (0.3025-1)\frac{1}{2}mv_0^2$

The mass divides out, and solving gives $s = 1.3$ m.

EVALUATE: Friction does negative work and this reduces her kinetic energy.

6.91.   **IDENTIFY:** To lift a mass m a height h requires work $W = mgh$. To accelerate mass m from rest to speed v requires $W = K_2 - K_1 = \frac{1}{2}mv^2$. $P_{\text{av}} = \dfrac{\Delta W}{\Delta t}$.

SET UP: $t = 60$ s

EXECUTE: **(a)** $(800 \text{ kg})(9.80 \text{ m/s}^2)(14.0 \text{ m}) = 1.10 \times 10^5$ J

(b) $(1/2)(800 \text{ kg})(18.0 \text{ m/s}^2) = 1.30 \times 10^5$ J.

(c) $\dfrac{1.10 \times 10^5 \text{ J} + 1.30 \times 10^5 \text{ J}}{60 \text{ s}} = 3.99$ kW.

EVALUATE: Approximately the same amount of work is required to lift the water against gravity as to accelerate it to its final speed.

6.93.   **IDENTIFY and SET UP:** Energy is $P_{\text{av}}t$. The total energy expended in one day is the sum of the energy expended in each type of activity.

EXECUTE: 1 day $= 8.64 \times 10^4$ s

Let t_{walk} be the time she spends walking and t_{other} be the time she spends in other activities;

$t_{\text{other}} = 8.64 \times 10^4 \text{ s} - t_{\text{walk}}$.

The energy expended in each activity is the power output times the time, so

$E = Pt = (280 \text{ W})t_{\text{walk}} + (100 \text{ W})t_{\text{other}} = 1.1 \times 10^7$ J

$(280 \text{ W})t_{\text{walk}} + (100 \text{ W})(8.64 \times 10^4 \text{ s} - t_{\text{walk}}) = 1.1 \times 10^7$ J

$(180 \text{ W})t_{\text{walk}} = 2.36 \times 10^6$ J

$t_{\text{walk}} = 1.31 \times 10^4$ s $= 218$ min $= 3.6$ h.

EVALUATE: Her average power for one day is $(1.1 \times 10^7 \text{ J})/([24][3600 \text{ s}]) = 127$ W. This is much closer to her 100 W rate than to her 280 W rate, so most of her day is spent at the 100 W rate.

6.95.   **IDENTIFY and SET UP:** For part (a) calculate m from the volume of blood pumped by the heart in one day. For part (b) use W calculated in part (a) in Eq. (6.15).

EXECUTE: (a) $W = mgh$, as in Example 6.10. We need the mass of blood lifted; we are given the volume

$$V = (7500 \text{ L})\left(\frac{1\times10^{-3} \text{ m}^3}{1 \text{ L}}\right) = 7.50 \text{ m}^3.$$

$$m = \text{density}\times\text{volume} = (1.05\times10^3 \text{ kg/m}^3)(7.50 \text{ m}^3) = 7.875\times10^3 \text{ kg}$$

Then $W = mgh = (7.875\times10^3 \text{ kg})(9.80 \text{ m/s}^2)(1.63 \text{ m}) = 1.26\times10^5 \text{ J}.$

(b) $P_{\text{av}} = \dfrac{\Delta W}{\Delta t} = \dfrac{1.26\times10^5 \text{ J}}{(24 \text{ h})(3600 \text{ s/h})} = 1.46 \text{ W}.$

EVALUATE: Compared to light bulbs or common electrical devices, the power output of the heart is rather small.

6.97. **IDENTIFY:** $P = F_{\parallel}v$. The force required to give mass m an acceleration a is $F = ma$. For an incline at an angle α above the horizontal, the component of mg down the incline is $mg\sin\alpha$.

SET UP: For small α, $\sin\alpha \approx \tan\alpha$.

EXECUTE: (a) $P_0 = Fv = (53\times10^3 \text{ N})(45 \text{ m/s}) = 2.4 \text{ MW}.$

(b) $P_1 = mav = (9.1\times10^5 \text{ kg})(1.5 \text{ m/s}^2)(45 \text{ m/s}) = 61 \text{ MW}.$

(c) Approximating $\sin\alpha$, by $\tan\alpha$, and using the component of gravity down the incline as $mg\sin\alpha$,

$$P_2 = (mg\sin\alpha)v = (9.1\times10^5 \text{ kg})(9.80 \text{ m/s}^2)(0.015)(45 \text{ m/s}) = 6.0 \text{ MW}.$$

EVALUATE: From Problem 6.96, we would expect that a 0.15 m/s^2 acceleration and a 1.5% slope would require the same power. We found that a 1.5 m/s^2 acceleration requires ten times more power than a 1.5% slope, which is consistent.

6.99. **IDENTIFY** and **SET UP:** Use Eq. (6.18) to relate the forces to the power required. The air resistance force is $F_{\text{air}} = \frac{1}{2}CA\rho v^2$, where C is the drag coefficient.

EXECUTE: (a) $P = F_{\text{tot}}v$, with $F_{\text{tot}} = F_{\text{roll}} + F_{\text{air}}$

$$F_{\text{air}} = \tfrac{1}{2}CA\rho v^2 = \tfrac{1}{2}(1.0)(0.463 \text{ m}^3)(1.2 \text{ kg/m}^3)(12.0 \text{ m/s})^2 = 40.0 \text{ N}$$

$$F_{\text{roll}} = \mu_r n = \mu_r w = (0.0045)(490 \text{ N} + 118 \text{ N}) = 2.74 \text{ N}$$

$$P = (F_{\text{roll}} + F_{\text{air}})v = (2.74 \text{ N} + 40.0 \text{ N})(12.0 \text{ s}) = 513 \text{ W}$$

(b) $F_{\text{air}} = \tfrac{1}{2}CA\rho v^2 = \tfrac{1}{2}(0.88)(0.366 \text{ m}^3)(1.2 \text{ kg/m}^3)(12.0 \text{ m/s})^2 = 27.8 \text{ N}$

$$F_{\text{roll}} = \mu_r n = \mu_r w = (0.0030)(490 \text{ N} + 88 \text{ N}) = 1.73 \text{ N}$$

$$P = (F_{\text{roll}} + F_{\text{air}})v = (1.73 \text{ N} + 27.8 \text{ N})(12.0 \text{ s}) = 354 \text{ W}$$

(c) $F_{\text{air}} = \tfrac{1}{2}CA\rho v^2 = \tfrac{1}{2}(0.88)(0.366 \text{ m}^3)(1.2 \text{ kg/m}^3)(6.0 \text{ m/s})^2 = 6.96 \text{ N}$

$$F_{\text{roll}} = \mu_r n = 1.73 \text{ N} \text{ (unchanged)}$$

$$P = (F_{\text{roll}} + F_{\text{air}})v = (1.73 \text{ N} + 6.96 \text{ N})(6.0 \text{ m/s}) = 52.1 \text{ W}$$

EVALUATE: Since F_{air} is proportional to v^2 and $P = Fv$, reducing the speed greatly reduces the power required.

6.101. **IDENTIFY** and **SET UP:** Use Eq. (6.18) to relate F and P. In part (a), F is the retarding force. In parts (b) and (c), F includes gravity.

EXECUTE: (a) $P = Fv$, so $F = P/v$.

$$P = (8.00 \text{ hp})\left(\frac{746 \text{ W}}{1 \text{ hp}}\right) = 5968 \text{ W}$$

$$v = (60.0 \text{ km/h})\left(\frac{1000 \text{ m}}{1 \text{ km}}\right)\left(\frac{1 \text{ h}}{3600 \text{ s}}\right) = 16.67 \text{ m/s}$$

$$F = \frac{P}{v} = \frac{5968 \text{ W}}{16.67 \text{ m/s}} = 358 \text{ N}.$$

(b) The power required is the 8.00 hp of part (a) plus the power P_g required to lift the car against gravity. The situation is sketched in Figure 6.101.

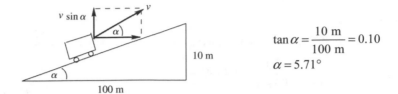

$$\tan\alpha = \frac{10 \text{ m}}{100 \text{ m}} = 0.10$$

$$\alpha = 5.71°$$

Figure 6.101

The vertical component of the velocity of the car is $v\sin\alpha = (16.67 \text{ m/s})\sin 5.71° = 1.658 \text{ m/s}$.

Then $P_g = F(v\sin a) = mgv\sin\alpha = (1800 \text{ kg})(9.80 \text{ m/s}^2)(1.658 \text{ m/s}) = 2.92\times10^4 \text{ W}$

$$P_g = 2.92\times10^4 \text{ W}\left(\frac{1 \text{ hp}}{746 \text{ W}}\right) = 39.1 \text{ hp}$$

The total power required is $8.00 \text{ hp} + 39.1 \text{ hp} = 47.1 \text{ hp}$.

(c) The power required from the engine is *reduced* by the rate at which gravity does positive work. The road incline angle α is given by $\tan\alpha = 0.0100$, so $\alpha = 0.5729°$.

$P_g = mg(v\sin\alpha) = (1800 \text{ kg})(9.80 \text{ m/s}^2)(16.67 \text{ m/s})\sin 0.5729° = 2.94\times10^3 \text{ W} = 3.94 \text{ hp}$.

The power required from the engine is then $8.00 \text{ hp} - 3.94 \text{ hp} = 4.06 \text{ hp}$.

(d) No power is needed from the engine if gravity does work at the rate of $P_g = 8.00 \text{ hp} = 5968 \text{ W}$.

$$P_g = mgv\sin\alpha, \text{ so } \sin\alpha = \frac{P_g}{mgv} = \frac{5968 \text{ W}}{(1800 \text{ kg})(9.80 \text{ m/s}^2)(16.67 \text{ m/s})} = 0.02030$$

$\alpha = 1.163°$ and $\tan\alpha = 0.0203$, a 2.03% grade.

EVALUATE: More power is required when the car goes uphill and less when it goes downhill. In part (d), at this angle the component of gravity down the incline is $mg\sin\alpha = 358 \text{ N}$ and this force cancels the retarding force and no force from the engine is required. The retarding force depends on the speed so it is the same in parts (a), (b) and (c).

POTENTIAL ENERGY AND ENERGY CONSERVATION

7.3. **IDENTIFY:** Use the free-body diagram for the bag and Newton's first law to find the force the worker applies. Since the bag starts and ends at rest, $K_2 - K_1 = 0$ and $W_{tot} = 0$.

SET UP: A sketch showing the initial and final positions of the bag is given in Figure 7.3a. $\sin\phi = \dfrac{2.0 \text{ m}}{3.5 \text{ m}}$ and $\phi = 34.85°$. The free-body diagram is given in Figure 7.3b. $\vec{F}$ is the horizontal force applied by the worker. In the calculation of U_{grav} take $+y$ upward and $y = 0$ at the initial position of the bag.

EXECUTE: **(a)** $\Sigma F_y = 0$ gives $T\cos\phi = mg$ and $\Sigma F_x = 0$ gives $F = T\sin\phi$. Combining these equations to eliminate T gives $F = mg\tan\phi = (120 \text{ kg})(9.80 \text{ m/s}^2)\tan 34.85° = 820 \text{ N}$.

(b) (i) The tension in the rope is radial and the displacement is tangential so there is no component of T in the direction of the displacement during the motion and the tension in the rope does no work.

(ii) $W_{tot} = 0$ so $W_{worker} = -W_{grav} = U_{grav,2} - U_{grav,1} = mg(y_2 - y_1) = (120 \text{ kg})(9.80 \text{ m/s}^2)(0.6277 \text{ m}) = 740 \text{ J}$.

EVALUATE: The force applied by the worker varies during the motion of the bag and it would be difficult to calculate W_{worker} directly.

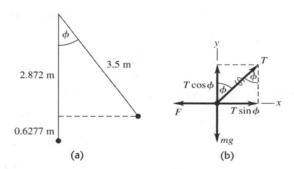

Figure 7.3

7.9. **IDENTIFY:** $W_{tot} = K_B - K_A$. The forces on the rock are gravity, the normal force and friction.

SET UP: Let $y = 0$ at point B and let $+y$ be upward. $y_A = R = 0.50 \text{ m}$. The work done by friction is negative; $W_f = -0.22 \text{ J}$. $K_A = 0$. The free-body diagram for the rock at point B is given in Figure 7.9. The acceleration of the rock at this point is $a_{rad} = v^2/R$, upward.

EXECUTE: **(a)** (i) The normal force is perpendicular to the displacement and does zero work.

(ii) $W_{grav} = U_{grav,A} - U_{grav,B} = mgy_A = (0.20 \text{ kg})(9.80 \text{ m/s}^2)(0.50 \text{ m}) = 0.98 \text{ J}$.

(b) $W_{tot} = W_n + W_f + W_{grav} = 0 + (-0.22 \text{ J}) + 0.98 \text{ J} = 0.76 \text{ J}$. $W_{tot} = K_B - K_A$ gives $\frac{1}{2}mv_B^2 = W_{tot}$.

$v_B = \sqrt{\dfrac{2W_{tot}}{m}} = \sqrt{\dfrac{2(0.76 \text{ J})}{0.20 \text{ kg}}} = 2.8 \text{ m/s}$.

(c) Gravity is constant and equal to mg. n is not constant; it is zero at A and not zero at B. Therefore, $f_k = \mu_k n$ is also not constant.

(d) $\Sigma F_y = ma_y$ applied to Figure 7.9 gives $n - mg = ma_{\text{rad}}$.

$$n = m\left(g + \frac{v^2}{R}\right) = (0.20 \text{ kg})\left(9.80 \text{ m/s}^2 + \frac{[2.8 \text{ m/s}]^2}{0.50 \text{ m}}\right) = 5.1 \text{ N}.$$

EVALUATE: In the absence of friction, the speed of the rock at point B would be $\sqrt{2gR} = 3.1$ m/s. As the rock slides through point B, the normal force is greater than the weight $mg = 2.0$ N of the rock.

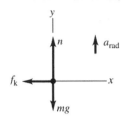

Figure 7.9

7.11. **IDENTIFY:** Apply Eq. (7.7) to the motion of the car.
SET UP: Take $y = 0$ at point A. Let point 1 be A and point 2 be B.

$$K_1 + U_1 + W_{\text{other}} = K_2 + U_2$$

EXECUTE: $U_1 = 0$, $U_2 = mg(2R) = 28{,}224$ J, $W_{\text{other}} = W_f$

$K_1 = \frac{1}{2}mv_1^2 = 37{,}500$ J, $K_2 = \frac{1}{2}mv_2^2 = 3840$ J

The work-energy relation then gives $W_f = K_2 + U_2 - K_1 = -5400$ J.

EVALUATE: Friction does negative work. The final mechanical energy $(K_2 + U_2 = 32{,}064$ J$)$ is less than the initial mechanical energy $(K_1 + U_1 = 37{,}500$ J$)$ because of the energy removed by friction work.

7.19. **IDENTIFY** and **SET UP:** Use energy methods. There are changes in both elastic and gravitational potential energy; elastic; $U = \frac{1}{2}kx^2$, gravitational: $U = mgy$.

EXECUTE: (a) $U = \frac{1}{2}kx^2$ so $x = \sqrt{\dfrac{2U}{k}} = \sqrt{\dfrac{2(3.20 \text{ J})}{1600 \text{ N/m}}} = 0.0632$ m $= 6.32$ cm

(b) Points 1 and 2 in the motion are sketched in Figure 7.19.

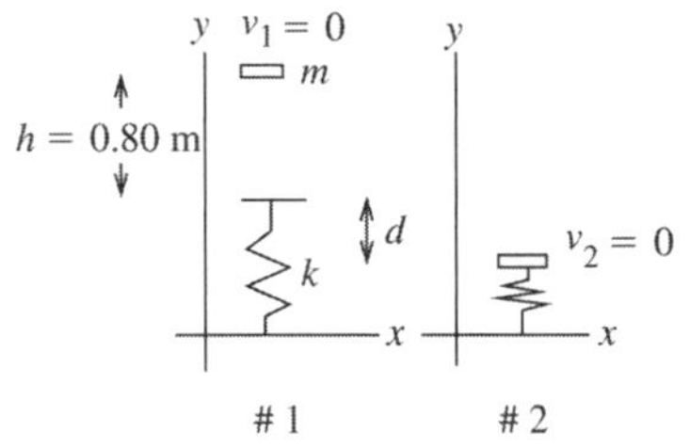

$$K_1 + U_1 + W_{\text{other}} = K_2 + U_2$$

$W_{\text{other}} = 0$ (Only work is that done by gravity and spring force)

$K_1 = 0$, $K_2 = 0$

$y = 0$ at final position of book

$U_1 = mg(h + d)$, $U_2 = \frac{1}{2}kd^2$

Figure 7.19

$$0 + mg(h + d) + 0 = \frac{1}{2}kd^2$$

The original gravitational potential energy of the system is converted into potential energy of the compressed spring.

$$\frac{1}{2}kd^2 - mgd - mgh = 0$$

$$d = \frac{1}{k}\left(mg \pm \sqrt{(mg)^2 + 4\left(\frac{1}{2}k\right)(mgh)}\right)$$

d must be positive, so $d = \dfrac{1}{k}\left(mg + \sqrt{(mg)^2 + 2kmgh}\right)$

$d = \dfrac{1}{1600 \text{ N/m}}(1.20 \text{ kg})(9.80 \text{ m/s}^2) +$

$$\sqrt{((1.20 \text{ kg})(9.80 \text{ m/s}^2))^2 + 2(1600 \text{ N/m})(1.20 \text{ kg})(9.80 \text{ m/s}^2)(0.80 \text{ m})}$$

$d = 0.0074 \text{ m} + 0.1087 \text{ m} = 0.12 \text{ m} = 12 \text{ cm}$

EVALUATE: It was important to recognize that the total displacement was $h + d$; gravity continues to do work as the book moves against the spring. Also note that with the spring compressed 0.12 m it exerts an upward force (192 N) greater than the weight of the book (11.8 N). The book will be accelerated upward from this position.

7.23. **IDENTIFY:** Only the spring does work and Eq. (7.11) applies. $a = \dfrac{F}{m} = \dfrac{-kx}{m}$, where F is the force the spring exerts on the mass.

SET UP: Let point 1 be the initial position of the mass against the compressed spring, so $K_1 = 0$ and $U_1 = 11.5$ J. Let point 2 be where the mass leaves the spring, so $U_{el,2} = 0$.

EXECUTE: (a) $K_1 + U_{el,1} = K_2 + U_{el,2}$ gives $U_{el,1} = K_2$. $\frac{1}{2}mv_2^2 = U_{el,1}$ and

$v_2 = \sqrt{\dfrac{2U_{el,1}}{m}} = \sqrt{\dfrac{2(11.5 \text{ J})}{2.50 \text{ kg}}} = 3.03 \text{ m/s}.$

K is largest when U_{el} is least and this is when the mass leaves the spring. The mass achieves its maximum speed of 3.03 m/s as it leaves the spring and then slides along the surface with constant speed.

(b) The acceleration is greatest when the force on the mass is the greatest, and this is when the spring has its maximum compression. $U_{el} = \frac{1}{2}kx^2$ so $x = -\sqrt{\dfrac{2U_{el}}{k}} = 2\sqrt{\dfrac{2(11.5 \text{ J})}{2500 \text{ N/m}}} = -0.0959 \text{ m}.$ The minus sign

indicates compression. $F = -kx = ma_x$ and $a_x = -\dfrac{kx}{m} = -\dfrac{(2500 \text{ N/m})(-0.0959 \text{ m})}{2.50 \text{ kg}} = 95.9 \text{ m/s}^2.$

EVALUATE: If the end of the spring is displaced to the left when the spring is compressed, then a_x in part (b) is to the right, and vice versa.

7.25. **IDENTIFY:** Apply Eq. (7.13) and $F = ma$.

SET UP: $W_{other} = 0.$ There is no change in U_{grav}. $K_1 = 0$, $U_2 = 0$.

EXECUTE: $\frac{1}{2}kx^2 = \frac{1}{2}mv_x^2$. The relations for m, v_x, k and x are $kx^2 = mv_x^2$ and $kx = 5mg$.

Dividing the first equation by the second gives $x = \dfrac{v_x^2}{5g}$, and substituting this into the second gives

$k = 25\dfrac{mg^2}{v_x^2}.$

(a) $k = 25\dfrac{(1160 \text{ kg})(9.80 \text{ m/s}^2)^2}{(2.50 \text{ m/s})^2} = 4.46 \times 10^5 \text{ N/m}$

(b) $x = \dfrac{(2.50 \text{ m/s})^2}{5(9.80 \text{ m/s}^2)} = 0.128 \text{ m}$

EVALUATE: Our results for k and x do give the required values for a_x and v_x:

$$a_x = \frac{kx}{m} = \frac{(4.46 \times 10^5 \text{ N/m})(0.128 \text{ m})}{1160 \text{ kg}} = 49.2 \text{ m/s}^2 = 5.0g \text{ and } v_x = x\sqrt{\frac{k}{m}} = 2.5 \text{ m/s}.$$

7.31. **IDENTIFY and SET UP:** The friction force is constant during each displacement and Eq. (6.2) can be used to calculate work, but the direction of the friction force can be different for different displacements.

$f = \mu_k mg = (0.25)(1.5 \text{ kg})(9.80 \text{ m/s}^2) = 3.675 \text{ N}$; direction of $\vec{f}$ is opposite to the motion.

EXECUTE: **(a)** The path of the book is sketched in Figure 7.31a.

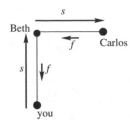

Figure 7.31a

For the motion from you to Beth the friction force is directed opposite to the displacement $\vec{s}$ and $W_1 = -fs = -(3.675 \text{ N})(8.0 \text{ m}) = -29.4 \text{ J}$.

For the motion from Beth to Carlos the friction force is again directed opposite to the displacement and $W_2 = -29.4 \text{ J}$.

$$W_{\text{tot}} = W_1 + W_2 = -29.4 \text{ J} - 29.4 \text{ J} = -59 \text{ J}$$

(b) The path of the book is sketched in Figure 7.31b.

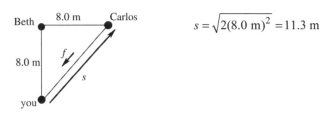

$$s = \sqrt{2(8.0 \text{ m})^2} = 11.3 \text{ m}$$

Figure 7.31b

$\vec{f}$ is opposite to $\vec{s}$, so $W = -fs = -(3.675 \text{ N})(11.3 \text{ m}) = -42 \text{ J}$

(c)

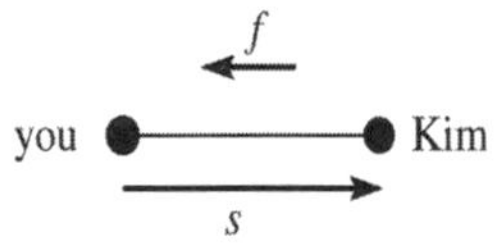

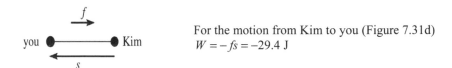

For the motion from you to Kim (Figure 7.31c)
$$W = -fs$$
$$W = -(3.675 \text{ N})(8.0 \text{ m}) = -29.4 \text{ J}$$

Figure 7.31c

For the motion from Kim to you (Figure 7.31d)
$$W = -fs = -29.4 \text{ J}$$

Figure 7.31d

The total work for the round trip is $-29.4 \text{ J} - 29.4 \text{ J} = -59 \text{ J}$.

(d) EVALUATE: Parts (a) and (b) show that for two different paths between you and Carlos, the work done by friction is different. Part (c) shows that when the starting and ending points are the same, the total work is not zero. Both these results show that the friction force is nonconservative.

7.33.   **IDENTIFY:** Some of the mechanical energy of the skier is converted to internal energy by the nonconservative force of friction on the rough patch.

SET UP: For part (a) use $E_{\text{mech},f} = E_{\text{mech},i} - f_k s$ where $f_k = \mu_k mg$. Let $y_f = 0$ at the bottom of the hill; then $y_i = 2.50$ m along the rough patch. The energy equation is thus $\frac{1}{2}mv_f^2 = \frac{1}{2}mv_i^2 + mgy_i - \mu_k mgs$.

Solving for her final speed gives $v_f = \sqrt{v_i^2 + 2gy_i - 2\mu_k gs}$. For part (b), the internal energy is calculated as the negative of the work done by friction: $-W_f = +f_k s = +\mu_k mgs$.

EXECUTE: **(a)** $v_f = \sqrt{(6.50\ \text{m/s})^2 + 2(9.80\ \text{m/s}^2)(2.50\ \text{m}) - 2(0.300)(9.80\ \text{m/s}^2)(3.50\ \text{m})} = 8.41$ m/s.

(b) Internal energy $= \mu_k mgs = (0.300)(62.0\ \text{kg})(9.80\ \text{m/s}^2)(3.50\ \text{m}) = 638$ J.

EVALUATE: Without friction the skier would be moving faster at the bottom of the hill than at the top, but in this case she is moving *slower* because friction converted some of her initial kinetic energy into internal energy.

7.35. **IDENTIFY:** Apply Eq. (7.16).

SET UP: The sign of F_x indicates its direction.

EXECUTE: $F_x = -\dfrac{dU}{dx} = -4\alpha x^3 = -(4.8\ \text{J/m}^4)x^3$. $F_x(-0.800\ \text{m}) = -(4.8\ \text{J/m}^4)(-0.80\ \text{m})^3 = 2.46$ N. The force is in the $+x$-direction.

EVALUATE: $F_x > 0$ when $x < 0$ and $F_x < 0$ when $x > 0$, so the force is always directed towards the origin.

7.39. **IDENTIFY** and **SET UP:** Use Eq. (7.17) to calculate the force from U. At equilibrium $F = 0$.

(a) EXECUTE: The graphs are sketched in Figure 7.39.

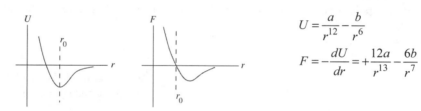

Figure 7.39

(b) At equilibrium $F = 0$, so $\dfrac{dU}{dr} = 0$

$$F = 0 \text{ implies } \frac{+12a}{r^{13}} - \frac{6b}{r^7} = 0$$

$6br^6 = 12a$; solution is the equilibrium distance $r_0 = (2a/b)^{1/6}$

U is a minimum at this r; the equilibrium is stable.

(c) At $r = (2a/b)^{1/6}$, $U = a/r^{12} - b/r^6 = a(b/2a)^2 - b(b/2a) = -b^2/4a$.

At $r \to \infty$, $U = 0$. The energy that must be added is $-\Delta U = b^2/4a$.

(d) $r_0 = (2a/b)^{1/6} = 1.13 \times 10^{-10}$ m gives that

$2a/b = 2.082 \times 10^{-60}$ m^6 and $b/4a = 2.402 \times 10^{59}$ m^{-6}

$b^2/4a = b(b/4a) = 1.54 \times 10^{-18}$ J

$b(2.402 \times 10^{59}\ \text{m}^{-6}) = 1.54 \times 10^{-18}$ J and $b = 6.41 \times 10^{-78}$ J·m^6.

Then $2a/b = 2.082 \times 10^{-60}$ m^6 gives $a = (b/2)(2.082 \times 10^{-60}\ \text{m}^6) =$

$\frac{1}{2}(6.41 \times 10^{-78}\ \text{J·m}^6)(2.082 \times 10^{-60}\ \text{m}^6) = 6.67 \times 10^{-138}$ J·m^{12}

EVALUATE: As the graphs in part (a) show, $F(r)$ is the slope of $U(r)$ at each r. $U(r)$ has a minimum where $F = 0$.

7.41.   **IDENTIFY:** Apply $\Sigma\vec{F} = m\vec{a}$ to the bag and to the box. Apply Eq. (7.7) to the motion of the system of the box and bucket after the bag is removed.

SET UP: Let $y = 0$ at the final height of the bucket, so $y_1 = 2.00$ m and $y_2 = 0$. $K_1 = 0$. The box and the bucket move with the same speed v, so $K_2 = \frac{1}{2}(m_{box} + m_{bucket})v^2$. $W_{other} = -f_k d$, with $d = 2.00$ m and $f_k = \mu_k m_{box} g$. Before the bag is removed, the maximum possible friction force the roof can exert on the box is $(0.700)(80.0 \text{ kg} + 50.0 \text{ kg})(9.80 \text{ m/s}^2) = 892$ N. This is larger than the weight of the bucket (637 N), so before the bag is removed the system is at rest.

EXECUTE: **(a)** The friction force on the bag of gravel is zero, since there is no other horizontal force on the bag for friction to oppose. The static friction force on the box equals the weight of the bucket, 637 N.

(b) Eq. (7.7) gives $m_{bucket}gy_1 - f_k d = \frac{1}{2}m_{tot}v^2$, with $m_{tot} = 145.0$ kg. $v = \sqrt{\dfrac{2}{m_{tot}}(m_{bucket}gy_1 - \mu_k m_{box} gd)}$.

$v = \sqrt{\dfrac{2}{145.0 \text{ kg}}[(65.0 \text{ kg})(9.80 \text{ m/s}^2)(2.00 \text{ m}) - (0.400)(80.0 \text{ kg})(9.80 \text{ m/s}^2)(2.00 \text{ m})]}$.

$v = 2.99$ m/s.

EVALUATE: If we apply $\Sigma\vec{F} = m\vec{a}$ to the box and to the bucket we can calculate their common acceleration a. Then a constant acceleration equation applied to either object gives $v = 2.99$ m/s, in agreement with our result obtained using energy methods.

7.43.   **IDENTIFY:** Use the work-energy theorem, Eq. (7.7). The target variable μ_k will be a factor in the work done by friction.

SET UP: Let point 1 be where the block is released and let point 2 be where the block stops, as shown in Figure 7.43.

$K_1 + U_1 + W_{other} = K_2 + U_2$

Work is done on the block by the spring and by friction, so $W_{other} = W_f$ and $U = U_{el}$.

Figure 7.43

EXECUTE: $K_1 = K_2 = 0$

$U_1 = U_{1,el} = \frac{1}{2}kx_1^2 = \frac{1}{2}(100 \text{ N/m})(0.200 \text{ m})^2 = 2.00$ J

$U_2 = U_{2,el} = 0$, since after the block leaves the spring has given up all its stored energy

$W_{other} = W_f = (f_k\cos\phi)s = \mu_k mg(\cos\phi)s = -\mu_k mgs$, since $\phi = 180°$ (The friction force is directed opposite to the displacement and does negative work.)

Putting all this into $K_1 + U_1 + W_{other} = K_2 + U_2$ gives

$U_{1,el} + W_f = 0$

$\mu_k mgs - U_{1,el}$

$\mu_k = \dfrac{U_{1,el}}{mgs} = \dfrac{2.00 \text{ J}}{(0.50 \text{ kg})(9.80 \text{ m/s}^2)(1.00 \text{ m})} = 0.41$.

EVALUATE: $U_{1,el} + W_f = 0$ says that the potential energy originally stored in the spring is taken out of the system by the negative work done by friction.

7.45.   **IDENTIFY:** The mechanical energy of the roller coaster is conserved since there is no friction with the track. We must also apply Newton's second law for the circular motion.

SET UP: For part (a), apply conservation of energy to the motion from point A to point B:

$K_B + U_{grav,B} = K_A + U_{grav,A}$ with $K_A = 0$. Defining $y_B = 0$ and $y_A = 13.0$ m, conservation of energy

becomes $\frac{1}{2}mv_B^2 = mgy_A$ or $v_B = \sqrt{2gy_A}$. In part (b), the free-body diagram for the roller coaster car at

point B is shown in Figure 7.45. $\Sigma F_y = ma_y$ gives $mg + n = ma_{rad}$, where $a_{rad} = v^2/r$. Solving for the

normal force gives $n = m\left(\dfrac{v^2}{r} - g\right)$.

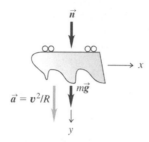

Figure 7.45

EXECUTE: **(a)** $v_B = \sqrt{2(9.80 \text{ m/s}^2)(13.0 \text{ m})} = 16.0$ m/s.

(b) $n = (350 \text{ kg})\left[\dfrac{(16.0 \text{ m/s})^2}{6.0 \text{ m}} - 9.80 \text{ m/s}^2\right] = 1.15 \times 10^4$ N.

EVALUATE: The normal force n is the force that the tracks exert on the roller coaster car. The car exerts a force of equal magnitude and opposite direction on the tracks.

7.47.   **(a) IDENTIFY:**   Use work-energy relation to find the kinetic energy of the wood as it enters the rough bottom.

SET UP: Let point 1 be where the piece of wood is released and point 2 be just before it enters the rough bottom. Let $y = 0$ be at point 2.

EXECUTE: $U_1 = K_2$ gives $K_2 = mgy_1 = 78.4$ J.

IDENTIFY: Now apply work-energy relation to the motion along the rough bottom.

SET UP: Let point 1 be where it enters the rough bottom and point 2 be where it stops.

$$K_1 + U_1 + W_{other} = K_2 + U_2$$

EXECUTE: $W_{other} = W_f = -\mu_k mgs$, $K_2 = U_1 = U_2 = 0$; $K_1 = 78.4$ J

$78.4 \text{ J} - \mu_k mgs = 0$; solving for s gives $s = 20.0$ m.

The wood stops after traveling 20.0 m along the rough bottom.

(b) Friction does -78.4 J of work.

EVALUATE: The piece of wood stops before it makes one trip across the rough bottom. The final mechanical energy is zero. The negative friction work takes away all the mechanical energy initially in the system.

7.49.   **IDENTIFY:**   Apply Eq. (7.7) to the motion of the stone.

SET UP: $K_1 + U_1 + W_{other} = K_2 + U_2$

Let point 1 be point A and point 2 be point B. Take $y = 0$ at point B.

EXECUTE: $mgy_1 + \frac{1}{2}mv_1^2 = \frac{1}{2}mv_2^2$, with $h = 20.0$ m and $v_1 = 10.0$ m/s

$v_2 = \sqrt{v_1^2 + 2gh} = 22.2$ m/s

EVALUATE: The loss of gravitational potential energy equals the gain of kinetic energy.

(b) IDENTIFY: Apply Eq. (7.8) to the motion of the stone from point B to where it comes to rest against the spring.

SET UP: Use $K_1 + U_1 + W_{other} = K_2 + U_2$, with point 1 at B and point 2 where the spring has its maximum compression x.

EXECUTE: $U_1 = U_2 = K_2 = 0$; $K_1 = \frac{1}{2}mv_1^2$ with $v_1 = 22.2$ m/s

$W_{\text{other}} = W_f + W_{\text{el}} = -\mu_k mgs - \frac{1}{2}kx^2$, with $s = 100$ m $+ x$

The work-energy relation gives $K_1 + W_{\text{other}} = 0$.

$\frac{1}{2}mv_1^2 - \mu_k mgs - \frac{1}{2}kx^2 = 0$

Putting in the numerical values gives $x^2 + 29.4x - 750 = 0$. The positive root to this equation is $x = 16.4$ m.

EVALUATE: Part of the initial mechanical (kinetic) energy is removed by friction work and the rest goes into the potential energy stored in the spring.

(c) IDENTIFY and **SET UP:** Consider the forces.

EXECUTE: When the spring is compressed $x = 16.4$ m the force it exerts on the stone is

$F_{\text{el}} = kx = 32.8$ N. The maximum possible static friction force is

$\max f_s = \mu_s mg = (0.80)(15.0 \text{ kg})(9.80 \text{ m/s}^2) = 118$ N.

EVALUATE: The spring force is less than the maximum possible static friction force so the stone remains at rest.

7.51. **IDENTIFY:** Apply $K_1 + U_1 + W_{\text{other}} = K_2 + U_2$ to the motion of the person.

SET UP: Point 1 is where he steps off the platform and point 2 is where he is stopped by the cord. Let $y = 0$ at point 2. $y_1 = 41.0$ m. $W_{\text{other}} = -\frac{1}{2}kx^2$, where $x = 11.0$ m is the amount the cord is stretched at point 2. The cord does negative work.

EXECUTE: $K_1 = K_2 = U_2 = 0$, so $mgy_1 - \frac{1}{2}kx^2 = 0$ and $k = 631$ N/m.

Now apply $F = kx$ to the test pulls:

$F = kx$ so $x = F/k = 0.602$ m.

EVALUATE: All his initial gravitational potential energy is taken away by the negative work done by the force exerted by the cord, and this amount of energy is stored as elastic potential energy in the stretched cord.

7.53. **IDENTIFY:** Use the work-energy theorem, Eq. (7.7). Solve for K_2 and then for v_2.

SET UP: Let point 1 be at his initial position against the compressed spring and let point 2 be at the end of the barrel, as shown in Figure 7.53. Use $F = kx$ to find the amount the spring is initially compressed by the 4400 N force.

$K_1 + U_1 + W_{\text{other}} = K_2 + U_2$

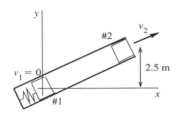

Take $y = 0$ at his initial position.

EXECUTE: $K_1 = 0$, $K_2 = \frac{1}{2}mv_2^2$

$W_{\text{other}} = W_{\text{fric}} = -fs$

$W_{\text{other}} = -(40 \text{ N})(4.0 \text{ m}) = -160$ J

Figure 7.53

$U_{1,\text{grav}} = 0$, $U_{1,\text{el}} = \frac{1}{2}kd^2$, where d is the distance the spring is initially compressed.

$F = kd$ so $d = \dfrac{F}{k} = \dfrac{4400 \text{ N}}{1100 \text{ N/m}} = 4.00$ m

and $U_{1,\text{el}} = \frac{1}{2}(1100 \text{ N/m})(4.00 \text{ m})^2 = 8800$ J

$U_{2,\text{grav}} = mgy_2 = (60 \text{ kg})(9.80 \text{ m/s}^2)(2.5 \text{ m}) = 1470$ J, $U_{2,\text{el}} = 0$

Then $K_1 + U_1 + W_{\text{other}} = K_2 + U_2$ gives

$8800 \text{ J} - 160 \text{ J} = \frac{1}{2}mv_2^2 + 1470$ J

$\frac{1}{2}mv_2^2 = 7170$ J and $v_2 = \sqrt{\dfrac{2(7170 \text{ J})}{60 \text{ kg}}} = 15.5$ m/s

EVALUATE: Some of the potential energy stored in the compressed spring is taken away by the work done by friction. The rest goes partly into gravitational potential energy and partly into kinetic energy.

7.55. **IDENTIFY:** Apply Eq. (7.7) to the system consisting of the two buckets. If we ignore the inertia of the pulley we ignore the kinetic energy it has.

SET UP: $K_1 + U_1 + W_{other} = K_2 + U_2$. Points 1 and 2 in the motion are sketched in Figure 7.55.

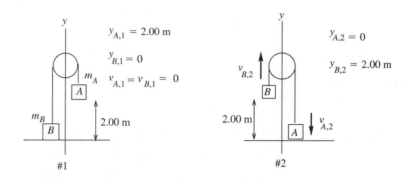

Figure 7.55

The tension force does positive work on the 4.0 kg bucket and an equal amount of negative work on the 12.0 kg bucket, so the net work done by the tension is zero.

Work is done on the system only by gravity, so $W_{other} = 0$ and $U = U_{grav}$

EXECUTE: $K_1 = 0$

$K_2 = \frac{1}{2} m_A v_{A,2}^2 + \frac{1}{2} m_B v_{B,2}^2$ But since the two buckets are connected by a rope they move together and have the same speed: $v_{A,2} = v_{B,2} = v_2$.

Thus $K_2 = \frac{1}{2}(m_A + m_B)v_2^2 = (8.00 \text{ kg})v_2^2$.

$U_1 = m_A g y_{A,1} = (12.0 \text{ kg})(9.80 \text{ m/s}^2)(2.00 \text{ m}) = 235.2 \text{ J}$.

$U_2 = m_B g y_{B,2} = (4.0 \text{ kg})(9.80 \text{ m/s}^2)(2.00 \text{ m}) = 78.4 \text{ J}$.

Putting all this into $K_1 + U_1 + W_{other} = K_2 + U_2$ gives

$U_1 = K_2 + U_2$

$235.2 \text{ J} = (8.00 \text{ kg})v_2^2 + 78.4 \text{ J}$

$v_2 = \sqrt{\dfrac{235.2 \text{ J} - 78.4 \text{ J}}{8.00 \text{ kg}}} = 4.4 \text{ m/s}$

EVALUATE: The gravitational potential energy decreases and the kinetic energy increases by the same amount. We could apply Eq. (7.7) to one bucket, but then we would have to include in W_{other} the work done on the bucket by the tension T.

7.57. **IDENTIFY:** Apply $K_1 + U_1 + W_{other} = K_2 + U_2$

SET UP: $U_1 = U_2 = K_2 = 0$. $W_{other} = W_f = -\mu_k mgs$, with $s = 280 \text{ ft} = 85.3 \text{ m}$

EXECUTE: **(a)** The work-energy expression gives $\frac{1}{2} mv_1^2 - \mu_k mgs = 0$.

$v_1 = \sqrt{2\mu_k gs} = 22.4 \text{ m/s} = 50 \text{ mph}$; the driver was speeding.

(b) 15 mph over speed limit so $150 ticket.

EVALUATE: The negative work done by friction removes the kinetic energy of the object.

7.59. **(a) IDENTIFY and SET UP:** Apply Eq. (7.7) to the motion of the potato. Let point 1 be where the potato is released and point 2 be at the lowest point in its motion, as shown in Figure 7.59a.

$K_1 + U_1 + W_{other} = K_2 + U_2$

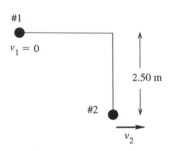

$y_1 = 2.50$ m

$y_2 = 0$

The tension in the string is at all points in the motion perpendicular to the displacement, so $W_r = 0$

The only force that does work on the potato is gravity, so $W_{\text{other}} = 0$.

Figure 7.59a

EXECUTE: $K_1 = 0$, $K_2 = \frac{1}{2}mv_2^2$, $U_1 = mgy_1$, $U_2 = 0$. Thus $U_1 = K_2$. $mgy_1 = \frac{1}{2}mv_2^2$, which gives

$v_2 = \sqrt{2gy_1} = \sqrt{2(9.80 \text{ m/s}^2)(2.50 \text{ m})} = 7.00$ m/s.

EVALUATE: The speed v_2 is the same as if the potato fell through 2.50 m.

(b) IDENTIFY: Apply $\Sigma \vec{F} = m\vec{a}$ to the potato. The potato moves in an arc of a circle so its acceleration is $\vec{a}_{\text{rad}}$, where $a_{\text{rad}} = v^2/R$ and is directed toward the center of the circle. Solve for one of the forces, the tension T in the string.

SET UP: The free-body diagram for the potato as it swings through its lowest point is given in Figure 7.59b.

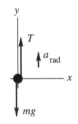

The acceleration $\vec{a}_{\text{rad}}$ is directed in toward the center of the circular path, so at this point it is upward.

Figure 7.59b

EXECUTE: $\Sigma F_y = ma_y$ gives $T - mg = ma_{\text{rad}}$. Solving for T gives $T = m(g + a_{\text{rad}}) = m\left(g + \dfrac{v_2^2}{R}\right)$, where

the radius R for the circular motion is the length L of the string. It is instructive to use the algebraic expression for v_2 from part (a) rather than just putting in the numerical value: $v_2 = \sqrt{2gy_1} = \sqrt{2gL}$, so

$v_2^2 = 2gL$. Then $T = m\left(g + \dfrac{v_2^2}{L}\right) = m\left(g + \dfrac{2gL}{L}\right) = 3mg$. The tension at this point is three times the weight

of the potato, so $T = 3mg = 3(0.300 \text{ kg})(9.80 \text{ m/s}^2) = 8.82$ N.

EVALUATE: The tension is greater than the weight; the acceleration is upward so the net force must be upward.

7.61.   **IDENTIFY** and **SET UP:**   There are two situations to compare: stepping off a platform and sliding down a pole. Apply the work-energy theorem to each.

(a) EXECUTE: Speed at ground if steps off platform at height h:

$K_1 + U_1 + W_{\text{other}} = K_2 + U_2$

$mgh = \frac{1}{2}mv_2^2$, so $v_2^2 = 2gh$

Motion from top to bottom of pole: (take $y = 0$ at bottom)

$K_1 + U_1 + W_{\text{other}} = K_2 + U_2$

$mgd - fd = \frac{1}{2}mv_2^2$

Use $v_2^2 = 2gh$ and get $mgd - fd = mgh$

$fd = mg(d - h)$

$f = mg(d - h)/d = mg(1 - h/d)$

EVALUATE: For $h = d$ this gives $f = 0$ as it should (friction has no effect).

For $h = 0$, $v_2 = 0$ (no motion). The equation for f gives $f = mg$ in this special case. When $f = mg$ the forces on him cancel and he doesn't accelerate down the pole, which agrees with $v_2 = 0$.

(b) EXECUTE: $f = mg(1 - h/d) = (75 \text{ kg})(9.80 \text{ m/s}^2)(1 - 1.0 \text{ m}/2.5 \text{ m}) = 441 \text{ N}$.

(c) Take $y = 0$ at bottom of pole, so $y_1 = d$ and $y_2 = y$.

$K_1 + U_1 + W_{\text{other}} = K_2 + U_2$

$0 + mgd - f(d - y) = \frac{1}{2}mv^2 + mgy$

$\frac{1}{2}mv^2 = mg(d - y) - f(d - y)$

Using $f = mg(1 - h/d)$ gives $\frac{1}{2}mv^2 = mg(d - y) - mg(1 - h/d)(d - y)$

$\frac{1}{2}mv^2 = mg(h/d)(d - y)$ and $v = \sqrt{2gh(1 - y/d)}$

EVALUATE: This gives the correct results for $y = 0$ and for $y = d$.

7.63.   **IDENTIFY** and **SET UP:** First apply $\Sigma \vec{F} = m\vec{a}$ to the skier.

Find the angle α where the normal force becomes zero, in terms of the speed v_2 at this point. Then apply the work-energy theorem to the motion of the skier to obtain another equation that relates v_2 and α. Solve these two equations for α.

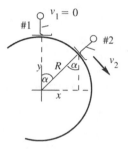

Let point 2 be where the skier loses contact with the snowball, as sketched in Figure 7.63a
Loses contact implies $n \to 0$.

$y_1 = R$, $y_2 = R\cos\alpha$

Figure 7.63a

First, analyze the forces on the skier when she is at point 2. The free-body diagram is given in Figure 7.63b. For this use coordinates that are in the tangential and radial directions. The skier moves in an arc of a circle, so her acceleration is $a_{\text{rad}} = v^2/R$, directed in towards the center of the snowball.

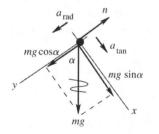

EXECUTE: $\Sigma F_y = ma_y$

$mg\cos\alpha - n = mv_2^2/R$

But $n = 0$ so $mg\cos\alpha = mv_2^2/R$

$v_2^2 = Rg\cos\alpha$

Figure 7.63b

Now use conservation of energy to get another equation relating v_2 to α:

$K_1 + U_1 + W_{\text{other}} = K_2 + U_2$

The only force that does work on the skier is gravity, so $W_{\text{other}} = 0$.

$K_1 = 0$, $K_2 = \frac{1}{2}mv_2^2$

$U_1 = mgy_1 = mgR$, $U_2 = mgy_2 = mgR\cos\alpha$

Then $mgR = \frac{1}{2}mv_2^2 + mgR\cos\alpha$

$v_2^2 = 2gR(1-\cos\alpha)$

Combine this with the $\Sigma F_y = ma_y$ equation:

$Rg\cos\alpha = 2gR(1-\cos\alpha)$

$\cos\alpha = 2 - 2\cos\alpha$

$3\cos\alpha = 2$ so $\cos\alpha = 2/3$ and $\alpha = 48.2°$

EVALUATE: She speeds up and her a_{rad} increases as she loses gravitational potential energy. She loses contact when she is going so fast that the radially inward component of her weight isn't large enough to keep her in the circular path. Note that α where she loses contact does not depend on her mass or on the radius of the snowball.

7.65.   **IDENTIFY** and **SET UP:**

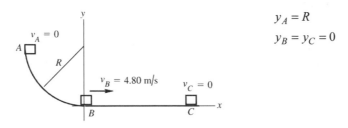

$y_A = R$

$y_B = y_C = 0$

Figure 7.65

(a) Apply conservation of energy to the motion from B to C:

$K_B + U_B + W_{other} = K_C + U_C$. The motion is described in Figure 7.65.

EXECUTE: The only force that does work on the package during this part of the motion is friction, so

$W_{other} = W_f = f_k(\cos\phi)s = \mu_k mg(\cos 180°)s = -\mu_k mgs$

$K_B = \frac{1}{2}mv_B^2, \quad K_C = 0$

$U_B = 0, \quad U_C = 0$

Thus $K_B + W_f = 0$

$\frac{1}{2}mv_B^2 - \mu_k mgs = 0$

$\mu_k = \dfrac{\mu_B^2}{2gs} = \dfrac{(4.80 \text{ m/s})^2}{2(9.80 \text{ m/s}^2)(3.00 \text{ m})} = 0.392$

EVALUATE: The negative friction work takes away all the kinetic energy.

(b) IDENTIFY and **SET UP:** Apply conservation of energy to the motion from A to B:

$$K_A + U_A + W_{other} = K_B + U_B$$

EXECUTE: Work is done by gravity and by friction, so $W_{other} = W_f$.

$K_A = 0, \quad K_B = \frac{1}{2}mv_B^2 = \frac{1}{2}(0.200 \text{ kg})(4.80 \text{ m/s})^2 = 2.304 \text{ J}$

$U_A = mgy_A = mgR = (0.200 \text{ kg})(9.80 \text{ m/s}^2)(1.60 \text{ m}) = 3.136 \text{ J}, \quad U_B - 0$

Thus $U_A + W_f = K_B$

$W_f = K_B - U_A = 2.304 \text{ J} - 3.136 \text{ J} = -0.83 \text{ J}$

EVALUATE: W_f is negative as expected; the friction force does negative work since it is directed opposite to the displacement.

7.67. $F_x = -\alpha x - \beta x^2, \quad \alpha = 60.0 \text{ N/m}$ and $\beta = 18.0 \text{ N/m}^2$

(a) IDENTIFY: Use Eq. (6.7) to calculate W and then use $W = -\Delta U$ to identify the potential energy function $U(x)$.

SET UP: $W_{F_x} = U_1 - U_2 = \int_{x_1}^{x_2} F_x(x)\, dx$

Let $x_1 = 0$ and $U_1 = 0$. Let x_2 be some arbitrary point x, so $U_2 = U(x)$.

EXECUTE: $U(x) = -\int_0^x F_x(x)\, dx = -\int_0^x (-\alpha x - \beta x^2)\, dx = \int_0^x (\alpha x + \beta x^2)\, dx = \frac{1}{2}\alpha x^2 + \frac{1}{3}\beta x^3.$

EVALUATE: If $\beta = 0$, the spring does obey Hooke's law, with $k = \alpha$, and our result reduces to $\frac{1}{2}kx^2$.

(b) IDENTIFY: Apply Eq. (7.15) to the motion of the object.

SET UP: The system at points 1 and 2 is sketched in Figure 7.67.

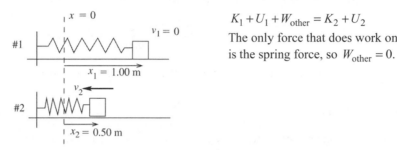

$K_1 + U_1 + W_{\text{other}} = K_2 + U_2$

The only force that does work on the object is the spring force, so $W_{\text{other}} = 0$.

Figure 7.67

EXECUTE: $K_1 = 0, \quad K_2 = \frac{1}{2}mv_2^2$

$U_1 = U(x_1) = \frac{1}{2}\alpha x_1^2 + \frac{1}{3}\beta x_1^3 = \frac{1}{2}(60.0 \text{ N/m})(1.00 \text{ m})^2 + \frac{1}{3}(18.0 \text{ N/m}^2)(1.00 \text{ m})^3 = 36.0 \text{ J}$

$U_2 = U(x_2) = \frac{1}{2}\alpha x_2^2 + \frac{1}{3}\beta x_2^3 = \frac{1}{2}(60.0 \text{ N/m})(0.500 \text{ m})^2 + \frac{1}{3}(18.0 \text{ N/m}^2)(0.500 \text{ m})^3 = 8.25 \text{ J}$

Thus $36.0 \text{ J} = \frac{1}{2}mv_2^2 + 8.25 \text{ J}$

$v_2 = \sqrt{\dfrac{2(36.0 \text{ J} - 8.25 \text{ J})}{0.900 \text{ kg}}} = 7.85 \text{ m/s}$

EVALUATE: The elastic potential energy stored in the spring decreases and the kinetic energy of the object increases.

7.71.   **IDENTIFY:**   Apply conservation of energy to relate x and h. Apply $\Sigma \vec{F} = m\vec{a}$ to relate a and x.

SET UP: The first condition, that the maximum height above the release point is h, is expressed as $\frac{1}{2}kx^2 = mgh$. The magnitude of the acceleration is largest when the spring is compressed to a distance x; at this point the net upward force is $kx - mg = ma$, so the second condition is expressed as $x = (m/k)(g + a)$.

EXECUTE: **(a)** Substituting the second expression into the first gives

$$\frac{1}{2}k\left(\frac{m}{k}\right)^2 (g+a)^2 = mgh, \text{ or } k = \frac{m(g+a)^2}{2gh}.$$

(b) Substituting this into the expression for x gives $x = \dfrac{2gh}{g+a}$.

EVALUATE: When $a \to 0$, our results become $k = \dfrac{mg}{2h}$ and $x = 2h$. The initial spring force is $kx = mg$ and the net upward force approaches zero. But $\frac{1}{2}kx^2 = mgh$ and sufficient potential energy is stored in the spring to move the mass to height h.

7.73.   **IDENTIFY:**   Only conservative forces (gravity and the spring) act on the fish, so its mechanical energy is conserved.

SET UP: Energy conservation tells us $K_1 + U_1 + W_{\text{other}} = K_2 + U_2$, where $W_{\text{other}} = 0$. $U_g = mgy$, $K = \frac{1}{2}mv^2$, and $U_{spring} = \frac{1}{2}ky^2$.

EXECUTE: (a) $K_1 + U_1 + W_{other} = K_2 + U_2$. Let y be the distance the fish has descended, so $y = 0.0500$ m.

$K_1 = 0$, $W_{other} = 0$, $U_1 = mgy$, $K_2 = \frac{1}{2}mv_2^2$, and $U_2 = \frac{1}{2}ky^2$. Solving for K_2 gives

$$K_2 = U_1 - U_2 = mgy - \frac{1}{2}ky^2 = (3.00 \text{ kg})(9.8 \text{ m/s}^2)(0.0500 \text{ m}) - \frac{1}{2}(900 \text{ N/m})(0.0500 \text{ m})^2$$

$K_2 = 1.47 \text{ J} - 1.125 \text{ J} = 0.345 \text{ J}$. Solving for v_2 gives $v_2 = \sqrt{\dfrac{2K_2}{m}} = \sqrt{\dfrac{2(0.345 \text{ J})}{3.00 \text{ kg}}} = 0.480$ m/s.

(b) The maximum speed is when K_2 is maximum, which is when $dK_2/dy = 0$. Using $K_2 = mgy - \frac{1}{2}ky^2$

gives $\dfrac{dK_2}{dy} = mg - ky = 0$. Solving for y gives $y = \dfrac{mg}{k} = \dfrac{(3.00 \text{ kg})(9.8 \text{ m/s}^2)}{900 \text{ N/m}} = 0.03267$ m. At this y,

$K_2 = (3.00 \text{ kg})(9.8 \text{ m/s}^2)(0.03267 \text{ m}) - \frac{1}{2}(900 \text{ N/m})(0.03267 \text{ m})^2$. $K_2 = 0.9604 \text{ J} - 0.4803 \text{ J} = 0.4801 \text{ J}$,

so $v_2 = \sqrt{\dfrac{2K_2}{m}} = 0.566$ m/s.

EVALUATE: The speed in part (b) is greater than the speed in part (a), as it should be since it is the maximum speed.

7.77. **IDENTIFY:** We can apply Newton's second law to the block. The only forces acting on the block are gravity downward and the normal force from the track pointing toward the center of the circle. The mechanical energy of the block is conserved since only gravity does work on it. The normal force does no work since it is perpendicular to the displacement of the block. The target variable is the normal force at the top of the track.

SET UP: For circular motion $\Sigma F = m\dfrac{v^2}{R}$. Energy conservation tells us that $K_A + U_A + W_{other} = K_B + U_B$, where $W_{other} = 0$. $U_g = mgy$ and $K = \frac{1}{2}mv^2$.

EXECUTE: Let point A be at the bottom of the path and point B be at the top of the path. At the bottom of the path, $n_A - mg = m\dfrac{v^2}{R}$ (from Newton's second law).

$v_A = \sqrt{\dfrac{R}{m}(n_A - mg)} = \sqrt{\dfrac{0.800 \text{ m}}{0.0500 \text{ kg}}(3.40 \text{ N} - 0.49 \text{ N})} = 6.82$ m/s. Use energy conservation to find the speed

at point B. $K_A + U_A + W_{other} = K_B + U_B$, giving $\frac{1}{2}mv_A^2 = \frac{1}{2}mv_B^2 + mg(2R)$. Solving for v_B gives

$v_B = \sqrt{v_A^2 - 4Rg} = \sqrt{(6.82 \text{ m/s})^2 - 4(0.800 \text{ M})(9.8 \text{ m/s}^2)} = 3.89$ m/s. Then at point B, Newton's second law

gives $n_B + mg = m\dfrac{v_B^2}{R}$. Solving for n_B gives

$$n_B = m\dfrac{v_B^2}{R} - mg = (0.0500 \text{ kg})\left(\dfrac{(3.89 \text{ m/s})^2}{0.800 \text{ m}} - 9.8 \text{ m/s}^2\right) = 0.456 \text{ N}.$$

EVALUATE: The normal force at the top is considerably less than it is at the bottom for two reasons: the block is moving slower at the top and the downward force of gravity at the top aids the normal force in keeping the block moving in a circle.

7.79. **IDENTIFY:** $U = mgh$. Use $h = 150$ m for all the water that passes through the dam.

SET UP: $m = \rho V$ and $V = A\Delta h$ is the volume of water in a height Δh of water in the lake.

EXECUTE: (a) Stored energy $= mgh = (\rho V)gh = \rho A(1 \text{ m})gh$.

stored energy $= (1000 \text{ kg/m}^3)(3.0 \times 10^6 \text{ m}^2)(1 \text{ m})(9.8 \text{ m/s}^2)(150 \text{ m}) = 4.4 \times 10^{12}$ J.

(b) 90% of the stored energy is converted to electrical energy, so $(0.90)(mgh) = 1000$ kWh.

$(0.90)\rho Vgh = 1000$ kWh. $V = \dfrac{(1000\ \text{kWh})((3600\ \text{s})/(1\ \text{h}))}{(0.90)(1000\ \text{kg/m}^3)(150\ \text{m})(9.8\ \text{m/s}^2)} = 2.7 \times 10^3\ \text{m}^3.$

Change in level of the lake: $A\Delta h = V_{\text{water}}.$ $\Delta h = \dfrac{V}{A} = \dfrac{2.7 \times 10^3\,\text{m}^3}{3.0 \times 10^6\,\text{m}^2} = 9.0 \times 10^{-4}\text{m}.$

EVALUATE: Δh is much less than 150 m, so using $h = 150$ m for all the water that passed through the dam was a very good approximation.

7.83. $\vec{F} = -\alpha xy^2 \hat{j},\ \ \alpha = 2.50\ \text{N/m}^3$

IDENTIFY: $\vec{F}$ is not constant so use Eq. (6.14) to calculate W. $\vec{F}$ must be evaluated along the path.
(a) SET UP: The path is sketched in Figure 7.83a.

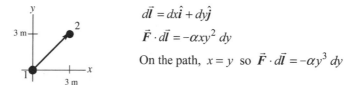

$d\vec{l} = dx\hat{i} + dy\hat{j}$

$\vec{F} \cdot d\vec{l} = -\alpha xy^2\ dy$

On the path, $x = y$ so $\vec{F} \cdot d\vec{l} = -\alpha y^3\ dy$

Figure 7.83a

EXECUTE: $W = \displaystyle\int_1^2 \vec{F} \cdot d\vec{l} = \int_{y_1}^{y_2} (-\alpha y^3)\,dy = -(\alpha/4)\left(y^4\, \Big|_{y_1}^{y_2} \right) = -(\alpha/4)(y_2^4 - y_1^4)$

$y_1 = 0,\ \ y_2 = 3.00$ m, so $W = -\tfrac{1}{4}(2.50\ \text{N/m}^3)(3.00\ \text{m})^4 = -50.6$ J

(b) SET UP: The path is sketched in Figure 7.83b.

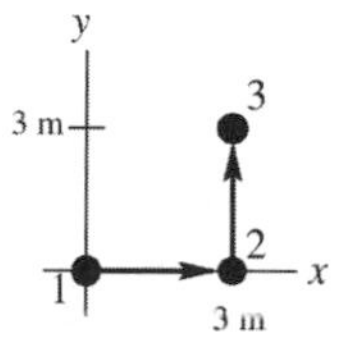

Figure 7.83b

For the displacement from point 1 to point 2, $d\vec{l} = dx\hat{i}$, so $\vec{F} \cdot d\vec{l} = 0$ and $W = 0$. (The force is perpendicular to the displacement at each point along the path, so $W = 0$.)

For the displacement from point 2 to point 3, $d\vec{l} = dy\hat{j}$, so $\vec{F} \cdot d\vec{l} = -\alpha xy^2\ dy$. On this path, $x = 3.00$ m, so

$$\vec{F} \cdot d\vec{l} = -(2.50\ \text{N/m}^3)(3.00\ \text{m})y^2\ dy = -(7.50\ \text{N/m}^2)y^2\ dy.$$

EXECUTE: $W = \displaystyle\int_2^3 \vec{F} \cdot d\vec{l} = -(7.50\ \text{N/m}^2)\int_{y_2}^{y_3} y^2\ dy = -(7.50\ \text{N/m}^2)\tfrac{1}{3}(y_3^3 - y_2^3)$

$W = -(7.50\ \text{N/m}^2)\left(\tfrac{1}{3}\right)(3.00\ \text{m})^3 = -67.5$ J

(c) EVALUATE: For these two paths between the same starting and ending points the work is different, so the force is nonconservative.

MOMENTUM, IMPULSE, AND COLLISIONS

8.5. **IDENTIFY:** For each object, $\vec{p} = m\vec{v}$ and $K = \frac{1}{2}mv^2$. The total momentum is the vector sum of the momenta of each object. The total kinetic energy is the scalar sum of the kinetic energies of each object.

SET UP: Let object A be the 110 kg lineman and object B the 125 kg lineman. Let $+x$ be to the right, so $v_{Ax} = +2.75$ m/s and $v_{Bx} = -2.60$ m/s.

EXECUTE: **(a)** $P_x = m_A v_{Ax} + m_B v_{Bx} = (110 \text{ kg})(2.75 \text{ m/s}) + (125 \text{ kg})(-2.60 \text{ m/s}) = -22.5 \text{ kg} \cdot \text{m/s}$. The net momentum has magnitude $22.5 \text{ kg} \cdot \text{m/s}$ and is directed to the left.

(b) $K = \frac{1}{2} m_A v_A^2 + \frac{1}{2} m_B v_B^2 = \frac{1}{2}(110 \text{ kg})(2.75 \text{ m/s})^2 + \frac{1}{2}(125 \text{ kg})(2.60 \text{ m/s})^2 = 838 \text{ J}$

EVALUATE: The kinetic energy of an object is a scalar and is never negative. It depends only on the magnitude of the velocity of the object, not on its direction. The momentum of an object is a vector and has both magnitude and direction. When two objects are in motion, their total kinetic energy is greater than the kinetic energy of either one. But if they are moving in opposite directions, the net momentum of the system has a smaller magnitude than the magnitude of the momentum of either object.

8.9. **IDENTIFY:** Use Eq. 8.9. We know the initial momentum and the impulse so can solve for the final momentum and then the final velocity.

SET UP: Take the x-axis to be toward the right, so $v_{1x} = +3.00$ m/s. Use Eq. 8.5 to calculate the impulse, since the force is constant.

EXECUTE: **(a)** $J_x = p_{2x} - p_{1x}$

$$J_x = F_x(t_2 - t_1) = (+25.0 \text{ N})(0.050 \text{ s}) = +1.25 \text{ kg} \cdot \text{m/s}$$

Thus $p_{2x} = J_x + p_{1x} = +1.25 \text{ kg} \cdot \text{m/s} + (0.160 \text{ kg})(+3.00 \text{ m/s}) = +1.73 \text{ kg} \cdot \text{m/s}$

$$v_{2x} = \frac{p_{2x}}{m} = \frac{1.73 \text{ kg} \cdot \text{m/s}}{0.160 \text{ kg}} = +10.8 \text{ m/s} \; (\text{to the right})$$

(b) $J_x = F_x(t_2 - t_1) = (-12.0 \text{ N})(0.050 \text{ s}) = -0.600 \text{ kg} \cdot \text{m/s}$ (negative since force is to left)

$$p_{2x} = J_x + p_{1x} = -0.600 \text{ kg} \cdot \text{m/s} + (0.160 \text{ kg})(+3.00 \text{ m/s}) = -0.120 \text{ kg} \cdot \text{m/s}$$

$$v_{2x} = \frac{p_{2x}}{m} = \frac{-0.120 \text{ kg} \cdot \text{m/s}}{0.160 \text{ kg}} = -0.75 \text{ m/s (to the left)}$$

EVALUATE: In part (a) the impulse and initial momentum are in the same direction and v_x increases. In part (b) the impulse and initial momentum are in opposite directions and the velocity decreases.

8.11. **IDENTIFY:** The force is not constant so $\vec{J} = \int_{t_1}^{t_2} \vec{F} dt$. The impulse is related to the change in velocity by Eq. 8.9.

SET UP: Only the x component of the force is nonzero, so $J_x = \int_{t_1}^{t_2} F_x dt$ is the only nonzero component of $\vec{J}$.

$$J_x = m(v_{2x} - v_{1x}). \; t_1 = 2.00 \text{ s}, \; t_2 = 3.50 \text{ s}.$$

EXECUTE: **(a)** $A = \dfrac{F_x}{t^2} = \dfrac{781.25 \text{ N}}{(1.25 \text{ s})^2} = 500 \text{ N/s}^2.$

(b) $J_x = \int_{t_1}^{t_2} At^2 dt = \frac{1}{3}A(t_2^3 - t_1^3) = \frac{1}{3}(500 \text{ N/s}^2)([3.50 \text{ s}]^3 - [2.00 \text{ s}]^3) = 5.81\times10^3 \text{ N}\cdot\text{s}$.

(c) $\Delta v_x = v_{2x} - v_{1x} = \dfrac{J_x}{m} = \dfrac{5.81\times10^3 \text{ N}\cdot\text{s}}{2150 \text{ kg}} = 2.70 \text{ m/s}$. The x component of the velocity of the rocket

increases by 2.70 m/s.

EVALUATE: The change in velocity is in the same direction as the impulse, which in turn is in the direction of the net force. In this problem the net force equals the force applied by the engine, since that is the only force on the rocket.

8.13. **IDENTIFY:** The force is constant during the 1.0 ms interval that it acts, so $\vec{J} = \vec{F}\Delta t$.

$\vec{J} = \vec{p}_2 - \vec{p}_1 = m(\vec{v}_2 - \vec{v}_1)$.

SET UP: Let $+x$ be to the right, so $v_{1x} = +5.00$ m/s. Only the x component of $\vec{J}$ is nonzero, and

$J_x = m(v_{2x} - v_{1x})$.

EXECUTE: (a) The magnitude of the impulse is $J = F\Delta t = (2.50\times10^3 \text{ N})(1.00\times10^{-3} \text{ s}) = 2.50 \text{ N}\cdot\text{s}$. The

direction of the impulse is the direction of the force.

(b) (i) $v_{2x} = \dfrac{J_x}{m} + v_{1x}$. $J_x = +2.50 \text{ N}\cdot\text{s}$. $v_{2x} = \dfrac{+2.50 \text{ N}\cdot\text{s}}{2.00 \text{ kg}} + 5.00 \text{ m/s} = 6.25 \text{ m/s}$. The stone's velocity has

magnitude 6.25 m/s and is directed to the right. (ii) Now $J_x = -2.50 \text{ N}\cdot\text{s}$ and

$v_{2x} = \dfrac{-2.50 \text{ N}\cdot\text{s}}{2.00 \text{ kg}} + 5.00 \text{ m/s} = 3.75 \text{ m/s}$. The stone's velocity has magnitude 3.75 m/s and is directed to the

right.

EVALUATE: When the force and initial velocity are in the same direction the speed increases and when they are in opposite directions the speed decreases.

8.15. **IDENTIFY:** The player imparts an impulse to the ball which gives it momentum, causing it to go upward.

SET UP: Take $+y$ to be upward. Use the motion of the ball after it leaves the racket to find its speed just after it is hit. After it leaves the racket $a_y = -g$. At the maximum height $v_y = 0$. Use $J_y = \Delta p_y$ and the

kinematics equation $v_y^2 = v_{0y}^2 + 2a_y(y - y_0)$ for constant acceleration.

EXECUTE: $v_y^2 = v_{0y}^2 + 2a_y(y - y_0)$ gives $v_{0y} = \sqrt{-2a_y(y - y_0)} = \sqrt{-2(-9.80 \text{ m/s}^2)(5.50 \text{ m})} = 10.4 \text{ m/s}$.

For the interaction with the racket $v_{1y} = 0$ and $v_{2y} = 10.4$ m/s.

$J_y = mv_{2y} - mv_{1y} = (57\times10^{-3} \text{ kg})(10.4 \text{ m/s} - 0) = 0.593 \text{ kg}\cdot\text{m/s}$.

EVALUATE: We could have found the initial velocity using energy conservation instead of free-fall kinematics.

8.17. **IDENTIFY:** Since the rifle is loosely held there is no net external force on the system consisting of the rifle, bullet and propellant gases and the momentum of this system is conserved. Before the rifle is fired everything in the system is at rest and the initial momentum of the system is zero.

SET UP: Let $+x$ be in the direction of the bullet's motion. The bullet has speed

601 m/s $-$ 1.85 m/s $= 599$ m/s relative to the earth. $P_{2x} = p_{rx} + p_{bx} + p_{gx}$, the momenta of the rifle, bullet

and gases. $v_{rx} = -1.85$ m/s and $v_{bx} = +599$ m/s.

EXECUTE: $P_{2x} = P_{1x} = 0$. $p_{rx} + p_{bx} + p_{gx} = 0$.

$p_{gx} = -p_{rx} - p_{bx} = -(2.80 \text{ kg})(-1.85 \text{ m/s}) - (0.00720 \text{ kg})(599 \text{ m/s})$ and

$p_{gx} = +5.18 \text{ kg}\cdot\text{m/s} - 4.31 \text{ kg}\cdot\text{m/s} = 0.87 \text{ kg}\cdot\text{m/s}$. The propellant gases have momentum 0.87 kg$\cdot$m/s, in

the same direction as the bullet is traveling.

EVALUATE: The magnitude of the momentum of the recoiling rifle equals the magnitude of the momentum of the bullet plus that of the gases as both exit the muzzle.

8.21.   **IDENTIFY:**   Apply conservation of momentum to the system of the two pucks.

SET UP: Let $+x$ be to the right.

EXECUTE: **(a)** $P_{1x} = P_{2x}$ says $(0.250 \text{ kg})v_{A1} = (0.250 \text{ kg})(-0.120 \text{ m/s}) + (0.350 \text{ kg})(0.650 \text{ m/s})$ and $v_{A1} = 0.790 \text{ m/s}$.

(b) $K_1 = \frac{1}{2}(0.250 \text{ kg})(0.790 \text{ m/s})^2 = 0.0780 \text{ J}$.

$K_2 = \frac{1}{2}(0.250 \text{ kg})(0.120 \text{ m/s})^2 + \frac{1}{2}(0.350 \text{ kg})(0.650 \text{ m/s})^2 = 0.0757 \text{ J}$ and $\Delta K = K_2 - K_1 = -0.0023 \text{ J}$.

EVALUATE: The total momentum of the system is conserved but the total kinetic energy decreases.

8.23.   **IDENTIFY:**   The momentum and the mechanical energy of the system are both conserved. The mechanical energy consists of the kinetic energy of the masses and the elastic potential energy of the spring. The potential energy stored in the spring is transformed into the kinetic energy of the two masses.

SET UP: Let the system be the two masses and the spring. The system is sketched in Figure 8.23, in its initial and final situations. Use coordinates where $+x$ is to the right. Call the masses A and B.

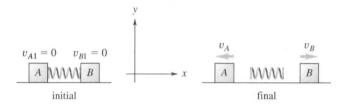

Figure 8.23

EXECUTE: $P_{1x} = P_{2x}$ so $0 = (1.50 \text{ kg})(-v_A) + (1.50 \text{ kg})(v_B)$ and, since the masses are equal, $v_A = v_B$. Energy conservation says the potential energy originally stored in the spring is all converted into kinetic energy of the masses, so $\frac{1}{2}kx_1^2 = \frac{1}{2}mv_A^2 + \frac{1}{2}mv_B^2$. Since $v_A = v_B$, this equation gives

$$v_A = x_1\sqrt{\frac{k}{2m}} = (0.200 \text{ m})\sqrt{\frac{175 \text{ N/m}}{2(1.50 \text{ kg})}} = 1.53 \text{ m/s}.$$

EVALUATE: If the objects have different masses they will end up with different speeds. The lighter one will have the greater speed, since they end up with equal magnitudes of momentum.

8.25.   **IDENTIFY:**   Since friction at the pond surface is neglected, there is no net external horizontal force and the horizontal component of the momentum of the system of hunter plus bullet is conserved. Both objects are initially at rest, so the initial momentum of the system is zero. Gravity and the normal force exerted by the ice together produce a net vertical force while the rifle is firing, so the vertical component of momentum is not conserved.

SET UP: Let object A be the hunter and object B be the bullet. Let $+x$ be the direction of the horizontal component of velocity of the bullet. Solve for v_{A2x}.

EXECUTE: **(a)** $v_{B2x} = +965 \text{ m/s}$. $P_{1x} = P_{2x} = 0$. $0 = m_A v_{A2x} + m_B v_{B2x}$ and

$$v_{A2x} = -\frac{m_B}{m_A}v_{B2x} = -\left(\frac{4.20 \times 10^{-3} \text{ kg}}{72.5 \text{ kg}}\right)(965 \text{ m/s}) = -0.0559 \text{ m/s}.$$

(b) $v_{B2x} = v_{B2}\cos\theta = (965 \text{ m/s})\cos 56.0° = 540 \text{ m/s}$. $v_{A2x} = -\left(\frac{4.20 \times 10^{-3} \text{ kg}}{72.5 \text{ kg}}\right)(540 \text{ m/s}) = -0.0313 \text{ m/s}$.

EVALUATE: The mass of the bullet is much less than the mass of the hunter, so the final mass of the hunter plus gun is still 72.5 kg, to three significant figures. Since the hunter has much larger mass, his final speed is much less than the speed of the bullet.

8.27.   **IDENTIFY:**   Each horizontal component of momentum is conserved. $K = \frac{1}{2}mv^2$.

SET UP: Let $+x$ be the direction of Rebecca's initial velocity and let the $+y$ axis make an angle of $36.9°$ with respect to the direction of her final velocity. $v_{D1x} = v_{D1y} = 0$. $v_{R1x} = 13.0 \text{ m/s}$; $v_{R1y} = 0$.

$v_{R2x} = (8.00 \text{ m/s})\cos 53.1° = 4.80 \text{ m/s}$; $v_{R2y} = (8.00 \text{ m/s})\sin 53.1° = 6.40 \text{ m/s}$. Solve for v_{D2x} and v_{D2y}.

EXECUTE: (a) $P_{1x} = P_{2x}$ gives $m_R v_{R1x} = m_R v_{R2x} + m_D v_{D2x}$.

$$v_{D2x} = \frac{m_R(v_{R1x} - v_{R2x})}{m_D} = \frac{(45.0 \text{ kg})(13.0 \text{ m/s} - 4.80 \text{ m/s})}{65.0 \text{ kg}} = 5.68 \text{ m/s}.$$

$P_{1y} = P_{2y}$ gives $0 = m_R v_{R2y} + m_D v_{D2y}$. $v_{D2y} = -\frac{m_R}{m_D} v_{R2y} = -\left(\frac{45.0 \text{ kg}}{65.0 \text{ kg}}\right)(6.40 \text{ m/s}) = -4.43 \text{ m/s}$.

The directions of $\vec{v}_{R1}$, $\vec{v}_{R2}$ and $\vec{v}_{D2}$ are sketched in Figure 8.27. $\tan\theta = \left|\frac{v_{D2y}}{v_{D2x}}\right| = \frac{4.43 \text{ m/s}}{5.68 \text{ m/s}}$ and

$\theta = 38.0°$. $v_D = \sqrt{v_{D2x}^2 + v_{D2y}^2} = 7.20 \text{ m/s}$.

(b) $K_1 = \frac{1}{2} m_R v_{R1}^2 = \frac{1}{2}(45.0 \text{ kg})(13.0 \text{ m/s})^2 = 3.80 \times 10^3$ J.

$K_2 = \frac{1}{2} m_R v_{R2}^2 + \frac{1}{2} m_D v_{D2}^2 = \frac{1}{2}(45.0 \text{ kg})(8.00 \text{ m/s})^2 + \frac{1}{2}(65.0 \text{ kg})(7.20 \text{ m/s})^2 = 3.12 \times 10^3$ J.

$\Delta K = K_2 - K_1 = -680$ J.

EVALUATE: Each component of momentum is separately conserved. The kinetic energy of the system decreases.

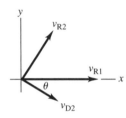

Figure 8.27

8.31.   **IDENTIFY:** The x and y components of the momentum of the system of the two asteroids are separately conserved.

SET UP: The before and after diagrams are given in Figure 8.31 and the choice of coordinates is indicated. Each asteroid has mass m.

EXECUTE: (a) $P_{1x} = P_{2x}$ gives $mv_{A1} = mv_{A2}\cos 30.0° + mv_{B2}\cos 45.0°$. $40.0 \text{ m/s} = 0.866 v_{A2} + 0.707 v_{B2}$ and $0.707 v_{B2} = 40.0 \text{ m/s} - 0.866 v_{A2}$.

$P_{2y} = P_{2y}$ gives $0 = mv_{A2}\sin 30.0° - mv_{B2}\sin 45.0°$ and $0.500 v_{A2} = 0.707 v_{B2}$.

Combining these two equations gives $0.500 v_{A2} = 40.0 \text{ m/s} - 0.866 v_{A2}$ and $v_{A2} = 29.3 \text{ m/s}$. Then

$$v_{B2} = \left(\frac{0.500}{0.707}\right)(29.3 \text{ m/s}) = 20.7 \text{ m/s}.$$

(b) $K_1 = \frac{1}{2} mv_{A1}^2$. $K_2 = \frac{1}{2} mv_{A2}^2 + \frac{1}{2} mv_{B2}^2$. $\frac{K_2}{K_1} = \frac{v_{A2}^2 + v_{B2}^2}{v_{A1}^2} = \frac{(29.3 \text{ m/s})^2 + (20.7 \text{ m/s})^2}{(40.0 \text{ m/s})^2} = 0.804$.

$$\frac{\Delta K}{K_1} = \frac{K_2 - K_1}{K_1} = \frac{K_2}{K_1} - 1 = -0.196.$$

19.6% of the original kinetic energy is dissipated during the collision.

EVALUATE: We could use any directions we wish for the x and y coordinate directions, but the particular choice we have made is especially convenient.

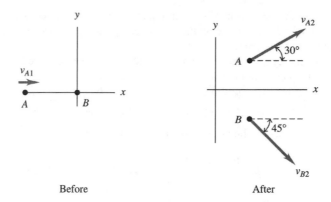

Before   After

Figure 8.31

8.37.   IDENTIFY: The forces the two players exert on each other during the collision are much larger than the horizontal forces exerted by the slippery ground and it is a good approximation to assume momentum conservation. Each component of momentum is separately conserved.

SET UP: Let $+x$ be east and $+y$ be north. After the collision the two players have velocity $\vec{v}_2$. Let the linebacker be object A and the halfback be object B, so $v_{A1x} = 0$, $v_{A1y} = 8.8$ m/s, $v_{B1x} = 7.2$ m/s and $v_{B1y} = 0$. Solve for v_{2x} and v_{2y}.

EXECUTE: $P_{1x} = P_{2x}$ gives $m_A v_{A1x} + m_B v_{B1x} = (m_A + m_B)v_{2x}$.

$$v_{2x} = \frac{m_A v_{A1x} + m_B v_{B1x}}{m_A + m_B} = \frac{(85 \text{ kg})(7.2 \text{ m/s})}{110 \text{ kg} + 85 \text{ kg}} = 3.14 \text{ m/s}.$$

$P_{1y} = P_{2y}$ gives $m_A v_{A1y} + m_B v_{B1y} = (m_A + m_B)v_{2y}$.

$$v_{2y} = \frac{m_A v_{A1y} + m_B v_{B1y}}{m_A + m_B} = \frac{(110 \text{ kg})(8.8 \text{ m/s})}{110 \text{ kg} + 85 \text{ kg}} = 4.96 \text{ m/s}.$$

$v = \sqrt{v_{2x}^2 + v_{2y}^2} = 5.9$ m/s.

$$\tan \theta = \frac{v_{2y}}{v_{2x}} = \frac{4.96 \text{ m/s}}{3.14 \text{ m/s}} \text{ and } \theta = 58°.$$

The players move with a speed of 5.9 m/s and in a direction 58° north of east.

EVALUATE: Each component of momentum is separately conserved.

8.41.   IDENTIFY: Since friction forces from the road are ignored, the x and y components of momentum are conserved.

SET UP: Let object A be the subcompact and object B be the truck. After the collision the two objects move together with velocity $\vec{v}_2$. Use the x and y coordinates given in the problem. $v_{A1y} = v_{B1x} = 0$.

$v_{2x} = (16.0 \text{ m/s})\sin 24.0° = 6.5 \text{ m/s}; \quad v_{2y} = (16.0 \text{ m/s})\cos 24.0° = 14.6 \text{ m/s}.$

EXECUTE: $P_{1x} = P_{2x}$ gives $m_A v_{A1x} = (m_A + m_B)v_{2x}$.

$$v_{A1x} = \left(\frac{m_A + m_B}{m_A}\right)v_{2x} = \left(\frac{950 \text{ kg} + 1900 \text{ kg}}{950 \text{ kg}}\right)(6.5 \text{ m/s}) = 19.5 \text{ m/s}.$$

$P_{1y} = P_{2y}$ gives $m_A v_{B1y} = (m_A + m_B)v_{2y}$.

$$v_{B1y} = \left(\frac{m_A + m_B}{m_A}\right)v_{2y} = \left(\frac{950 \text{ kg} + 1900 \text{ kg}}{1900 \text{ kg}}\right)(14.6 \text{ m/s}) = 21.9 \text{ m/s}.$$

Before the collision the subcompact car has speed 19.5 m/s and the truck has speed 21.9 m/s.

EVALUATE: Each component of momentum is independently conserved.

8.43.   **IDENTIFY:**   Apply conservation of momentum to the collision and conservation of energy to the motion after the collision. After the collision the kinetic energy of the combined object is converted to gravitational potential energy.

SET UP: Immediately after the collision the combined object has speed V. Let h be the vertical height through which the pendulum rises.

EXECUTE: (a) Conservation of momentum applied to the collision gives

$(12.0\times10^{-3}\text{ kg})(380\text{ m/s}) = (6.00\text{ kg} + 12.0\times10^{-3}\text{ kg})V$ and $V = 0.758$ m/s.

Conservation of energy applied to the motion after the collision gives $\frac{1}{2}m_{tot}V^2 = m_{tot}gh$ and

$$h = \frac{V^2}{2g} = \frac{(0.758\text{ m/s})^2}{2(9.80\text{ m/s}^2)} = 0.0293\text{ m } = 2.93\text{ cm.}$$

(b) $K = \frac{1}{2}m_b v_b^2 = \frac{1}{2}(12.0\times10^{-3}\text{ kg})(380\text{ m/s})^2 = 866$ J.

(c) $K = \frac{1}{2}m_{tot}V^2 = \frac{1}{2}(6.00\text{ kg} + 12.0\times10^{-3}\text{ kg})(0.758\text{ m/s})^2 = 1.73$ J.

EVALUATE: Most of the initial kinetic energy of the bullet is dissipated in the collision.

8.45.   **IDENTIFY:**   The missile gives momentum to the ornament causing it to swing in a circular arc and thereby be accelerated toward the center of the circle.

SET UP: After the collision the ornament moves in an arc of a circle and has acceleration $a_{rad} = \dfrac{v^2}{r}$.

During the collision, momentum is conserved, so $P_{1x} = P_{2x}$. The free-body diagram for the ornament plus missile is given in Figure 8.45. Take $+y$ to be upward, since that is the direction of the acceleration. Take the $+x$ direction to be the initial direction of motion of the missile.

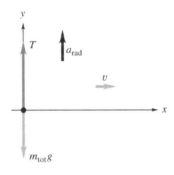

Figure 8.45

EXECUTE: Apply conservation of momentum to the collision. Using $P_{1x} = P_{2x}$, we get

$(3.00\text{ kg})(12.0\text{ m/s}) = (8.00\text{ kg})V$, which gives $V = 4.50$ m/s, the speed of the ornament immediately after

the collision. Then $\Sigma F_y = ma_y$ gives $T - m_{tot}g = m_{tot}\dfrac{v^2}{r}$. Solving for T gives

$$T = m_{tot}\left(g + \frac{v^2}{r}\right) = (8.00\text{ kg})\left(9.80\text{ m/s}^2 + \frac{(4.50\text{ m/s})^2}{1.50\text{ m}}\right) = 186\text{ N.}$$

EVALUATE: We cannot use energy conservation during the collision because it is an inelastic collision (the objects stick together).

8.47.   **IDENTIFY:**   When the spring is compressed the maximum amount the two blocks aren't moving relative to each other and have the same velocity $\vec{V}$ relative to the surface. Apply conservation of momentum to find V and conservation of energy to find the energy stored in the spring. Since the collision is elastic, Eqs. 8.24 and 8.25 give the final velocity of each block after the collision.

SET UP: Let $+x$ be the direction of the initial motion of A.

EXECUTE: (a) Momentum conservation gives $(2.00 \text{ kg})(2.00 \text{ m/s}) = (12.0 \text{ kg})V$ and $V = 0.333 \text{ m/s}$. Both blocks are moving at 0.333 m/s, in the direction of the initial motion of block A. Conservation of energy says the initial kinetic energy of A equals the total kinetic energy at maximum compression plus the potential energy U_b stored in the bumpers: $\frac{1}{2}(2.00 \text{ kg})(2.00 \text{ m/s})^2 = U_b + \frac{1}{2}(12.0 \text{ kg})(0.333 \text{ m/s})^2$ and $U_b = 3.33 \text{ J}$.

(b) $v_{A2x} = \left(\dfrac{m_A - m_B}{m_A + m_B}\right)v_{A1x} = \left(\dfrac{2.00 \text{ kg} - 10.0 \text{ kg}}{12.0 \text{ kg}}\right)(2.00 \text{ m/s}) = -1.33 \text{ m/s}$. Block A is moving in the

$-x$ direction at 1.33 m/s.

$v_{B2x} = \left(\dfrac{2m_A}{m_A + m_B}\right)v_{A1x} = \dfrac{2(2.00 \text{ kg})}{12.0 \text{ kg}}(2.00 \text{ m/s}) = +0.667 \text{ m/s}$. Block B is moving in the $+x$ direction at 0.667 m/s.

EVALUATE: When the spring is compressed the maximum amount the system must still be moving in order to conserve momentum.

8.49. **IDENTIFY:** Eqs. 8.24 and 8.25 apply, with object A being the neutron.

SET UP: Let $+x$ be the direction of the initial momentum of the neutron. The mass of a neutron is $m_n = 1.0 \text{ u}$.

EXECUTE: (a) $v_{A2x} = \left(\dfrac{m_A - m_B}{m_A + m_B}\right)v_{A1x} = \dfrac{1.0 \text{ u} - 2.0 \text{ u}}{1.0 \text{ u} + 2.0 \text{ u}}v_{A1x} = -v_{A1x}/3.0$. The speed of the neutron after the

collision is one-third its initial speed.

(b) $K_2 = \frac{1}{2}m_n v_n^2 = \frac{1}{2}m_n(v_{A1}/3.0)^2 = \dfrac{1}{9.0}K_1$.

(c) After n collisions, $v_{A2} = \left(\dfrac{1}{3.0}\right)^n v_{A1} \cdot \left(\dfrac{1}{3.0}\right)^n = \dfrac{1}{59,000}$, so $3.0^n = 59,000$. $n\log 3.0 = \log 59,000$ and $n = 10$.

EVALUATE: Since the collision is elastic, in each collision the kinetic energy lost by the neutron equals the kinetic energy gained by the deuteron.

8.51. **IDENTIFY:** Apply Eq. 8.28.

SET UP: $m_A = 0.300 \text{ kg}$, $m_B = 0.400 \text{ kg}$, $m_C = 0.200 \text{ kg}$.

EXECUTE: $x_{cm} = \dfrac{m_A x_A + m_B x_B + m_C x_C}{m_A + m_B + m_C}$.

$x_{cm} = \dfrac{(0.300 \text{ kg})(0.200 \text{ m}) + (0.400 \text{ kg})(0.100 \text{ m}) + (0.200 \text{ kg})(-0.300 \text{ m})}{0.300 \text{ kg} + 0.400 \text{ kg} + 0.200 \text{ kg}} = 0.0444 \text{ m}$.

$y_{cm} = \dfrac{m_A y_A + m_B y_B + m_C y_C}{m_A + m_B + m_C}$.

$y_{cm} = \dfrac{(0.300 \text{ kg})(0.300 \text{ m}) + (0.400 \text{ kg})(-0.400 \text{ m}) + (0.200 \text{ kg})(0.600 \text{ m})}{0.300 \text{ kg} + 0.400 \text{ kg} + 0.200 \text{ kg}} = 0.0556 \text{ m}$.

EVALUATE: There is mass at both positive and negative x and at positive and negative y and therefore the center of mass is close to the origin.

8.53. **IDENTIFY:** The location of the center of mass is given by Eq. 8.48. The mass can be expressed in terms of the diameter. Each object can be replaced by a point mass at its center.

SET UP: Use coordinates with the origin at the center of Pluto and the $+x$ direction toward Charon, so

$x_P = 0$, $x_C = 19,700 \text{ km}$. $m = \rho V = \rho \frac{4}{3}\pi r^3 = \frac{1}{6}\rho\pi d^3$.

EXECUTE: $x_{cm} = \dfrac{m_P x_P + m_C x_C}{m_P + m_C} = \left(\dfrac{m_C}{m_P + m_C}\right)x_C = \left(\dfrac{\frac{1}{6}\rho\pi d_C^3}{\frac{1}{6}\rho\pi d_P^3 + \frac{1}{6}\rho\pi d_C^3}\right)x_C = \left(\dfrac{d_C^3}{d_P^3 + d_C^3}\right)x_C$.

$x_{cm} = \left(\dfrac{[1250 \text{ km}]^3}{[2370 \text{ km}]^3 + [1250 \text{ km}]^3}\right)(19,700 \text{ km}) = 2.52 \times 10^3 \text{ km}$.

The center of mass of the system is $2.52 \times 10^3 \text{ km}$ from the center of Pluto.

EVALUATE: The center of mass is closer to Pluto because Pluto has more mass than Charon.

8.55.   **IDENTIFY:**   Use Eq. 8.28 to find the x and y coordinates of the center of mass of the machine part for each configuration of the part. In calculating the center of mass of the machine part, each uniform bar can be represented by a point mass at its geometrical center.

SET UP: Use coordinates with the axis at the hinge and the $+x$ and $+y$ axes along the horizontal and vertical bars in the figure in the problem. Let (x_i, y_i) and (x_f, y_f) be the coordinates of the bar before and after the vertical bar is pivoted. Let object 1 be the horizontal bar, object 2 be the vertical bar and 3 be the ball.

EXECUTE: $x_i = \dfrac{m_1 x_1 + m_2 x_2 + m_3 x_3}{m_1 + m_2 + m_3} = \dfrac{(4.00\ \text{kg})(0.750\ \text{m}) + 0 + 0}{4.00\ \text{kg} + 3.00\ \text{kg} + 2.00\ \text{kg}} = 0.333\ \text{m.}$

$y_i = \dfrac{m_1 y_1 + m_2 y_2 + m_3 y_3}{m_1 + m_2 + m_3} = \dfrac{0 + (3.00\ \text{kg})(0.900\ \text{m}) + (2.00\ \text{kg})(1.80\ \text{m})}{9.00\ \text{kg}} = 0.700\ \text{m.}$

$x_f = \dfrac{(4.00\ \text{kg})(0.750\ \text{m}) + (3.00\ \text{kg})(-0.900\ \text{m}) + (2.00\ \text{kg})(-1.80\ \text{m})}{9.00\ \text{kg}} = -0.366\ \text{m.}$

$y_f = 0.$ $x_f - x_i = -0.700\ \text{m}$ and $y_f - y_i = -0.700\ \text{m.}$ The center of mass moves 0.700 m to the right and 0.700 m upward.

EVALUATE: The vertical bar moves upward and to the right so it is sensible for the center of mass of the machine part to move in these directions.

8.61.   **IDENTIFY:** $a = -\dfrac{v_{\text{ex}}}{m}\dfrac{dm}{dt}$. Assume that dm/dt is constant over the 5.0 s interval, since m doesn't change much during that interval. The thrust is $F = -v_{\text{ex}}\dfrac{dm}{dt}$.

SET UP: Take m to have the constant value $110\ \text{kg} + 70\ \text{kg} = 180\ \text{kg}$. dm/dt is negative since the mass of the MMU decreases as gas is ejected.

EXECUTE: **(a)** $\dfrac{dm}{dt} = -\dfrac{m}{v_{\text{ex}}}a = -\left(\dfrac{180\ \text{kg}}{490\ \text{m/s}}\right)(0.029\ \text{m/s}^2) = -0.0106\ \text{kg/s.}$ In 5.0 s the mass that is ejected is $(0.0106\ \text{kg/s})(5.0\ \text{s}) = 0.053\ \text{kg.}$

(b) $F = -v_{\text{ex}}\dfrac{dm}{dt} = -(490\ \text{m/s})(-0.0106\ \text{kg/s}) = 5.19\ \text{N.}$

EVALUATE: The mass change in the 5.0 s is a very small fraction of the total mass m, so it is accurate to take m to be constant.

8.65.   **IDENTIFY:** $v - v_0 = v_{\text{ex}} \ln\left(\dfrac{m_0}{m}\right)$.

SET UP: $v_0 = 0$.

EXECUTE: $\ln\left(\dfrac{m_0}{m}\right) = \dfrac{v}{v_{\text{ex}}} = \dfrac{8.00 \times 10^3\ \text{m/s}}{2100\ \text{m/s}} = 3.81$ and $\dfrac{m_0}{m} = e^{3.81} = 45.2$.

EVALUATE: Note that the final speed of the rocket is greater than the relative speed of the exhaust gas.

8.67.   **IDENTIFY:**   Use the heights to find v_{1y} and v_{2y}, the velocity of the ball just before and just after it strikes the slab. Then apply $J_y = F_y \Delta t = \Delta p_y$.

SET UP: Let $+y$ be downward.

EXECUTE: **(a)** $\frac{1}{2}mv^2 = mgh$ so $v = \pm\sqrt{2gh}$.

$v_{1y} = +\sqrt{2(9.80\ \text{m/s}^2)(2.00\ \text{m})} = 6.26\ \text{m/s.}$ $v_{2y} = -\sqrt{2(9.80\ \text{m/s}^2)(1.60\ \text{m})} = -5.60\ \text{m/s.}$

$J_y = \Delta p_y = m(v_{2y} - v_{1y}) = (40.0 \times 10^{-3}\ \text{kg})(-5.60\ \text{m/s} - 6.26\ \text{m/s}) = -0.474\ \text{kg} \cdot \text{m/s.}$

The impulse is $0.474\ \text{kg} \cdot \text{m/s}$, upward.

(b) $F_y = \dfrac{J_y}{\Delta t} = \dfrac{-0.474\ \text{kg} \cdot \text{m/s}}{2.00 \times 10^{-3}\ \text{s}} = -237\ \text{N.}$ The average force on the ball is 237 N, upward.

EVALUATE: The upward force on the ball changes the direction of its momentum.

8.73. **IDENTIFY:** The x and y components of the momentum of the system are conserved.

SET UP: After the collision the combined object with mass $m_{tot} = 0.100$ kg moves with velocity $\vec{v}_2$.

Solve for v_{Cx} and v_{Cy}.

EXECUTE: **(a)** $P_{1x} = P_{2x}$ gives $m_A v_{Ax} + m_B v_{Bx} + m_C v_{Cx} = m_{tot} v_{2x}$.

$$v_{Cx} = -\frac{m_A v_{Ax} + m_B v_{Bx} - m_{tot} v_{2x}}{m_C}$$

$$v_{Cx} = -\frac{(0.020 \text{ kg})(-1.50 \text{ m/s}) + (0.030 \text{ kg})(-0.50 \text{ m/s})\cos 60° - (0.100 \text{ kg})(0.50 \text{ m/s})}{0.050 \text{ kg}}.$$

$$v_{Cx} = 1.75 \text{ m/s}.$$

$P_{1y} = P_{2y}$ gives $m_A v_{Ay} + m_B v_{By} + m_C v_{Cy} = m_{tot} v_{2y}$.

$$v_{Cy} = -\frac{m_A v_{Ay} + m_B v_{By} - m_{tot} v_{2y}}{m_C} = -\frac{(0.030 \text{ kg})(-0.50 \text{ m/s})\sin 60°}{0.050 \text{ kg}} = +0.260 \text{ m/s}.$$

(b) $v_C = \sqrt{v_{Cx}^2 + v_{Cy}^2} = 1.77$ m/s. $\Delta K = K_2 - K_1$.

$$\Delta K = \tfrac{1}{2}(0.100 \text{ kg})(0.50 \text{ m/s})^2 - [\tfrac{1}{2}(0.020 \text{ kg})(1.50 \text{ m/s})^2 + \tfrac{1}{2}(0.030 \text{ kg})(0.50 \text{ m/s})^2 + \tfrac{1}{2}(0.050 \text{ kg})(1.77 \text{ m/s})^2]$$

$\Delta K = -0.092$ J.

EVALUATE: Since there is no horizontal external force the vector momentum of the system is conserved. The forces the spheres exert on each other do negative work during the collision and this reduces the kinetic energy of the system.

8.75. **IDENTIFY:** Apply conservation of momentum to the nucleus and its fragments. The initial momentum is zero. The ^{214}Po nucleus has mass $214(1.67 \times 10^{-27} \text{ kg}) = 3.57 \times 10^{-25}$ kg, where 1.67×10^{-27} kg is the mass of a nucleon (proton or neutron). $K = \tfrac{1}{2}mv^2$.

SET UP: Let $+x$ be the direction in which the alpha particle is emitted. The nucleus that is left after the decay has mass $m_n = 3.75 \times 10^{-25}$ kg $- m_\alpha = 3.57 \times 10^{-25}$ kg $- 6.65 \times 10^{-27}$ kg $= 3.50 \times 10^{-25}$ kg.

EXECUTE: $P_{2x} = P_{1x} = 0$ gives $m_\alpha v_\alpha + m_n v_n = 0$. $v_n - \frac{m_\alpha}{m_n} v_\alpha$.

$$v_\alpha = \sqrt{\frac{2K_\alpha}{m_\alpha}} = \sqrt{\frac{2(1.23 \times 10^{-12} \text{ J})}{6.65 \times 10^{-27} \text{ kg}}} = 1.92 \times 10^7 \text{ m/s}. \quad v_n = \left(\frac{6.65 \times 10^{-27} \text{ kg}}{3.50 \times 10^{-25} \text{ kg}}\right)(1.92 \times 10^7 \text{ m/s}) = 3.65 \times 10^5 \text{ m/s}.$$

EVALUATE: The recoil velocity of the more massive nucleus is much less than the speed of the emitted alpha particle.

8.77. **IDENTIFY:** Momentum is conserved during the collision, and the wood (with the clay attached) is in free fall as it falls since only gravity acts on it.

SET UP: Apply conservation of momentum to the collision to find the velocity V of the combined object just after the collision. After the collision, the wood's downward acceleration is g and it has no horizontal acceleration, so we can use the standard kinematics equations: $y - y_0 = v_{0y}t + \frac{1}{2}a_y t^2$ and $x - x_0 = v_{0x}t + \frac{1}{2}a_x t^2$.

EXECUTE: Momentum conservation gives $(0.500 \text{ kg})(24.0 \text{ m/s}) = (8.50 \text{ kg})V$, so $V = 1.412$ m/s. Consider the projectile motion after the collision: $a_y = +9.8 \text{ m/s}^2$, $v_{0y} = 0$, $y - y_0 = +2.20$ m, and t is unknown.

$y - y_0 = v_{0y}t + \frac{1}{2}a_y t^2$ gives $t = \sqrt{\frac{2(y - y_0)}{a_y}} = \sqrt{\frac{2(2.20 \text{ m})}{9.8 \text{ m/s}^2}} = 0.6701$ s. The horizontal acceleration is zero

so $x - x_0 = v_{0x}t + \frac{1}{2}a_x t^2 = (1.412 \text{ m/s})(0.6701 \text{ s}) = 0.946$ m.

EVALUATE: The momentum is *not* conserved after the collision because an external force (gravity) acts on the system. Mechanical energy is *not* conserved during the collision because the clay and block stick together, making it an inelastic collision.

8.79.   **IDENTIFY:**   During the collision, momentum is conserved, but after the collision mechanical energy is conserved. We cannot solve this problem in a single step because the collision and the motion after the collision involve different conservation laws.

SET UP: Use coordinates where $+x$ is to the right and $+y$ is upward. Momentum is conserved during the collision, so $P_{1x} = P_{2x}$. Energy is conserved after the collision, so $K_1 = U_2$, where $K = \frac{1}{2}mv^2$ and $U = mgh$.

EXECUTE: *Collision*: There is no external horizontal force during the collision so $P_{1x} = P_{2x}$. This gives $(5.00 \text{ kg})(12.0 \text{ m/s}) = (10.0 \text{ kg})v_2$ and $v_2 = 6.0 \text{ m/s}$.

Motion after the collision: Only gravity does work and the initial kinetic energy of the combined chunks is converted entirely to gravitational potential energy when the chunk reaches its maximum height h above the valley floor. Conservation of energy gives $\frac{1}{2}m_{\text{tot}}v^2 = m_{\text{tot}}gh$ and $h = \dfrac{v^2}{2g} = \dfrac{(6.0 \text{ m/s})^2}{2(9.8 \text{ m/s}^2)} = 1.8 \text{ m}$.

EVALUATE: After the collision the energy of the system is $\frac{1}{2}m_{\text{tot}}v^2 = \frac{1}{2}(10.0 \text{ kg})(6.0 \text{ m/s})^2 = 180 \text{ J}$ when it is all kinetic energy and the energy is $m_{\text{tot}}gh = (10.0 \text{ kg})(9.8 \text{ m/s}^2)(1.8 \text{ m}) = 180 \text{ J}$ when it is all gravitational potential energy. Mechanical energy is conserved during the motion after the collision. But before the collision the total energy of the system is $\frac{1}{2}(5.0 \text{ kg})(12.0 \text{ m/s})^2 = 360 \text{ J}$; 50% of the mechanical energy is dissipated during the inelastic collision of the two chunks.

8.81.   **IDENTIFY:**   During the inelastic collision, momentum is conserved (in two dimensions), but after the collision we must use energy principles.

SET UP: The friction force is $\mu_k m_{\text{tot}} g$. Use energy considerations to find the velocity of the combined object immediately after the collision. Apply conservation of momentum to the collision. Use coordinates where $+x$ is west and $+y$ is south. For momentum conservation, we have $P_{1x} = P_{2x}$ and $P_{1y} = P_{2y}$.

EXECUTE: *Motion after collision*: The negative work done by friction takes away all the kinetic energy that the combined object has just after the collision. Calling ϕ the angle south of west at which the enmeshed cars slid, we have $\tan\phi = \dfrac{6.43 \text{ m}}{5.39 \text{ m}}$ and $\phi = 50.0°$. The wreckage slides 8.39 m in a direction 50.0° south of west. Energy conservation gives $\frac{1}{2}m_{\text{tot}}V^2 = \mu_k m_{\text{tot}}gd$, so

$V = \sqrt{2\mu_k gd} = \sqrt{2(0.75)(9.80 \text{ m/s}^2)(8.39 \text{ m})} = 11.1 \text{ m/s}$.   The velocity components are
$V_x = V\cos\phi = 7.13 \text{ m/s};\ V_y = V\sin\phi = 8.50 \text{ m/s}$.

Collision: $P_{1x} = P_{2x}$ gives $(2200 \text{ kg})v_{\text{SUV}} = (1500 \text{ kg} + 2200 \text{ kg})V_x$ and $v_{\text{SUV}} = 12 \text{ m/s}$. $P_{1y} = P_{2y}$ gives $(1500 \text{ kg})v_{\text{sedan}} = (1500 \text{ kg} + 2200 \text{ kg})V_y$ and $v_{\text{sedan}} = 21 \text{ m/s}$.

EVALUATE: We cannot solve this problem in a single step because the collision and the motion after the collision involve different principles (momentum conservation and energy conservation).

8.85.   **IDENTIFY:**   Apply conservation of momentum to the collision between the two people. Apply conservation of energy to the motion of the stuntman before the collision and to the entwined people after the collision.

SET UP: For the motion of the stuntman, $y_1 - y_2 = 5.0 \text{ m}$. Let v_S be the magnitude of his horizontal velocity just before the collision. Let V be the speed of the entwined people just after the collision. Let d be the distance they slide along the floor.

EXECUTE: **(a)** Motion before the collision: $K_1 + U_1 = K_2 + U_2$. $K_1 = 0$ and $\frac{1}{2}mv_S^2 = mg(y_1 - y_2)$.

$v_S = \sqrt{2g(y_1 - y_2)} = \sqrt{2(9.80 \text{ m/s}^2)(5.0 \text{ m})} = 9.90 \text{ m/s}$.

Collision: $m_S v_S = m_{\text{tot}}V$. $V = \dfrac{m_S}{m_{\text{tot}}}v_S = \left(\dfrac{80.0 \text{ kg}}{150.0 \text{ kg}}\right)(9.90 \text{ m/s}) = 5.28 \text{ m/s}$.

(b) Motion after the collision: $K_1 + U_1 + W_{\text{other}} = K_2 + U_2$ gives $\frac{1}{2}m_{\text{tot}}V^2 - \mu_k m_{\text{tot}}gd = 0$.

$$d = \frac{V^2}{2\mu_k g} = \frac{(5.28\text{ m/s})^2}{2(0.250)(9.80\text{ m/s}^2)} = 5.7\text{ m}.$$

EVALUATE: Mechanical energy is dissipated in the inelastic collision, so the kinetic energy just after the collision is less than the initial potential energy of the stuntman.

8.87. **IDENTIFY:** Eqs. 8.24 and 8.25 give the outcome of the elastic collision. Apply conservation of energy to the motion of the block after the collision.

SET UP: Object B is the block, initially at rest. If L is the length of the wire and θ is the angle it makes with the vertical, the height of the block is $y = L(1 - \cos\theta)$. Initially, $y_1 = 0$.

EXECUTE: Eq. 8.25 gives $v_B = \left(\frac{2m_A}{m_A + m_B}\right)v_A = \left(\frac{2M}{M + 3M}\right)(4.00\text{ m/s}) = 2.00\text{ m/s}$.

Conservation of energy gives $\frac{1}{2}m_B v_B^2 = m_B gL(1 - \cos\theta)$.

$$\cos\theta = 1 - \frac{v_B^2}{2gL} = 1 - \frac{(2.00\text{ m/s})^2}{2(9.80\text{ m/s}^2)(0.500\text{ m})} = 0.5918, \text{ which gives } \theta = 53.7°.$$

EVALUATE: Only a portion of the initial kinetic energy of the ball is transferred to the block in the collision.

8.91. **IDENTIFY:** Apply conservation of momentum to the collision between the bullet and the block and apply conservation of energy to the motion of the block after the collision.

(a) SET UP: For the collision between the bullet and the block, let object A be the bullet and object B be the block. Apply momentum conservation to find the speed v_{B2} of the block just after the collision (see Figure 8.91a).

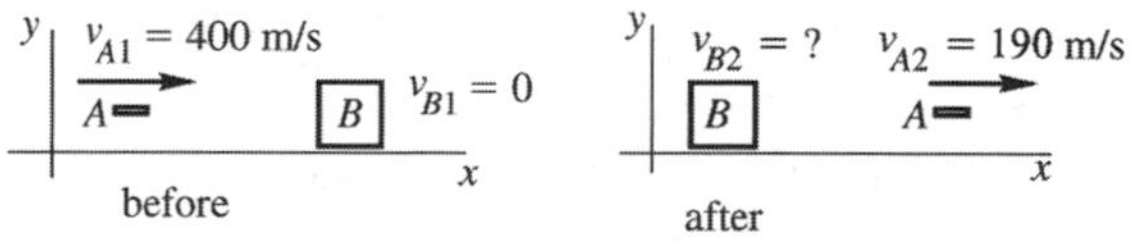

Figure 8.91a

EXECUTE: P_x is conserved so $m_A v_{A1x} + m_B v_{B1x} = m_A v_{A2x} + m_B v_{B2x}$. $m_A v_{A1} = m_A v_{A2} + m_B v_{B2x}$.

$$v_{B2x} = \frac{m_A(v_{A1} - v_{A2})}{m_B} = \frac{4.00\times10^{-3}\text{ kg}(400\text{ m/s} - 190\text{ m/s})}{0.800\text{ kg}} = 1.05\text{ m/s}.$$

SET UP: For the motion of the block after the collision, let point 1 in the motion be just after the collision, where the block has the speed 1.05 m/s calculated above, and let point 2 be where the block has come to rest (see Figure 8.91b).
$K_1 + U_1 + W_{\text{other}} = K_2 + U_2$.

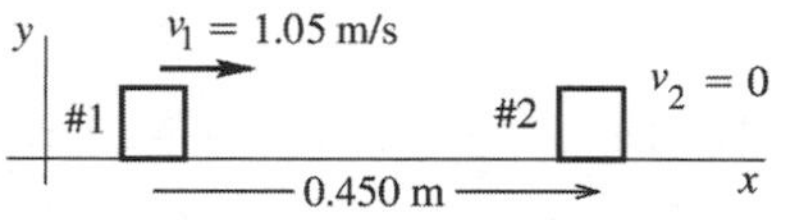

Figure 8.91b

EXECUTE: Work is done on the block by friction, so $W_{\text{other}} = W_f$.

$W_{\text{other}} = W_f = (f_k \cos\phi)s = -f_k s = -\mu_k mgs$, where $s = 0.450$ m. $U_1 = 0$, $U_2 = 0$, $K_1 = \frac{1}{2}mv_1^2$, $K_2 = 0$ (the

block has come to rest). Thus $\frac{1}{2}mv_1^2 - \mu_k mgs = 0$. Therefore $\mu_k = \frac{v_1^2}{2gs} = \frac{(1.05\text{ m/s})^2}{2(9.80\text{ m/s}^2)(0.450\text{ m})} = 0.125$.

(b) For the bullet, $K_1 = \frac{1}{2}mv_1^2 = \frac{1}{2}(4.00 \times 10^{-3} \text{ kg})(400 \text{ m/s})^2 = 320 \text{ J}$ and

$K_2 = \frac{1}{2}mv_2^2 = \frac{1}{2}(4.00 \times 10^{-3} \text{ kg})(190 \text{ m/s})^2 = 72.2 \text{ J}$. $\Delta K = K_2 - K_1 = 72.2 \text{ J} - 320 \text{ J} = -248 \text{ J}$. The kinetic energy of the bullet decreases by 248 J.

(c) Immediately after the collision the speed of the block is 1.05 m/s, so its kinetic energy is

$K = \frac{1}{2}mv^2 = \frac{1}{2}(0.800 \text{ kg})(1.05 \text{ m/s})^2 = 0.441 \text{ J}$.

EVALUATE: The collision is highly inelastic. The bullet loses 248 J of kinetic energy but only 0.441 J is gained by the block. But momentum is conserved in the collision. All the momentum lost by the bullet is gained by the block.

8.95.   **IDENTIFY:** Apply conservation of energy to the motion of the package before the collision and apply conservation of the horizontal component of momentum to the collision.

(a) SET UP: Apply conservation of energy to the motion of the package from point 1 as it leaves the chute to point 2 just before it lands in the cart. Take $y = 0$ at point 2, so $y_1 = 4.00$ m. Only gravity does work, so

$$K_1 + U_1 = K_2 + U_2.$$

EXECUTE: $\frac{1}{2}mv_1^2 + mgy_1 = \frac{1}{2}mv_2^2$.

$v_2 = \sqrt{v_1^2 + 2gy_1} = 9.35 \text{ m/s}$.

(b) SET UP: In the collision between the package and the cart, momentum is conserved in the horizontal direction. (But not in the vertical direction, due to the vertical force the floor exerts on the cart.) Take $+x$ to be to the right. Let A be the package and B be the cart.

EXECUTE: P_x is constant gives $m_A v_{A1x} + m_B v_{B1x} = (m_A + m_B)v_{2x}$.

$v_{B1x} = -5.00 \text{ m/s}$.

$v_{A1x} = (3.00 \text{ m/s})\cos 37.0°$. (The horizontal velocity of the package is constant during its free fall.)

Solving for v_{2x} gives $v_{2x} = -3.29$ m/s. The cart is moving to the left at 3.29 m/s after the package lands in it.

EVALUATE: The cart is slowed by its collision with the package, whose horizontal component of momentum is in the opposite direction to the motion of the cart.

8.99.   **IDENTIFY and SET UP:**

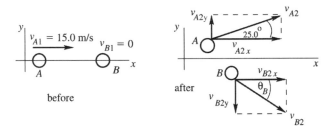

Figure 8.99

P_x and P_y are conserved in the collision since there is no external horizontal force.

The collision is elastic, so $25.0° + \theta_B = 90°$, so that $\theta_B = 65.0°$. (A and B move off in perpendicular directions.)

EXECUTE: P_x is conserved so $m_A v_{A1x} + m_B v_{B1x} = m_A v_{A2x} + m_B v_{B2x}$.

But $m_A = m_B$ so $v_{A1} = v_{A2}\cos 25.0° + v_{B2}\cos 65.0°$.

$\qquad\qquad\qquad P_y$ is conserved so $m_A v_{A1y} + m_B v_{B1y} = m_A v_{A2y} + m_B v_{B2y}$.

$0 = v_{A2y} + v_{B2y}$.

$0 = v_{A2}\sin 25.0° - v_{B2}\sin 65.0°$.

$v_{B2} = (\sin 25.0° / \sin 65.0°)v_{A2}$.

This result in the first equation gives $v_{A1} = v_{A2}\cos 25.0° + \left(\dfrac{\sin 25.0°\cos 65.0°}{\sin 65.0°}\right)v_{A2}.$

$v_{A1} = 1.103 v_{A2}.$

$v_{A2} = v_{A1}/1.103 = (15.0\ \text{m/s})/1.103 = 13.6\ \text{m/s}.$

And then $v_{B2} = (\sin 25.0°/\sin 65.0°)(13.6\ \text{m/s}) = 6.34\ \text{m/s}.$

EVALUATE: We can use our numerical results to show that $K_1 = K_2$ and that $P_{1x} = P_{2x}$ and $P_{1y} = P_{2y}.$

8.101. **IDENTIFY:** Apply conservation of momentum to the system of the neutron and its decay products.
SET UP: Let the proton be moving in the $+x$ direction with speed v_p after the decay. The initial momentum of the neutron is zero, so to conserve momentum the electron must be moving in the $-x$ direction after the decay. Let the speed of the electron be v_e.

EXECUTE: $P_{1x} = P_{2x}$ gives $0 = m_p v_p - m_e v_e$ and $v_e = \left(\dfrac{m_p}{m_e}\right)v_p.$ The total kinetic energy after the decay is

$$K_{\text{tot}} = \tfrac{1}{2}m_e v_e^2 + \tfrac{1}{2}m_p v_p^2 = \tfrac{1}{2}m_e\left(\dfrac{m_p}{m_e}\right)^2 v_p^2 + \tfrac{1}{2}m_p v_p^2 = \tfrac{1}{2}m_p v_p^2\left(1 + \dfrac{m_p}{m_e}\right).$$

Thus, $\dfrac{K_p}{K_{\text{tot}}} = \dfrac{1}{1 + m_p/m_e} = \dfrac{1}{1 + 1836} = 5.44\times10^{-4} = 0.0544\%.$

EVALUATE: Most of the released energy goes to the electron, since it is much lighter than the proton.

8.105. **IDENTIFY:** No net external force acts on the Burt-Ernie-log system, so the center of mass of the system does not move.
SET UP: $x_{\text{cm}} = \dfrac{m_1 x_1 + m_2 x_2 + m_3 x_3}{m_1 + m_2 + m_3}.$

EXECUTE: Use coordinates where the origin is at Burt's end of the log and where $+x$ is toward Ernie, which makes $x_1 = 0$ for Burt initially. The initial coordinate of the center of mass is

$x_{\text{cm},1} = \dfrac{(20.0\ \text{kg})(1.5\ \text{m}) + (40.0\ \text{kg})(3.0\ \text{m})}{90.0\ \text{kg}}.$ Let d be the distance the log moves toward Ernie's original

position. The final location of the center of mass is $x_{\text{cm},2} = \dfrac{(30.0\ \text{kg})d + (1.5\ \text{kg} + d)(20.0\ \text{kg}) + (40.0\ \text{kg})d}{90.0\ \text{kg}}.$

The center of mass does not move, so $x_{\text{cm},1} = x_{\text{cm},2}$, which gives

$(20.0\ \text{kg})(1.5\ \text{m}) + (40.0\ \text{kg})(3.0\ \text{m}) = (30.0\ \text{kg})d + (20.0\ \text{kg})(1.5\ \text{m} + d) + (40.0\ \text{kg})d.$ Solving for d gives

$d = 1.33\ \text{m}.$

EVALUATE: Burt, Ernie and the log all move, but the center of mass of the system does not move.

8.109. **IDENTIFY:** The rocket moves in projectile motion before the explosion and its fragments move in projectile motion after the explosion. Apply conservation of energy and conservation of momentum to the explosion.
(a) SET UP: Apply conservation of energy to the explosion. Just before the explosion the rocket is at its maximum height and has zero kinetic energy. Let A be the piece with mass 1.40 kg and B be the piece with mass 0.28 kg. Let v_A and v_B be the speeds of the two pieces immediately after the collision.

EXECUTE: $\tfrac{1}{2}m_A v_A^2 + \tfrac{1}{2}m_B v_B^2 = 860\ \text{J}$

SET UP: Since the two fragments reach the ground at the same time, their velocities just after the explosion must be horizontal. The initial momentum of the rocket before the explosion is zero, so after the explosion the pieces must be moving in opposite horizontal directions and have equal magnitude of momentum: $m_A v_A = m_B v_B.$

EXECUTE: Use this to eliminate v_A in the first equation and solve for v_B:

$$\tfrac{1}{2}m_B v_B^2(1 + m_B/m_A) = 860\ \text{J} \quad \text{and} \quad v_B = 71.6\ \text{m/s}.$$

Then $v_A = (m_B/m_A)v_B = 14.3\ \text{m/s}.$

(b) SET UP: Use the vertical motion from the maximum height to the ground to find the time it takes the pieces to fall to the ground after the explosion. Take $+y$ downward.

$$v_{0y} = 0, \quad a_y = +9.80 \text{ m/s}^2, \quad y - y_0 = 80.0 \text{ m}, \quad t = ?$$

EXECUTE: $y - y_0 = v_{0y}t + \frac{1}{2}a_y t^2$ gives $t = 4.04$ s.

During this time the horizontal distance each piece moves is $x_A = v_A t = 57.8$ m and $x_B = v_B t = 289.1$ m. They move in opposite directions, so they are $x_A + x_B = 347$ m apart when they land.

EVALUATE: Fragment A has more mass so it is moving slower right after the collision, and it travels horizontally a smaller distance as it falls to the ground.

8.113.   **IDENTIFY** and **SET UP:** Apply Eq. 8.40 to the single-stage rocket and to each stage of the two-stage rocket.

(a) EXECUTE: $v - v_0 = v_{ex} \ln(m_0/m); \quad v_0 = 0$ so $v = v_{ex} \ln(m_0/m)$

The total initial mass of the rocket is $m_0 = 12{,}000 \text{ kg} + 1000 \text{ kg} = 13{,}000 \text{ kg}$. Of this, $9000 \text{ kg} + 700 \text{ kg} = 9700 \text{ kg}$ is fuel, so the mass m left after all the fuel is burned is $13{,}000 \text{ kg} - 9700 \text{ kg} = 3300 \text{ kg}$.

$$v = v_{ex} \ln(13{,}000 \text{ kg}/3300 \text{ kg}) = 1.37 v_{ex}.$$

(b) First stage: $v = v_{ex} \ln(m_0/m)$

$m_0 = 13{,}000$ kg

The first stage has 9000 kg of fuel, so the mass left after the first stage fuel has burned is $13{,}000 \text{ kg} - 9000 \text{ kg} = 4000 \text{ kg}$.

$$v = v_{ex} \ln(13{,}000 \text{ kg}/4000 \text{ kg}) = 1.18 v_{ex}.$$

(c) Second stage: $m_0 = 1000 \text{ kg}, \quad m = 1000 \text{ kg} - 700 \text{ kg} = 300 \text{ kg}$.

$$v = v_0 + v_{ex} \ln(m_0/m) = 1.18 v_{ex} + v_{ex} \ln(1000 \text{ kg}/300 \text{ kg}) = 2.38 v_{ex}.$$

(d) $v = 7.00$ km/s

$$v_{ex} = v/2.38 = (7.00 \text{ km/s})/2.38 = 2.94 \text{ km/s}.$$

EVALUATE: The two-stage rocket achieves a greater final speed because it jettisons the left-over mass of the first stage before the second-state fires and this reduces the final m and increases m_0/m.

ROTATION OF RIGID BODIES

9

9.3. **IDENTIFY** $\alpha_z(t) = \dfrac{d\omega_z}{dt}$. Writing Eq. (2.16) in terms of angular quantities gives $\theta - \theta_0 = \int_{t_1}^{t_2} \omega_z\, dt$.

SET UP: $\dfrac{d}{dt} t^n = nt^{n-1}$ and $\int t^n dt = \dfrac{1}{n+1} t^{n+1}$

EXECUTE: (a) A must have units of rad/s and B must have units of rad/s^3.

(b) $\alpha_z(t) = 2Bt = (3.00 \text{ rad/s}^3)t$. (i) For $t = 0$, $\alpha_z = 0$. (ii) For $t = 5.00$ s, $\alpha_z = 15.0$ rad/s^2.

(c) $\theta_2 - \theta_1 = \int_{t_1}^{t_2} (A + Bt^2)\, dt = A(t_2 - t_1) + \frac{1}{3} B(t_2^3 - t_1^3)$. For $t_1 = 0$ and $t_2 = 2.00$ s,

$\theta_2 - \theta_1 = (2.75 \text{ rad/s})(2.00 \text{ s}) + \frac{1}{3}(1.50 \text{ rad/s}^3)(2.00 \text{ s})^3 = 9.50$ rad.

EVALUATE: Both α_z and ω_z are positive and the angular speed is increasing.

9.7. **IDENTIFY:** $\omega_z(t) = \dfrac{d\theta}{dt}$. $\alpha_z(t) = \dfrac{d\omega_z}{dt}$. Use the values of θ and ω_z at $t = 0$ and α_z at 1.50 s to calculate a, b, and c.

SET UP: $\dfrac{d}{dt} t^n = nt^{n-1}$

EXECUTE: (a) $\omega_z(t) = b - 3ct^2$. $\alpha_z(t) = -6ct$. At $t = 0$, $\theta = a = \pi/4$ rad and $\omega_z = b = 2.00$ rad/s. At $t = 1.50$ s, $\alpha_z = -6c(1.50 \text{ s}) = 1.25$ rad/s^2 and $c = -0.139$ rad/s^3.

(b) $\theta = \pi/4$ rad and $\alpha_z = 0$ at $t = 0$.

(c) $\alpha_z = 3.50$ rad/s^2 at $t = -\dfrac{\alpha_z}{6c} = -\dfrac{3.50 \text{ rad/s}^2}{6(-0.139 \text{ rad/s}^3)} = 4.20$ s. At $t = 4.20$ s,

$\theta = \dfrac{\pi}{4} \text{ rad} + (2.00 \text{ rad/s})(4.20 \text{ s}) - (-0.139 \text{ rad/s}^3)(4.20 \text{ s})^3 = 19.5$ rad.

$\omega_z = 2.00 \text{ rad/s} - 3(-0.139 \text{ rad/s}^3)(4.20 \text{ s})^2 = 9.36$ rad/s.

EVALUATE: θ, ω_z and α_z all increase as t increases.

9.11. **IDENTIFY:** Apply the constant angular acceleration equations to the motion. The target variables are t and $\theta - \theta_0$.

SET UP: (a) $\alpha_z = 1.50$ rad/s^2; $\omega_{0z} = 0$ (starts from rest); $\omega_z = 36.0$ rad/s; $t = ?$

$\omega_z = \omega_{0z} + \alpha_z t$

EXECUTE: $t = \dfrac{\omega_z - \omega_{0z}}{\alpha_z} = \dfrac{36.0 \text{ rad/s} - 0}{1.50 \text{ rad/s}^2} = 24.0$ s

(b) $\theta - \theta_0 = ?$

$\theta - \theta_0 = \omega_{0z} t + \frac{1}{2} \alpha_z t^2 = 0 + \frac{1}{2}(1.50 \text{ rad/s}^2)(24.0 \text{ s})^2 = 432$ rad

$\theta - \theta_0 = 432 \text{ rad}(1 \text{ rev}/2\pi \text{ rad}) = 68.8$ rev

EVALUATE: We could use $\theta - \theta_0 = \frac{1}{2}(\omega_z + \omega_{0z})t$ to calculate $\theta - \theta_0 = \frac{1}{2}(0 + 36.0 \text{ rad/s})(24.0 \text{ s}) = 432 \text{ rad}$, which checks.

9.15.   **IDENTIFY:**   Apply constant angular acceleration equations.
SET UP: Let the direction the flywheel is rotating be positive.
$\theta - \theta_0 = 200 \text{ rev}$, $\omega_{0z} = 500 \text{ rev/min} = 8.333 \text{ rev/s}$, $t = 30.0 \text{ s}$.

EXECUTE: (a) $\theta - \theta_0 = \left(\dfrac{\omega_{0z} + \omega_z}{2}\right)t$ gives $\omega_z = 5.00 \text{ rev/s} = 300 \text{ rpm}$

(b) Use the information in part (a) to find α_z: $\omega_z = \omega_{0z} + \alpha_z t$ gives $\alpha_z = -0.1111 \text{ rev/s}^2$. Then $\omega_z = 0$,

$\alpha_z = -0.1111 \text{ rev/s}^2$, $\omega_{0z} = 8.333 \text{ rev/s}$ in $\omega_z = \omega_{0z} + \alpha_z t$ gives $t = 75.0 \text{ s}$ and $\theta - \theta_0 = \left(\dfrac{\omega_{0z} + \omega_z}{2}\right)t$

gives $\theta - \theta_0 = 312 \text{ rev}$.

EVALUATE: The mass and diameter of the flywheel are not used in the calculation.

9.17.   **IDENTIFY:**   Apply Eq. (9.12) to relate ω_z to $\theta - \theta_0$.
SET UP: Establish a proportionality.
EXECUTE: From Eq. (9.12), with $\omega_{0z} = 0$, the number of revolutions is proportional to the square of the initial angular velocity, so tripling the initial angular velocity increases the number of revolutions by 9, to 9.00 rev.
EVALUATE: We don't have enough information to calculate α_z; all we need to know is that it is constant.

9.23.   **IDENTIFY and SET UP:**   Use constant acceleration equations to find ω and α after each displacement. Then use Eqs. (9.14) and (9.15) to find the components of the linear acceleration.
EXECUTE: (a) at the start $t = 0$

flywheel starts from rest so $\omega = \omega_{0z} = 0$

$a_{\text{tan}} = r\alpha = (0.300 \text{ m})(0.600 \text{ rad/s}^2) = 0.180 \text{ m/s}^2$

$a_{\text{rad}} = r\omega^2 = 0$

$a = \sqrt{a_{\text{rad}}^2 + a_{\text{tan}}^2} = 0.180 \text{ m/s}^2$

(b) $\theta - \theta_0 = 60°$

$a_{\text{tan}} = r\alpha = 0.180 \text{ m/s}^2$

Calculate ω:

$\theta - \theta_0 = 60°(\pi \text{ rad}/180°) = 1.047 \text{ rad}$; $\omega_{0z} = 0$; $\alpha_z = 0.600 \text{ rad/s}^2$; $\omega_z = ?$

$\omega_z^2 = \omega_{0z}^2 + 2\alpha_z(\theta - \theta_0)$

$\omega_z = \sqrt{2\alpha_z(\theta - \theta_0)} = \sqrt{2(0.600 \text{ rad/s}^2)(1.047 \text{ rad})} = 1.121 \text{ rad/s}$ and $\omega = \omega_z$.

Then $a_{\text{rad}} = r\omega^2 = (0.300 \text{ m})(1.121 \text{ rad/s})^2 = 0.377 \text{ m/s}^2$.

$a = \sqrt{a_{\text{rad}}^2 + a_{\text{tan}}^2} = \sqrt{(0.377 \text{ m/s}^2)^2 + (0.180 \text{ m/s}^2)^2} = 0.418 \text{ m/s}^2$

(c) $\theta - \theta_0 = 120°$

$a_{\text{tan}} = r\alpha = 0.180 \text{ m/s}^2$

Calculate ω:

$\theta - \theta_0 = 120°(\pi \text{ rad}/180°) = 2.094 \text{ rad}$; $\omega_{0z} = 0$; $\alpha_z = 0.600 \text{ rad/s}^2$; $\omega_z = ?$

$\omega_z^2 = \omega_{0z}^2 + 2\alpha_z(\theta - \theta_0)$

$\omega_z = \sqrt{2\alpha_z(\theta - \theta_0)} = \sqrt{2(0.600 \text{ rad/s}^2)(2.094 \text{ rad})} = 1.585 \text{ rad/s}$ and $\omega = \omega_z$.

Then $a_{\text{rad}} = r\omega^2 = (0.300 \text{ m})(1.585 \text{ rad/s})^2 = 0.754 \text{ m/s}^2$.

$a = \sqrt{a_{\text{rad}}^2 + a_{\text{tan}}^2} = \sqrt{(0.754 \text{ m/s}^2)^2 + (0.180 \text{ m/s}^2)^2} = 0.775 \text{ m/s}^2$.

EVALUATE: α is constant so α_{tan} is constant. ω increases so a_{rad} increases.

9.25.   **IDENTIFY:**   Use Eq. (9.15) and solve for r.

SET UP: $a_{rad} = r\omega^2$ so $r = a_{rad}/\omega^2$, where ω must be in rad/s

EXECUTE: $a_{rad} = 3000g = 3000(9.80 \text{ m/s}^2) = 29,400 \text{ m/s}^2$

$$\omega = (5000 \text{ rev/min})\left(\frac{1 \text{ min}}{60 \text{ s}}\right)\left(\frac{2\pi \text{ rad}}{1 \text{ rev}}\right) = 523.6 \text{ rad/s}$$

Then $r = \dfrac{a_{rad}}{\omega^2} = \dfrac{29,400 \text{ m/s}^2}{(523.6 \text{ rad/s})^2} = 0.107 \text{ m}.$

EVALUATE: The diameter is then 0.214 m, which is larger than 0.127 m, so the claim is *not* realistic.

9.29.   **IDENTIFY and SET UP:**   Use Eq. (9.15) to relate ω to a_{rad} and $\sum \vec{F} = m\vec{a}$ to relate a_{rad} to F_{rad}. Use Eq. (9.13) to relate ω and v, where v is the tangential speed.

EXECUTE: **(a)** $a_{rad} = r\omega^2$ and $F_{rad} = ma_{rad} = mr\omega^2$

$$\frac{F_{rad,2}}{F_{rad,1}} = \left(\frac{\omega_2}{\omega_1}\right)^2 = \left(\frac{640 \text{ rev/min}}{423 \text{ rev/min}}\right)^2 = 2.29$$

(b) $v = r\omega$

$$\frac{v_2}{v_1} = \frac{\omega_2}{\omega_1} = \frac{640 \text{ rev/min}}{423 \text{ rev/min}} = 1.51$$

(c) $v = r\omega$

$$\omega = (640 \text{ rev/min})\left(\frac{1 \text{ min}}{60 \text{ s}}\right)\left(\frac{2\pi \text{ rad}}{1 \text{ rev}}\right) = 67.0 \text{ rad/s}$$

Then $v = r\omega = (0.235 \text{ m})(67.0 \text{ rad/s}) = 15.7 \text{ m/s}.$

$a_{rad} = r\omega^2 = (0.235 \text{ m})(67.0 \text{ rad/s})^2 = 1060 \text{ m/s}^2$

$\dfrac{a_{rad}}{g} = \dfrac{1060 \text{ m/s}^2}{9.80 \text{ m/s}^2} = 108; \quad a = 108g$

EVALUATE: In parts (a) and (b), since a ratio is used the units cancel and there is no need to convert ω to rad/s. In part (c), v and a_{rad} are calculated from ω, and ω must be in rad/s.

9.33.   **IDENTIFY:** I for the object is the sum of the values of I for each part.

SET UP: For the bar, for an axis perpendicular to the bar, use the appropriate expression from Table 9.2. For a point mass, $I = mr^2$, where r is the distance of the mass from the axis.

EXECUTE: **(a)** $I = I_{bar} + I_{balls} = \frac{1}{12}M_{bar}L^2 + 2m_{balls}\left(\frac{L}{2}\right)^2.$

$$I = \frac{1}{12}(4.00 \text{ kg})(2.00 \text{ m})^2 + 2(0.500 \text{ kg})(1.00 \text{ m})^2 = 2.33 \text{ kg} \cdot \text{m}^2$$

(b) $I = \frac{1}{3}m_{bar}L^2 + m_{ball}L^2 = \frac{1}{3}(4.00 \text{ kg})(2.00 \text{ m})^2 + (0.500 \text{ kg})(2.00 \text{ m})^2 = 7.33 \text{ kg} \cdot \text{m}^2$

(c) $I = 0$ because all masses are on the axis.

(d) All the mass is a distance $d = 0.500$ m from the axis and

$I = m_{bar}d^2 + 2m_{ball}d^2 = M_{Total}d^2 = (5.00 \text{ kg})(0.500 \text{ m})^2 = 1.25 \text{ kg} \cdot \text{m}^2.$

EVALUATE: I for an object depends on the location and direction of the axis.

9.37.   **IDENTIFY:** I for the compound disk is the sum of I of the solid disk and of the ring.

SET UP: For the solid disk, $I = \frac{1}{2}m_d r_d^2$. For the ring, $I_r = \frac{1}{2}m_r(r_1^2 + r_2^2)$, where

$r_1 = 50.0$ cm, $r_2 = 70.0$ cm. The mass of the disk and ring is their area times their area density.

EXECUTE: $I = I_d + I_r.$

Disk: $m_d = (3.00 \text{ g/cm}^2)\pi r_d^2 = 23.56 \text{ kg}. \quad I_d = \frac{1}{2}m_d r_d^2 = 2.945 \text{ kg} \cdot \text{m}^2.$

Ring: $m_r = (2.00 \text{ g/cm}^2)\pi(r_2^2 - r_1^2) = 15.08$ kg. $I_r = \frac{1}{2}m_r(r_1^2 + r_2^2) = 5.580 \text{ kg} \cdot \text{m}^2$.

$I = I_d + I_r = 8.52 \text{ kg} \cdot \text{m}^2$.

EVALUATE: Even though $m_r < m_d$, $I_r > I_d$ since the mass of the ring is farther from the axis.

9.39.   **IDENTIFY:** Knowing the kinetic energy, mass and radius of the sphere, we can find its angular velocity. From this we can find the tangential velocity (the target variable) of a point on the rim.

SET UP: $K = \frac{1}{2}I\omega^2$ and $I = \frac{2}{5}MR^2$ for a solid uniform sphere. The tagential velocity is $v = r\omega$.

EXECUTE: $I = \frac{2}{5}MR^2 = \frac{2}{5}(28.0 \text{ kg})(0.380 \text{ m})^2 = 1.617 \text{ kg} \cdot \text{m}^2$. $K = \frac{1}{2}I\omega^2$ so

$$\omega = \sqrt{\frac{2K}{I}} = \sqrt{\frac{2(176 \text{ J})}{1.617 \text{ kg} \cdot \text{m}^2}} = 14.75 \text{ rad/s}.$$

$v = r\omega = (0.380 \text{ m})(14.75 \text{ rad/s}) = 5.61 \text{ m/s}$.

EVALUATE: This is the speed of a point on the surface of the sphere that is farthest from the axis of rotation (the "equator" of the sphere). Points off the "equator" would have smaller tangential velocity but the same angular velocity.

9.43.   **IDENTIFY:** $K = \frac{1}{2}I\omega^2$, with ω in rad/s. Solve for I.

SET UP: 1 rev/min $= (2\pi/60)$ rad/s. $\Delta K = -500$ J

EXECUTE: $\omega_i = 650$ rev/min $= 68.1$ rad/s. $\omega_f = 520$ rev/min $= 54.5$ rad/s. $\Delta K = K_f - K_i = \frac{1}{2}I(\omega_f^2 - \omega_i^2)$

and $I = \dfrac{2(\Delta K)}{\omega_f^2 - \omega_i^2} = \dfrac{2(-500 \text{ J})}{(54.5 \text{ rad/s})^2 - (68.1 \text{ rad/s})^2} = 0.600 \text{ kg} \cdot \text{m}^2$.

EVALUATE: In $K = \frac{1}{2}I\omega^2$, ω must be in rad/s.

9.45.   **IDENTIFY and SET UP:** Combine Eqs. (9.17) and (9.15) to solve for K. Use Table 9.2 to get I.

EXECUTE: $K = \frac{1}{2}I\omega^2$

$a_{rad} = R\omega^2$, so $\omega = \sqrt{a_{rad}/R} = \sqrt{(3500 \text{ m/s}^2)/1.20 \text{ m}} = 54.0$ rad/s

For a disk, $I = \frac{1}{2}MR^2 = \frac{1}{2}(70.0 \text{ kg})(1.20 \text{ m})^2 = 50.4 \text{ kg} \cdot \text{m}^2$

Thus $K = \frac{1}{2}I\omega^2 = \frac{1}{2}(50.4 \text{ kg} \cdot \text{m}^2)(54.0 \text{ rad/s})^2 = 7.35 \times 10^4$ J

EVALUATE: The limit on a_{rad} limits ω which in turn limits K.

9.47.   **IDENTIFY:** Apply conservation of energy to the system of stone plus pulley. $v = r\omega$ relates the motion of the stone to the rotation of the pulley.

SET UP: For a uniform solid disk, $I = \frac{1}{2}MR^2$. Let point 1 be when the stone is at its initial position and point 2 be when it has descended the desired distance. Let $+y$ be upward and take $y = 0$ at the initial position of the stone, so $y_1 = 0$ and $y_2 = -h$, where h is the distance the stone descends.

EXECUTE: **(a)** $K_p = \frac{1}{2}I_p\omega^2$. $I_p = \frac{1}{2}M_pR^2 = \frac{1}{2}(2.50 \text{ kg})(0.200 \text{ m})^2 = 0.0500 \text{ kg} \cdot \text{m}^2$.

$\omega = \sqrt{\dfrac{2K_p}{I_p}} = \sqrt{\dfrac{2(4.50 \text{ J})}{0.0500 \text{ kg} \cdot \text{m}^2}} = 13.4$ rad/s. The stone has speed $v = R\omega = (0.200 \text{ m})(13.4 \text{ rad/s}) = 2.68$ m/s.

The stone has kinetic energy $K_s = \frac{1}{2}mv^2 = \frac{1}{2}(1.50 \text{ kg})(2.68 \text{ m/s})^2 = 5.39$ J. $K_1 + U_1 = K_2 + U_2$ gives

$0 = K_2 + U_2$. $0 = 4.50 \text{ J} + 5.39 \text{ J} + mg(-h)$. $h = \dfrac{9.89 \text{ J}}{(1.50 \text{ kg})(9.80 \text{ m/s}^2)} = 0.673$ m.

(b) $K_{tot} = K_p + K_s = 9.89$ J. $\dfrac{K_p}{K_{tot}} = \dfrac{4.50 \text{ J}}{9.89 \text{ J}} = 45.5\%$.

EVALUATE: The gravitational potential energy of the pulley doesn't change as it rotates. The tension in the wire does positive work on the pulley and negative work of the same magnitude on the stone, so no net work on the system.

9.51. **IDENTIFY:** The general expression for I is Eq. (9.16). $K = \frac{1}{2} I \omega^2$.

SET UP: R will be multiplied by f.

EXECUTE: **(a)** In the expression of Eq. (9.16), each term will have the mass multiplied by f^3 and the distance multiplied by f, and so the moment of inertia is multiplied by $f^3(f)^2 = f^5$.

(b) $(2.5 \text{ J})(48)^5 = 6.37 \times 10^8 \text{ J}$.

EVALUATE: Mass and volume are proportional to each other so both scale by the same factor.

9.53. **IDENTIFY:** Use Eq. (9.19) to relate I for the wood sphere about the desired axis to I for an axis along a diameter.

SET UP: For a thin-walled hollow sphere, axis along a diameter, $I = \frac{2}{3} MR^2$.

For a solid sphere with mass M and radius R, $I_{cm} = \frac{2}{5} MR^2$, for an axis along a diameter.

EXECUTE: Find d such that $I_P = I_{cm} + Md^2$ with $I_P = \frac{2}{3} MR^2$:

$$\frac{2}{3} MR^2 = \frac{2}{5} MR^2 + Md^2$$

The factors of M divide out and the equation becomes $(\frac{2}{3} - \frac{2}{5})R^2 = d^2$

$$d = \sqrt{(10-6)/15} R = 2R/\sqrt{15} = 0.516R.$$

The axis is parallel to a diameter and is $0.516R$ from the center.

EVALUATE: $I_{cm}(\text{lead}) > I_{cm}(\text{wood})$ even though M and R are the same since for a hollow sphere all the mass is a distance R from the axis. Eq. (9.19) says $I_P > I_{cm}$, so there must be a d where $I_P(\text{wood}) = I_{cm}(\text{lead})$.

9.55. **IDENTIFY and SET UP:** Use Eq. (9.19). The cm of the sheet is at its geometrical center. The object is sketched in Figure 9.55.

EXECUTE: $I_P = I_{cm} + Md^2$.

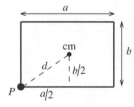

From part (c) of Table 9.2,
$I_{cm} = \frac{1}{12} M(a^2 + b^2)$.

The distance d of P from the cm is
$d = \sqrt{(a/2)^2 + (b/2)^2}$.

Figure 9.55

Thus $I_P = I_{cm} + Md^2 = \frac{1}{12} M(a^2 + b^2) + M(\frac{1}{4} a^2 + \frac{1}{4} b^2) = (\frac{1}{12} + \frac{1}{4})M(a^2 + b^2) = \frac{1}{3} M(a^2 + b^2)$

EVALUATE: $I_P = 4I_{cm}$. For an axis through P mass is farther from the axis.

9.57. **IDENTIFY:** Use the equations in Table 9.2. I for the rod is the sum of I for each segment. The parallel-axis theorem says $I_p = I_{cm} + Md^2$.

SET UP: The bent rod and axes a and b are shown in Figure 9.57. Each segment has length $L/2$ and mass $M/2$.

EXECUTE: **(a)** For each segment the moment of inertia is for a rod with mass $M/2$, length $L/2$ and the axis through one end. For one segment, $I_s = \frac{1}{3}\left(\frac{M}{2}\right)\left(\frac{L}{2}\right)^2 = \frac{1}{24} ML^2$. For the rod, $I_a = 2I_s = \frac{1}{12} ML^2$.

(b) The center of mass of each segment is at the center of the segment, a distance of $L/4$ from each end.

For each segment, $I_{cm} = \frac{1}{12}\left(\frac{M}{2}\right)\left(\frac{L}{2}\right)^2 = \frac{1}{96}ML^2$. Axis b is a distance $L/4$ from the cm of each segment,

so for each segment the parallel axis theorem gives I for axis b to be $I_s = \frac{1}{96}ML^2 + \frac{M}{2}\left(\frac{L}{4}\right)^2 = \frac{1}{24}ML^2$ and

$I_b = 2I_s = \frac{1}{12}ML^2$.

EVALUATE: I for these two axes are the same.

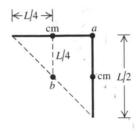

Figure 9.57

9.63.   **IDENTIFY:** The target variable is the horizontal distance the piece travels before hitting the floor. Using the angular acceleration of the blade, we can find its angular velocity when the piece breaks off. This will give us the linear horizontal speed of the piece. It is then in free fall, so we can use the linear kinematics equations.

SET UP: $\omega_z^2 = \omega_{0z}^2 + 2\alpha_z(\theta - \theta_0)$ for the blade, and $v = r\omega$ is the horizontal velocity of the piece.

$y - y_0 = v_{0y}t + \frac{1}{2}a_yt^2$ for the falling piece.

EXECUTE: Find the initial horizontal velocity of the piece just after it breaks off.

$\theta - \theta_0 = (155 \text{ rev})(2\pi \text{ rad}/1 \text{ rev}) = 973.9 \text{ rad}$.

$\alpha_z = (3.00 \text{ rev/s}^2)(2\pi \text{ rad}/1 \text{ rev}) = 18.85 \text{ rad/s}^2$. $\omega_z^2 = \omega_{0z}^2 + 2\alpha_z(\theta - \theta_0)$.

$\omega_z = \sqrt{2\alpha_z(\theta - \theta_0)} = \sqrt{2(18.85 \text{ rad/s}^2)(973.9 \text{ rad})} = 191.6 \text{ rad/s}$. The horizontal velocity of the piece is

$v = r\omega = (0.120 \text{ m})(191.6 \text{ rad/s}) = 23.0 \text{ m/s}$. Now consider the projectile motion of the piece. Take $+y$

downward and use the vertical motion to find t. Solving $y - y_0 = v_{0y}t + \frac{1}{2}a_yt^2$ for t gives

$t = \sqrt{\frac{2(y - y_0)}{a_y}} = \sqrt{\frac{2(0.820 \text{ m})}{9.8 \text{ m/s}^2}} = 0.4091 \text{ s}$. Then $x - x_0 = v_{0x}t + \frac{1}{2}a_xt^2 = (23.0 \text{ m/s})(0.4091 \text{ s}) = 9.41 \text{ m}$.

EVALUATE: Once the piece is free of the blade, the only force acting on it is gravity so its acceleration is g downward.

9.65.   **IDENTIFY:** The angular acceleration α of the disk is related to the linear acceleration a of the ball by

$a = R\alpha$. Since the acceleration is not constant, use $\omega_z - \omega_{0z} = \int_0^t \alpha_z \, dt$ and $\theta - \theta_0 = \int_0^t \omega_z \, dt$ to relate θ,

ω_z, α_z and t for the disk. $\omega_{0z} = 0$.

SET UP: $\int t^n dt = \frac{1}{n+1}t^{n+1}$. In $a = R\alpha$, α is in rad/s^2.

EXECUTE: **(a)** $A = \frac{a}{t} = \frac{1.80 \text{ m/s}^2}{3.00 \text{ s}} = 0.600 \text{ m/s}^3$

(b) $\alpha = \frac{a}{R} = \frac{(0.600 \text{ m/s}^3)t}{0.250 \text{ m}} = (2.40 \text{ rad/s}^3)t$

(c) $\omega_z = \int_0^t (2.40 \text{ rad/s}^3) t\, dt = (1.20 \text{ rad/s}^3) t^2.$ $\omega_z = 15.0 \text{ rad/s}$ for $t = \sqrt{\dfrac{15.0 \text{ rad/s}}{1.20 \text{ rad/s}^3}} = 3.54 \text{ s}.$

(d) $\theta - \theta_0 = \int_0^t \omega_z\, dt = \int_0^t (1.20 \text{ rad/s}^3) t^2\, dt = (0.400 \text{ rad/s}^3) t^3.$ For $t = 3.54 \text{ s},$ $\theta - \theta_0 = 17.7 \text{ rad}.$

EVALUATE: If the disk had turned at a constant angular velocity of 15.0 rad/s for 3.54 s it would have turned through an angle of 53.1 rad in 3.54 s. It actually turns through less than half this because the angular velocity is increasing in time and is less than 15.0 rad/s at all but the end of the interval.

9.71. **IDENTIFY:** Apply $v = r\omega.$

SET UP: Points on the chain all move at the same speed, so $r_r \omega_r = r_f \omega_f.$

EXECUTE: The angular velocity of the rear wheel is $\omega_r = \dfrac{v_r}{r} = \dfrac{5.00 \text{ m/s}}{0.330 \text{ m}} = 15.15 \text{ rad/s}.$

The angular velocity of the front wheel is $\omega_f = 0.600 \text{ rev/s} = 3.77 \text{ rad/s}.$ $r_r = r_f(\omega_f / \omega_r) = 2.99 \text{ cm}.$

EVALUATE: The rear sprocket and wheel have the same angular velocity and the front sprocket and wheel have the same angular velocity. $r\omega$ is the same for both, so the rear sprocket has a smaller radius since it has a larger angular velocity. The speed of a point on the chain is $v = r_r \omega_r = (2.99 \times 10^{-2} \text{ m})(15.15 \text{ rad/s}) = 0.453 \text{ m/s}.$ The linear speed of the bicycle is 5.00 m/s.

9.75. **IDENTIFY:** $K = \frac{1}{2} I \omega^2.$ $a_{\text{rad}} = r\omega^2.$ $m = \rho V.$

SET UP: For a disk with the axis at the center, $I = \frac{1}{2} m R^2.$ $V = t\pi R^2,$ where $t = 0.100 \text{ m}$ is the thickness of the flywheel. $\rho = 7800 \text{ kg/m}^3$ is the density of the iron.

EXECUTE: **(a)** $\omega = 90.0 \text{ rpm} = 9.425 \text{ rad/s}.$ $I = \dfrac{2K}{\omega^2} = \dfrac{2(10.0 \times 10^6 \text{ J})}{(9.425 \text{ rad/s})^2} = 2.252 \times 10^5 \text{ kg} \cdot \text{m}^2.$

$m = \rho V = \rho \pi R^2 t.$ $I = \dfrac{1}{2} m R^2 = \dfrac{1}{2} \rho \pi t R^4.$ This gives $R = (2I/\rho \pi t)^{1/4} = 3.68 \text{ m}$ and the diameter is 7.36 m.

(b) $a_{\text{rad}} = R\omega^2 = 327 \text{ m/s}^2$

EVALUATE: In $K = \frac{1}{2} I \omega^2,$ ω must be in rad/s. a_{rad} is about 33g; the flywheel material must have large cohesive strength to prevent the flywheel from flying apart.

9.79. **IDENTIFY:** Use Eq. (9.20) to calculate $I.$ Then use $K = \frac{1}{2} I \omega^2$ to calculate $K.$

(a) SET UP: The object is sketched in Figure 9.79.

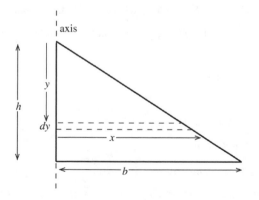

Consider a small strip of width dy and a distance y below the top of the triangle. The length of the strip is $x = (y/h)b.$

Figure 9.79

EXECUTE: The strip has area $x\, dy$ and the area of the sign is $\frac{1}{2} bh,$ so the mass of the strip is

$$dm = M \left(\frac{x\, dy}{\frac{1}{2} bh} \right) = M \left(\frac{yb}{h} \right) \left(\frac{2\, dy}{bh} \right) = \left(\frac{2M}{h^2} \right) y\, dy$$

$$dI = \tfrac{1}{3}(dm)x^2 = \left(\frac{2Mb^2}{3h^4}\right)y^3\,dy$$

$$I = \int_0^h dI = \frac{2Mb^2}{3h^4}\int_0^h y^3\,dy = \frac{2Mb^2}{3h^4}\left(\frac{1}{4}y^4\Big|_0^h\right) = \frac{1}{6}Mb^2$$

(b) $I = \tfrac{1}{6}Mb^2 = 2.304 \ \text{kg}\cdot\text{m}^2$

$\omega = 2.00 \ \text{rev/s} = 4.00\pi \ \text{rad/s}$

$K = \tfrac{1}{2}I\omega^2 = 182 \ \text{J}$

EVALUATE: From Table (9.2), if the sign were rectangular, with length b, then $I = \tfrac{1}{3}Mb^2$. Our result is one-half this, since mass is closer to the axis for the triangular than for the rectangular shape.

9.81. **IDENTIFY:** Use conservation of energy. The stick rotates about a fixed axis so $K = \tfrac{1}{2}I\omega^2$. Once we have ω use $v = r\omega$ to calculate v for the end of the stick.

SET UP: The object is sketched in Figure 9.81.

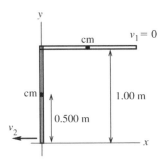

Take the origin of coordinates at the lowest point reached by the stick and take the positive y-direction to be upward.

Figure 9.81

EXECUTE: **(a)** Use Eq.(9.18): $U = Mgy_{cm}$. $\Delta U = U_2 - U_1 = Mg(y_{cm2} - y_{cm1})$. The center of mass of the meter stick is at its geometrical center, so $y_{cm1} = 1.00$ m and $y_{cm2} = 0.50$ m. Then

$\Delta U = (0.180 \ \text{kg})(9.80 \ \text{m/s}^2)(0.50 \ \text{m} - 1.00 \ \text{m}) = -0.882 \ \text{J}.$

(b) Use conservation of energy: $K_1 + U_1 + W_{other} = K_2 + U_2$. Gravity is the only force that does work on the meter stick, so $W_{other} = 0$. $K_1 = 0$. Thus $K_2 = U_1 - U_2 = -\Delta U$, where ΔU was calculated in part (a). $K_2 = \tfrac{1}{2}I\omega_2^2$ so $\tfrac{1}{2}I\omega_2^2 = -\Delta U$ and $\omega_2 = \sqrt{2(-\Delta U)/I}$. For stick pivoted about one end, $I = \tfrac{1}{3}ML^2$ where

$L = 1.00$ m, so $\omega_2 = \sqrt{\dfrac{6(-\Delta U)}{ML^2}} = \sqrt{\dfrac{6(0.882 \ \text{J})}{(0.180 \ \text{kg})(1.00 \ \text{m})^2}} = 5.42 \ \text{rad/s}.$

(c) $v = r\omega = (1.00 \ \text{m})(5.42 \ \text{rad/s}) = 5.42 \ \text{m/s}.$

(d) For a particle in free fall, with $+y$ upward, $v_{0y} = 0$; $y - y_0 = -1.00$ m; $a_y = -9.80 \ \text{m/s}^2$; and $v_y = ?$ Solving the equation $v_y^2 = v_{0y}^2 + 2a_y(y - y_0)$ for v_y gives

$v_y = -\sqrt{2a_y(y - y_0)} = -\sqrt{2(-9.80 \ \text{m/s}^2)(-1.00 \ \text{m})} = -4.43 \ \text{m/s}.$

EVALUATE: The magnitude of the answer in part (c) is larger. $U_{1,grav}$ is the same for the stick as for a particle falling from a height of 1.00 m. For the stick $K = \tfrac{1}{2}I\omega_2^2 = \tfrac{1}{2}(\tfrac{1}{3}ML^2)(v/L)^2 = \tfrac{1}{6}Mv^2$. For the stick and for the particle, K_2 is the same but the same K gives a larger v for the end of the stick than for the particle. The reason is that all the other points along the stick are moving slower than the end opposite the axis.

9.83. **IDENTIFY:** Apply conservation of energy to the system consisting of blocks A and B and the pulley.

SET UP: The system at points 1 and 2 of its motion is sketched in Figure 9.83.

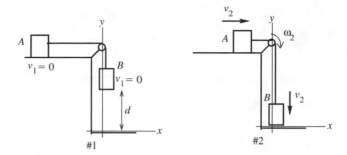

Figure 9.83

Use the work-energy relation $K_1 + U_1 + W_{\text{other}} = K_2 + U_2$. Use coordinates where $+y$ is upward and where the origin is at the position of block B after it has descended. The tension in the rope does positive work on block A and negative work of the same magnitude on block B, so the net work done by the tension in the rope is zero. Both blocks have the same speed.

EXECUTE: Gravity does work on block B and kinetic friction does work on block A. Therefore $W_{\text{other}} = W_f = -\mu_k m_A g d$.

$K_1 = 0$ (system is released from rest)

$U_1 = m_B g y_{B1} = m_B g d;\ \ U_2 = m_B g y_{B2} = 0$

$K_2 = \frac{1}{2} m_A v_2^2 + \frac{1}{2} m_B v_2^2 + \frac{1}{2} I \omega_2^2.$

But $v(\text{blocks}) = R\omega(\text{pulley})$, so $\omega_2 = v_2/R$ and

$K_2 = \frac{1}{2}(m_A + m_B)v_2^2 + \frac{1}{2} I (v_2/R)^2 = \frac{1}{2}(m_A + m_B + I/R^2)v_2^2$

Putting all this into the work-energy relation gives

$m_B g d - \mu_k m_A g d = \frac{1}{2}(m_A + m_B + I/R^2)v_2^2$

$(m_A + m_B + I/R^2)v_2^2 = 2gd(m_B - \mu_k m_A)$

$v_2 = \sqrt{\dfrac{2gd(m_B - \mu_k m_A)}{m_A + m_B + I/R^2}}$

EVALUATE: If $m_B \gg m_A$ and I/R^2, then $v_2 = \sqrt{2gd}$; block B falls freely. If I is very large, v_2 is very small. Must have $m_B > \mu_k m_A$ for motion, so the weight of B will be larger than the friction force on A.

I/R^2 has units of mass and is in a sense the "effective mass" of the pulley.

9.85. **IDENTIFY and SET UP:** Apply conservation of energy to the motion of the hoop. Use Eq. (9.18) to calculate U_{grav}. Use $K = \frac{1}{2} I \omega^2$ for the kinetic energy of the hoop. Solve for ω. The center of mass of the hoop is at its geometrical center.

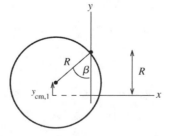

Take the origin to be at the original location of the center of the hoop, before it is rotated to one side, as shown in Figure 9.85.

Figure 9.85

$y_{\text{cm1}} = R - R\cos\beta = R(1 - \cos\beta)$

$y_{\text{cm2}} = 0$ (at equilibrium position hoop is at original position)

EXECUTE: $K_1 + U_1 + W_{other} = K_2 + U_2$

$W_{other} = 0$ (only gravity does work)

$K_1 = 0$ (released from rest), $K_2 = \frac{1}{2}I\omega_2^2$

For a hoop, $I_{cm} = MR^2$, so $I = Md^2 + MR^2$ with $d = R$ and $I = 2MR^2$, for an axis at the edge. Thus $K_2 = \frac{1}{2}(2MR^2)\omega_2^2 = MR^2\omega_2^2$.

$U_1 = Mgy_{cm1} = MgR(1-\cos\beta)$, $U_2 = mgy_{cm2} = 0$

Thus $K_1 + U_1 + W_{other} = K_2 + U_2$ gives

$MgR(1-\cos\beta) = MR^2\omega_2^2$ and $\omega_2 = \sqrt{g(1-\cos\beta)/R}$

EVALUATE: If $\beta = 0$, then $\omega_2 = 0$. As β increases, ω_2 increases.

9.87.   **IDENTIFY:** $I = I_1 + I_2$. Apply conservation of energy to the system. The calculation is similar to Example 9.8.

SET UP: $\omega = \dfrac{v}{R_1}$ for part (b) and $\omega = \dfrac{v}{R_2}$ for part (c).

EXECUTE: **(a)** $I = \frac{1}{2}M_1R_1^2 + \frac{1}{2}M_2R_2^2 = \frac{1}{2}((0.80\text{ kg})(2.50\times10^{-2}\text{ m})^2 + (1.60\text{ kg})(5.00\times10^{-2}\text{ m})^2)$

$I = 2.25\times10^{-3}\text{ kg}\cdot\text{m}^2$.

(b) The method of Example 9.8 yields $v = \sqrt{\dfrac{2gh}{1+(I/mR_1^2)}}$.

$v = \sqrt{\dfrac{2(9.80\text{ m/s}^2)(2.00\text{ m})}{(1+((2.25\times10^{-3}\text{ kg}\cdot\text{m}^2)/(1.50\text{ kg})(0.025\text{ m})^2))}} = 3.40\text{ m/s}.$

(c) The same calculation, with R_2 instead of R_1 gives $v = 4.95$ m/s.

EVALUATE: The final speed of the block is greater when the string is wrapped around the larger disk. $v = R\omega$, so when $R = R_2$ the factor that relates v to ω is larger. For $R = R_2$ a larger fraction of the total kinetic energy resides with the block. The total kinetic energy is the same in both cases (equal to mgh), so when $R = R_2$ the kinetic energy and speed of the block are greater.

9.89.   **IDENTIFY:**   Apply conservation of energy to relate the height of the mass to the kinetic energy of the cylinder.

SET UP: First use $K(\text{cylinder}) = 480$ J to find ω for the cylinder and v for the mass.

EXECUTE: $I = \frac{1}{2}MR^2 = \frac{1}{2}(10.0\text{ kg})(0.150\text{ m})^2 = 0.1125\text{ kg}\cdot\text{m}^2$. $K = \frac{1}{2}I\omega^2$ so $\omega = \sqrt{2K/I} = 92.38$ rad/s.

$v = R\omega = 13.86$ m/s.

SET UP: Use conservation of energy $K_1 + U_1 = K_2 + U_2$ to solve for the distance the mass descends. Take $y = 0$ at lowest point of the mass, so $y_2 = 0$ and $y_1 = h$, the distance the mass descends.

EXECUTE: $K_1 = U_2 = 0$ so $U_1 = K_2$. $mgh = \frac{1}{2}mv^2 + \frac{1}{2}I\omega^2$, where $m = 12.0$ kg. For the cylinder, $I = \frac{1}{2}MR^2$ and $\omega = v/R$, so $\frac{1}{2}I\omega^2 = \frac{1}{4}Mv^2$. Solving $mgh = \frac{1}{2}mv^2 + \frac{1}{4}Mv^2$ for h gives

$h = \dfrac{v^2}{2g}\left(1 + \dfrac{M}{2m}\right) = 13.9$ m.

EVALUATE: For the cylinder $K_{cyl} = \frac{1}{2}I\omega^2 = \frac{1}{2}(\frac{1}{2}MR^2)(v/R)^2 = \frac{1}{4}Mv^2$. $K_{mass} = \frac{1}{2}mv^2$, so

$K_{mass} = (2m/M)K_{cyl} = [2(12.0\text{ kg})/10.0\text{ kg}](480\text{ J}) = 1150$ J. The mass has 1150 J of kinetic energy when the cylinder has 480 J of kinetic energy and at this point the system has total energy 1630 J since $U_2 = 0$. Initially the total energy of the system is $U_1 = mgy_1 = mgh = 1630$ J, so the total energy is shown to be conserved.

9.91.   **IDENTIFY:** $I = I_{disk} - I_{hole}$, where I_{hole} is I for the piece punched from the disk. Apply the parallel-axis theorem to calculate the required moments of inertia.

SET UP: For a uniform disk, $I = \frac{1}{2}MR^2$.

EXECUTE: **(a)** The initial moment of inertia is $I_0 = \frac{1}{2}MR^2$. The piece punched has a mass of $\dfrac{M}{16}$ and a moment of inertia with respect to the axis of the original disk of

$$\frac{M}{16}\left[\frac{1}{2}\left(\frac{R}{4}\right)^2 + \left(\frac{R}{2}\right)^2\right] = \frac{9}{512}MR^2.$$

The moment of inertia of the remaining piece is then $I = \frac{1}{2}MR^2 - \frac{9}{512}MR^2 = \frac{247}{512}MR^2$.

(b) $I = \frac{1}{2}MR^2 + M(R/2)^2 - \frac{1}{2}(M/16)(R/4)^2 = \frac{383}{512}MR^2$.

EVALUATE: For a solid disk and an axis at a distance $R/2$ from the disk's center, the parallel-axis theorem gives $I = \frac{1}{2}MR^2 = \frac{3}{4}MR^2 = \frac{384}{512}MR^2$. For both choices of axes the presence of the hole reduces I, but the effect of the hole is greater in part (a), when it is farther from the axis.

9.93.   **IDENTIFY:** The total kinetic energy of a walker is the sum of his translational kinetic energy plus the rotational kinetic of his arms and legs. We can model these parts of the body as uniform bars.

SET UP: For a uniform bar pivoted about one end, $I = \frac{1}{3}mL^2$. $v = 5.0$ km/h $= 1.4$ m/s.

$K_{tran} = \frac{1}{2}mv^2$ and $K_{rot} = \frac{1}{2}I\omega^2$.

EXECUTE: **(a)** $60° = \left(\frac{1}{3}\right)$ rad. The average angular speed of each arm and leg is $\dfrac{\frac{1}{3}\text{ rad}}{1\text{ s}} = 1.05$ rad/s.

(b) Adding the moments of inertia gives

$I = \frac{1}{3}m_{arm}L_{arm}^2 + \frac{1}{3}m_{leg}L_{leg}^2 = \frac{1}{3}[(0.13)(75\text{ kg})(0.70\text{ m})^2 + (0.37)(75\text{ kg})(0.90\text{ m})^2].$ $I = 9.08$ kg·m².

$K_{rot} = \frac{1}{2}I\omega^2 = \frac{1}{2}(9.08\text{ kg·m}^2)(1.05\text{ rad/s})^2 = 5.0$ J.

(c) $K_{tran} = \frac{1}{2}mv^2 = \frac{1}{2}(75\text{ kg})(1.4\text{ m/s})^2 = 73.5$ J and $K_{tot} = K_{tran} + K_{rot} = 78.5$ J.

(d) $\dfrac{K_{rot}}{K_{tran}} = \dfrac{5.0\text{ J}}{78.5\text{ J}} = 6.4\%$.

EVALUATE: If you swing your arms more vigorously more of your energy input goes into the kinetic energy of walking and it is more effective exercise. Carrying weights in our hands would also be effective.

9.97.   **IDENTIFY:** Use Eq. (9.20) to calculate I.

(a) SET UP: Let L be the length of the cylinder. Divide the cylinder into thin cylindrical shells of inner radius r and outer radius $r + dr$. An end view is shown in Figure 9.97

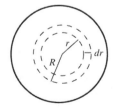

$\rho = \alpha r$

The mass of the thin cylindrical shell is

$dm = \rho \, dV = \rho(2\pi r \, dr)L = 2\pi\alpha Lr^2 \, dr$

Figure 9.97

EXECUTE: $I = \int r^2 \, dm = 2\pi\alpha L \int_0^R r^4 \, dr = 2\pi\alpha L\left(\frac{1}{5}R^5\right) = \frac{2}{5}\pi\alpha LR^5$

Relate M to α: $M = \int dm = 2\pi\alpha L \int_0^R r^2 \, dr = 2\alpha\pi L\left(\frac{1}{3}R^3\right) = \frac{2}{3}\pi\alpha LR^3$, so $\pi\alpha LR^3 = 3M/2$.

Using this in the above result for I gives $I = \frac{2}{5}(3M/2)R^2 = \frac{3}{5}MR^2$.

(b) Evaluate: For a cylinder of uniform density $I = \frac{1}{2}MR^2$. The answer in (a) is larger than this. Since the density increases with distance from the axis the cylinder in (a) has more mass farther from the axis than for a cylinder of uniform density.

9.99.   **Identify:**   The density depends on the distance from the center of the sphere, so it is a function of r. We need to integrate to find the mass and the moment of inertia.

Set Up: $M = \int dm = \int \rho \, dV$ and $I = \int dI$.

Execute: **(a)** Divide the sphere into thin spherical shells of radius r and thickness dr. The volume of each shell is $dV = 4\pi r^2 \, dr$. $\rho(r) = a - br$, with $a = 3.00 \times 10^3$ kg/m^3 and $b = 9.00 \times 10^3$ kg/m^4. Integrating gives $M = \int dm = \int \rho \, dV = \int_0^R (a - br)4\pi r^2 \, dr = \frac{4}{3}\pi R^3 \left(a - \frac{3}{4}bR \right)$.

$$M = \frac{4}{3}\pi (0.200)^3 \left(3.00 \times 10^3 \text{ kg/m}^3 - \frac{3}{4}(9.00 \times 10^3 \text{ kg/m}^4)(0.200 \text{ m}) \right) = 55.3 \text{ kg}.$$

(b) The moment of inertia of each thin spherical shell is

$$dI = \frac{2}{3}r^2 \, dm = \frac{2}{3}r^2 \rho \, dV = \frac{2}{3}r^2(a - br)4\pi r^2 \, dr = \frac{8\pi}{3}r^4(a - br)dr.$$

$$I = \int_0^R dI = \frac{8\pi}{3}\int_0^R r^4(a - br)dr = \frac{8\pi}{15}R^5 \left(a - \frac{5b}{6}R \right).$$

$$I = \frac{8\pi}{15}(0.200 \text{ m})^5 \left(3.00 \times 10^3 \text{ kg/m}^3 - \frac{5}{6}(9.00 \times 10^3 \text{ kg/m}^4)(0.200 \text{ m}) \right) = 0.804 \text{ kg} \cdot \text{m}^2.$$

Evaluate: We cannot use the formulas $M = \rho V$ and $I = \frac{1}{2}MR^2$ because this sphere is not uniform throughout. Its density increases toward the surface. For a uniform sphere with density 3.00×10^3 kg/m^3, the mass is $\frac{4}{3}\pi R^3 \rho = 100.5$ kg.   The mass of the sphere in this problem is less than this. For a uniform sphere with mass 55.3 kg and $R = 0.200$ m, $I = \frac{2}{5}MR^2 = 0.885$ kg$\cdot$m^2. The moment of inertia for the sphere in this problem is less than this, since the density decreases with distance from the center of the sphere.

DYNAMICS OF ROTATIONAL MOTION

10.3. **IDENTIFY** and **SET UP:** Use Eq. (10.2) to calculate the magnitude of each torque and use the right-hand rule (Figure 10.4 in the textbook) to determine the direction. Consider Figure 10.3.

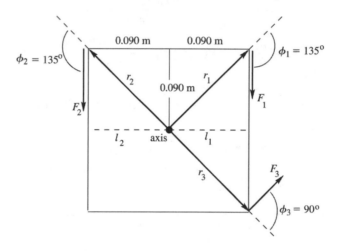

Figure 10.3

Let counterclockwise be the positive sense of rotation.

EXECUTE: $r_1 = r_2 = r_3 = \sqrt{(0.090 \text{ m})^2 + (0.090 \text{ m})^2} = 0.1273 \text{ m}$

$\tau_1 = -F_1 l_1$

$l_1 = r_1 \sin \phi_1 = (0.1273 \text{ m}) \sin 135° = 0.0900 \text{ m}$

$\tau_1 = -(18.0 \text{ N})(0.0900 \text{ m}) = -1.62 \text{ N} \cdot \text{m}$

$\vec{\tau}_1$ is directed into paper

$\tau_2 = +F_2 l_2$

$l_2 = r_2 \sin \phi_2 = (0.1273) \sin 135° = 0.0900 \text{ m}$

$\tau_2 = +(26.0 \text{ N})(0.0900 \text{ m}) = +2.34 \text{ N} \cdot \text{m}$

$\vec{\tau}_2$ is directed out of paper

$\tau_3 = +F_3 l_3$

$l_3 = r_3 \sin \phi_3 = (0.1273 \text{ m}) \sin 90° = 0.1273 \text{ m}$

$\tau_3 = +(14.0 \text{ N})(0.1273 \text{ m}) = +1.78 \text{ N} \cdot \text{m}$

$\vec{\tau}_3$ is directed out of paper

$\sum \tau = \tau_1 + \tau_2 + \tau_3 = -1.62 \text{ N} \cdot \text{m} + 2.34 \text{ N} \cdot \text{m} + 1.78 \text{ N} \cdot \text{m} = 2.50 \text{ N} \cdot \text{m}$

EVALUATE: The net torque is positive, which means it tends to produce a counterclockwise rotation; the vector torque is directed out of the plane of the paper. In summing the torques it is important to include + or − signs to show direction.

10.7. **IDENTIFY:** The total torque is the sum of the torques due to all the forces.
SET UP: The torque due to a force is the product of the force times its moment arm: $\tau = Fl$. Let counterclockwise torques be positive.
EXECUTE: (a) $\tau_A = +(50\ \text{N})(0.20\ \text{m})\sin 60° = +8.7\ \text{N}\cdot\text{m}$, counterclockwise. $\tau_B = 0$.
$\tau_C = -(50\ \text{N})(0.20\ \text{m})\sin 30° = -5.0\ \text{N}\cdot\text{m}$, clockwise. $\tau_D = -(50\ \text{N})(0.20\ \text{m})\sin 90° = -10.0\ \text{N}\cdot\text{m}$, clockwise.
(b) $\sum\tau = \tau_A + \tau_B + \tau_C + \tau_D = -6.3\ \text{N}\cdot\text{m}$, clockwise.
EVALUATE: In the above solution, we used the force component perpendicular to the 20-cm line. We could also have constructed the component of the 20-cm line perpendicular to each force, but that would have been a bit more intricate.

10.9. **IDENTIFY:** Apply $\sum\tau_z = I\alpha_z$.

SET UP: $\omega_{0z} = 0$. $\omega_z = (400\ \text{rev/min})\left(\dfrac{2\pi\ \text{rad/rev}}{60\ \text{s/min}}\right) = 41.9\ \text{rad/s}$

EXECUTE: $\tau_z = I\alpha_z = I\dfrac{\omega_z - \omega_{0z}}{t} = (2.50\ \text{kg}\cdot\text{m}^2)\dfrac{41.9\ \text{rad/s}}{8.00\ \text{s}} = 13.1\ \text{N}\cdot\text{m}$.

EVALUATE: In $\tau_z = I\alpha_z$, α_z must be in rad/s^2.

10.11. **IDENTIFY:** Use $\sum\tau_z = I\alpha_z$ to calculate α. Use a constant angular acceleration kinematic equation to relate α_z, ω_z and t.
SET UP: For a solid uniform sphere and an axis through its center, $I = \frac{2}{5}MR^2$. Let the direction the sphere is spinning be the positive sense of rotation. The moment arm for the friction force is $l = 0.0150\ \text{m}$ and the torque due to this force is negative.
EXECUTE: (a) $\alpha_z = \dfrac{\tau_z}{I} = \dfrac{-(0.0200\ \text{N})(0.0150\ \text{m})}{\frac{2}{5}(0.225\ \text{kg})(0.0150\ \text{m})^2} = -14.8\ \text{rad/s}^2$

(b) $\omega_z - \omega_{0z} = -22.5\ \text{rad/s}$. $\omega_z = \omega_{0z} + \alpha_z t$ gives $t = \dfrac{\omega_z - \omega_{0z}}{\alpha_z} = \dfrac{-22.5\ \text{rad/s}}{-14.8\ \text{rad/s}^2} = 1.52\ \text{s}$.

EVALUATE: The fact that α_z is negative means its direction is opposite to the direction of spin. The negative α_z causes ω_z to decrease.

10.13. **IDENTIFY:** Apply $\sum\vec{F} = m\vec{a}$ to each book and apply $\sum\tau_z = I\alpha_z$ to the pulley. Use a constant acceleration equation to find the common acceleration of the books.
SET UP: $m_1 = 2.00\ \text{kg}$, $m_2 = 3.00\ \text{kg}$. Let T_1 be the tension in the part of the cord attached to m_1 and T_2 be the tension in the part of the cord attached to m_2. Let the +x-direction be in the direction of the acceleration of each book. $a = R\alpha$.
EXECUTE: (a) $x - x_0 = v_{0x}t + \frac{1}{2}a_x t^2$ gives $a_x = \dfrac{2(x-x_0)}{t^2} = \dfrac{2(1.20\ \text{m})}{(0.800\ \text{s})^2} = 3.75\ \text{m/s}^2$. $a_1 = 3.75\ \text{m/s}^2$ so

$T_1 = m_1 a_1 = 7.50\ \text{N}$ and $T_2 = m_2(g - a_1) = 18.2\ \text{N}$.
(b) The torque on the pulley is $(T_2 - T_1)R = 0.803\ \text{N}\cdot\text{m}$, and the angular acceleration is
$\alpha = a_1/R = 50\ \text{rad/s}^2$, so $I = \tau/\alpha = 0.016\ \text{kg}\cdot\text{m}^2$.
EVALUATE: The tensions in the two parts of the cord must be different, so there will be a net torque on the pulley.

10.15. **IDENTIFY:** The constant force produces a torque which gives a constant angular acceleration to the wheel.
SET UP: $\omega_z = \omega_{0z} + \alpha_z t$ because the angular acceleration is constant, and $\sum\tau_z = I\alpha_z$ applies to the wheel.

EXECUTE: $\omega_{0z} = 0$ and $\omega_z = 12.0 \text{ rev/s} = 75.40 \text{ rad/s}.$ $\omega_z = \omega_{0z} + \alpha_z t,$ so

$$\alpha_z = \frac{\omega_z - \omega_{0z}}{t} = \frac{75.40 \text{ rad/s}}{2.00 \text{ s}} = 37.70 \text{ rad/s}^2. \quad \sum \tau_z = I\alpha_z \text{ gives}$$

$$I = \frac{Fr}{\alpha_z} = \frac{(80.0 \text{ N})(0.120 \text{ m})}{37.70 \text{ rad/s}^2} = 0.255 \text{ kg} \cdot \text{m}^2.$$

EVALUATE: The units of the answer are the proper ones for moment of inertia.

10.17.   **IDENTIFY:** Apply $\sum \vec{F} = m\vec{a}$ to each box and $\sum \tau_z = I\alpha_z$ to the pulley. The magnitude a of the acceleration of each box is related to the magnitude of the angular acceleration α of the pulley by $a = R\alpha.$

SET UP: The free-body diagrams for each object are shown in Figure 10.17a–c. For the pulley, $R = 0.250 \text{ m}$ and $I = \frac{1}{2}MR^2.$ T_1 and T_2 are the tensions in the wire on either side of the pulley.

$m_1 = 12.0 \text{ kg},$ $m_2 = 5.00 \text{ kg}$ and $M = 2.00 \text{ kg}.$ $\vec{F}$ is the force that the axle exerts on the pulley. For the pulley, let clockwise rotation be positive.

EXECUTE: **(a)** $\sum F_x = ma_x$ for the 12.0 kg box gives $T_1 = m_1 a.$ $\sum F_y = ma_y$ for the 5.00 kg weight gives

$m_2 g - T_2 = m_2 a.$ $\sum \tau_z = I\alpha_z$ for the pulley gives $(T_2 - T_1)R = \left(\frac{1}{2}MR^2\right)\alpha.$ $a = R\alpha$ and $T_2 - T_1 = \frac{1}{2}Ma.$

Adding these three equations gives $m_2 g = (m_1 + m_2 + \frac{1}{2}M)a$ and

$$a = \left(\frac{m_2}{m_1 + m_2 + \frac{1}{2}M}\right)g = \left(\frac{5.00 \text{ kg}}{12.0 \text{ kg} + 5.00 \text{ kg} + 1.00 \text{ kg}}\right)(9.80 \text{ m/s}^2) = 2.72 \text{ m/s}^2. \text{ Then}$$

$T_1 = m_1 a = (12.0 \text{ kg})(2.72 \text{ m/s}^2) = 32.6 \text{ N}.$ $m_2 g - T_2 = m_2 a$ gives

$T_2 = m_2(g - a) = (5.00 \text{ kg})(9.80 \text{ m/s}^2 - 2.72 \text{ m/s}^2) = 35.4 \text{ N}.$ The tension to the left of the pulley is 32.6 N and below the pulley it is 35.4 N.

(b) $a = 2.72 \text{ m/s}^2$

(c) For the pulley, $\sum F_x = ma_x$ gives $F_x = T_1 = 32.6 \text{ N}$ and $\sum F_y = ma_y$ gives

$F_y = Mg + T_2 = (2.00 \text{ kg})(9.80 \text{ m/s}^2) + 35.4 \text{ N} = 55.0 \text{ N}.$

EVALUATE: The equation $m_2 g - (m_1 + m_2 + \frac{1}{2}M)a$ says that the external force $m_2 g$ must accelerate all three objects.

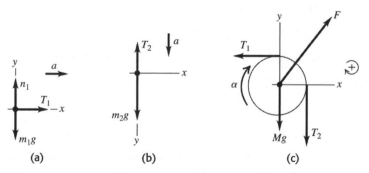

Figure 10.17

10.21.   **IDENTIFY:** Apply Eq. (10.8).

SET UP: For an object that is rolling without slipping, $v_{cm} = R\omega.$

EXECUTE: The fraction of the total kinetic energy that is rotational is

$$\frac{(1/2)I_{cm}\omega^2}{(1/2)Mv_{cm}^2 + (1/2)I_{cm}\omega^2} = \frac{1}{1 + (M/I_{cm})v_{cm}^2/\omega^2} = \frac{1}{1 + (MR^2/I_{cm})}$$

(a) $I_{cm} = (1/2)MR^2,$ so the above ratio is 1/3.

(b) $I_{cm} = (2/5)MR^2$ so the above ratio is 2/7.

(c) $I_{cm} = (2/3)MR^2$ so the ratio is $2/5$.

(d) $I_{cm} = (5/8)MR^2$ so the ratio is $5/13$.

EVALUATE: The moment of inertia of each object takes the form $I = \beta MR^2$. The ratio of rotational

kinetic energy to total kinetic energy can be written as $\dfrac{1}{1+1/\beta} = \dfrac{\beta}{1+\beta}$. The ratio increases as β increases.

10.23.   **IDENTIFY:** Apply $\sum \vec{F}_{ext} = m\vec{a}_{cm}$ and $\sum \tau_z = I_{cm}\alpha_z$ to the motion of the ball.

(a) **SET UP:** The free-body diagram is given in Figure 10.23a.

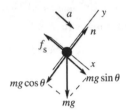

EXECUTE: $\sum F_y = ma_y$

$n = mg \cos\theta$ and $f_s = \mu_s mg \cos\theta$

$\sum F_x = ma_x$

$mg \sin\theta - \mu_s mg \cos\theta = ma$

$g(\sin\theta - \mu_s \cos\theta) = a$ (eq. 1)

Figure 10.23a

SET UP: Consider Figure 10.23b.

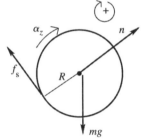

n and mg act at the center of the ball and provide no torque.

Figure 10.23b

EXECUTE: $\sum \tau = \tau_f = \mu_s mg \cos\theta R; \quad I = \frac{2}{5}mR^2$

$\sum \tau_z = I_{cm}\alpha_z$ gives $\mu_s mg \cos\theta R = \frac{2}{5}mR^2\alpha$

No slipping means $\alpha = a/R$, so $\mu_s g \cos\theta = \frac{2}{5}a$ (eq.2)

We have two equations in the two unknowns a and μ_s. Solving gives $a = \frac{5}{7}g \sin\theta$ and

$\mu_s = \frac{2}{7}\tan\theta = \frac{2}{7}\tan 65.0° = 0.613$.

(b) Repeat the calculation of part (a), but now $I = \frac{2}{3}mR^2$. $a = \frac{3}{5}g \sin\theta$ and

$\mu_s = \frac{2}{5}\tan\theta = \frac{2}{5}\tan 65.0° = 0.858$

The value of μ_s calculated in part (a) is not large enough to prevent slipping for the hollow ball.

(c) **EVALUATE:** There is no slipping at the point of contact. More friction is required for a hollow ball since for a given m and R it has a larger I and more torque is needed to provide the same α. Note that the required μ_s is independent of the mass or radius of the ball and only depends on how that mass is distributed.

10.27.   **IDENTIFY:** As the cylinder falls, its potential energy is transformed into both translational and rotational kinetic energy. Its mechanical energy is conserved.

SET UP: The hollow cylinder has $I = \frac{1}{2}m(R_a^2 + R_b^2)$, where $R_a = 0.200$ m and $R_b = 0.350$ m. Use

coordinates where $+y$ is upward and $y = 0$ at the initial position of the cylinder. Then $y_1 = 0$ and

$y_2 = -d$, where d is the distance it has fallen. $v_{cm} = R\omega$. $K_{cm} = \frac{1}{2}Mv_{cm}^2$ and $K_{rot} = \frac{1}{2}I_{cm}\omega^2$.

EXECUTE: (a) Conservation of energy gives $K_1 + U_1 = K_2 + U_2$. $K_1 = 0$, $U_1 = 0$. $0 = U_2 + K_2$ and

$0 = -mgd + \frac{1}{2}mv_{cm}^2 + \frac{1}{2}I_{cm}\omega^2$. $\frac{1}{2}I\omega^2 = \frac{1}{2}\left(\frac{1}{2}m[R_a^2 + R_b^2]\right)(v_{cm}/R_b)^2 = \frac{1}{4}m[1 + (R_a/R_b)^2]v_{cm}^2$, so

$\frac{1}{2}\left(1 + \frac{1}{2}[1 + (R_a/R_b)^2]\right)v_{cm}^2 = gd$ and $d = \dfrac{\left(1 + \frac{1}{2}[1 + (R_a/R_b)^2]\right)v_{cm}^2}{2g} = \dfrac{(1 + 0.663)(6.66 \text{ m/s})^2}{2(9.80 \text{ m/s}^2)} = 3.76 \text{ m}.$

(b) $K_2 = \frac{1}{2}mv_{cm}^2$ since there is no rotation. So $mgd = \frac{1}{2}mv_{cm}^2$ which gives

$v_{cm} = \sqrt{2gd} = \sqrt{2(9.80 \text{ m/s}^2)(3.76 \text{ m})} = 8.58 \text{ m/s}.$

(c) In part (a) the cylinder has rotational as well as translational kinetic energy and therefore less translational speed at a given kinetic energy. The kinetic energy comes from a decrease in gravitational potential energy and that is the same, so in (a) the translational speed is less.

EVALUATE: If part (a) were repeated for a solid cylinder, $R_a = 0$ and $d = 3.39$ m. For a thin-walled hollow cylinder, $R_a = R_b$ and $d = 4.52$ cm. Note that all of these answers are independent of the mass m of the cylinder.

10.29.   **IDENTIFY:**   As the ball rolls up the hill, its kinetic energy (translational and rotational) is transformed into gravitational potential energy. Since there is no slipping, its mechanical energy is conserved.

SET UP: The ball has moment of inertia $I_{cm} = \frac{2}{3}mR^2$. Rolling without slipping means $v_{cm} = R\omega$. Use coordinates where $+y$ is upward and $y = 0$ at the bottom of the hill, so $y_1 = 0$ and $y_2 = h = 5.00$ m. The ball's kinetic energy is $K = \frac{1}{2}mv_{cm}^2 + \frac{1}{2}I_{cm}\omega^2$ and its potential energy is $U = mgh$.

EXECUTE: (a) Conservation of energy gives $K_1 + U_1 = K_2 + U_2$. $U_1 = 0$, $K_2 = 0$ (the ball stops).

Therefore $K_1 = U_2$ and $\frac{1}{2}mv_{cm}^2 + \frac{1}{2}I_{cm}\omega^2 = mgh$. $\frac{1}{2}I_{cm}\omega^2 = \frac{1}{2}\left(\frac{2}{3}mR^2\right)\left(\dfrac{v_{cm}}{R}\right)^2 = \frac{1}{3}mv_{cm}^2$ so

$\frac{5}{6}mv_{cm}^2 = mgh$. Therefore $v_{cm} = \sqrt{\dfrac{6gh}{5}} = \sqrt{\dfrac{6(9.80 \text{ m/s}^2)(5.00 \text{ m})}{5}} = 7.67 \text{ m/s}$ and

$\omega = \dfrac{v_{cm}}{R} = \dfrac{7.67 \text{ m/s}}{0.113 \text{ m}} = 67.9 \text{ rad/s}.$

(b) $K_{rot} = \frac{1}{2}I\omega^2 = \frac{1}{3}mv_{cm}^2 = \frac{1}{3}(0.426 \text{ kg})(7.67 \text{ m/s})^2 = 8.35 \text{ J}.$

EVALUATE: Its translational kinetic energy at the base of the hill is $\frac{1}{2}mv_{cm}^2 = \frac{3}{2}K_{rot} = 12.52$ J. Its total kinetic energy is 20.9 J, which equals its final potential energy:

$mgh = (0.426 \text{ kg})(9.80 \text{ m/s}^2)(5.00 \text{ m}) = 20.9 \text{ J}.$

10.35.   (a) **IDENTIFY and SET UP:**   Use Eq. (10.23) and solve for τ_z. $P = \tau_z\omega_z$, where ω_z must be in rad/s

EXECUTE: $\omega_z = (4000 \text{ rev/min})(2\pi \text{ rad/1 rev})(1 \text{ min/60 s}) = 418.9 \text{ rad/s}$

$\tau_z = \dfrac{P}{\omega_z} = \dfrac{1.50 \times 10^5 \text{ W}}{418.9 \text{ rad/s}} = 358 \text{ N} \cdot \text{m}$

(b) **IDENTIFY and SET UP:**   Apply $\Sigma \vec{F} = m\vec{a}$ to the drum. Find the tension T in the rope using τ_z from part (a). The system is sketched in Figure 10.35.

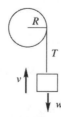

EXECUTE: v constant implies $a = 0$ and $T = w$

$\tau_z = TR$ implies

$T = \tau_z/R = 358 \text{ N} \cdot \text{m}/0.200 \text{ m} = 1790 \text{ N}$

Thus a weight $w = 1790$ N can be lifted.

Figure 10.35

(c) IDENTIFY and **SET UP:** Use $v = R\omega$.

EXECUTE: The drum has $\omega = 418.9$ rad/s, so $v = (0.200 \text{ m})(418.9 \text{ rad/s}) = 83.8$ m/s.

EVALUATE: The rate at which T is doing work on the drum is $P = Tv = (1790 \text{ N})(83.8 \text{ m/s}) = 150$ kW.
This agrees with the work output of the motor.

10.39.   **IDENTIFY** and **SET UP:** Use $L = I\omega$.

EXECUTE: The second hand makes 1 revolution in 1 minute, so
$\omega = (1.00 \text{ rev/min})(2\pi \text{ rad/1 rev})(1 \text{ min/60 s}) = 0.1047$ rad/s.

For a slender rod, with the axis about one end,

$I = \frac{1}{3}ML^2 = \frac{1}{3}(6.00 \times 10^{-3} \text{ kg})(0.150 \text{ m})^2 = 4.50 \times 10^{-5} \text{ kg} \cdot \text{m}^2$.

Then $L = I\omega = (4.50 \times 10^{-5} \text{ kg} \cdot \text{m}^2)(0.1047 \text{ rad/s}) = 4.71 \times 10^{-6} \text{ kg} \cdot \text{m}^2/\text{s}$.

EVALUATE: $\vec{L}$ is clockwise.

10.43.   **IDENTIFY:** Apply conservation of angular momentum to the motion of the skater.

SET UP: For a thin-walled hollow cylinder $I = mR^2$. For a slender rod rotating about an axis through its center, $I = \frac{1}{12}ml^2$.

EXECUTE: $L_i = L_f$ so $I_i\omega_i = I_f\omega_f$.

$I_i = 0.40 \text{ kg} \cdot \text{m}^2 + \frac{1}{12}(8.0 \text{ kg})(1.8 \text{ m})^2 = 2.56 \text{ kg} \cdot \text{m}^2$. $I_f = 0.40 \text{ kg} \cdot \text{m}^2 + (8.0 \text{ kg})(0.25 \text{ m})^2 = 0.90 \text{ kg} \cdot \text{m}^2$.

$\omega_f = \left(\dfrac{I_i}{I_f}\right)\omega_i = \left(\dfrac{2.56 \text{ kg} \cdot \text{m}^2}{0.90 \text{ kg} \cdot \text{m}^2}\right)(0.40 \text{ rev/s}) = 1.14$ rev/s.

EVALUATE: $K = \frac{1}{2}I\omega^2 = \frac{1}{2}L\omega$. ω increases and L is constant, so K increases. The increase in kinetic energy comes from the work done by the skater when he pulls in his hands.

10.45.   **IDENTIFY** and **SET UP:** There is no net external torque about the rotation axis so the angular momentum $L = I\omega$ is conserved.

EXECUTE: **(a)** $L_1 = L_2$ gives $I_1\omega_1 = I_2\omega_2$, so $\omega_2 = (I_1/I_2)\omega_1$

$I_1 = I_{tt} = \frac{1}{2}MR^2 = \frac{1}{2}(120 \text{ kg})(2.00 \text{ m})^2 = 240 \text{ kg} \cdot \text{m}^2$

$I_2 = I_{tt} + I_p = 240 \text{ kg} \cdot \text{m}^2 + mR^2 = 240 \text{ kg} \cdot \text{m}^2 + (70 \text{ kg})(2.00 \text{ m})^2 = 520 \text{ kg} \cdot \text{m}^2$

$\omega_2 = (I_1/I_2)\omega_1 = (240 \text{ kg} \cdot \text{m}^2/520 \text{ kg} \cdot \text{m}^2)(3.00 \text{ rad/s}) = 1.38$ rad/s

(b) $K_1 = \frac{1}{2}I_1\omega_1^2 = \frac{1}{2}(240 \text{ kg} \cdot \text{m}^2)(3.00 \text{ rad/s})^2 = 1080$ J

$K_2 = \frac{1}{2}I_2\omega_2^2 = \frac{1}{2}(520 \text{ kg} \cdot \text{m}^2)(1.38 \text{ rad/s})^2 = 495$ J

EVALUATE: The kinetic energy decreases because of the negative work done on the turntable and the parachutist by the friction force between these two objects.
The angular speed decreases because I increases when the parachutist is added to the system.

10.47.   **(a) IDENTIFY** and **SET UP:** Apply conservation of angular momentum $\vec{L}$, with the axis at the nail. Let object A be the bug and object B be the bar. Initially, all objects are at rest and $L_1 = 0$. Just after the bug jumps, it has angular momentum in one direction of rotation and the bar is rotating with angular velocity ω_B in the opposite direction.

EXECUTE: $L_2 = m_A v_A r - I_B \omega_B$ where $r = 1.00$ m and $I_B = \frac{1}{3}m_B r^2$

$L_1 = L_2$ gives $m_A v_A r = \frac{1}{3}m_B r^2 \omega_B$

$\omega_B = \dfrac{3m_A v_A}{m_B r} = 0.120$ rad/s

(b) $K_1 = 0$; $K_2 = \frac{1}{2}m_A v_A^2 + \frac{1}{2}I_B\omega_B^2 = $

$\frac{1}{2}(0.0100 \text{ kg})(0.200 \text{ m/s})^2 + \frac{1}{2}\left(\frac{1}{3}[0.0500 \text{ kg}][1.00 \text{ m}]^2\right)(0.120 \text{ rad/s})^2 = 3.2 \times 10^{-4}$ J.

(c) The increase in kinetic energy comes from work done by the bug when it pushes against the bar in order to jump.

EVALUATE: There is no external torque applied to the system and the total angular momentum of the system is constant. There are internal forces, forces the bug and bar exert on each other. The forces exert torques and change the angular momentum of the bug and the bar, but these changes are equal in magnitude and opposite in direction. These internal forces do positive work on the two objects and the kinetic energy of each object and of the system increases.

10.49.   **IDENTIFY:**   Apply conservation of angular momentum to the collision.

SET UP: The system before and after the collision is sketched in Figure 10.49. Let counterclockwise rotation be positive. The bar has $I = \frac{1}{3} m_2 L^2$.

EXECUTE: **(a)** Conservation of angular momentum: $m_1 v_0 d = -m_1 v d + \frac{1}{3} m_2 L^2 \omega.$

$$(3.00 \text{ kg})(10.0 \text{ m/s})(1.50 \text{ m}) = -(3.00 \text{ kg})(6.00 \text{ m/s})(1.50 \text{ m}) + \frac{1}{3}\left(\frac{90.0 \text{ N}}{9.80 \text{ m/s}^2}\right)(2.00 \text{ m})^2 \omega$$

$\omega = 5.88$ rad/s.

(b) There are no unbalanced torques about the pivot, so angular momentum is conserved. But the pivot exerts an unbalanced horizontal external force on the system, so the linear momentum is not conserved.

EVALUATE: Kinetic energy is not conserved in the collision.

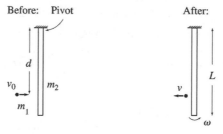

Figure 10.49

10.51.   **IDENTIFY:**   If we take the raven and the gate as a system, the torque about the pivot is zero, so the angular momentum of the system about the pivot is conserved.

SET UP: The system before and after the collision is sketched in Figure 10.51. The gate has $I = \frac{1}{3} ML^2$.

Take counterclockwise torques to be positive.

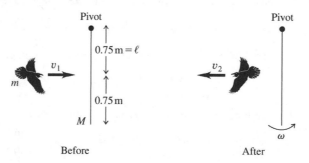

Figure 10.51

EXECUTE: **(a)** The gravity forces exert no torque at the moment of collision and angular momentum is conserved. $L_1 = L_2$. $m v_1 l = -m v_2 l + I_{\text{gate}} \omega$ with $l = L/2$.

$$\omega = \frac{m(v_1 + v_2)l}{\frac{1}{3}ML^2} = \frac{3m(v_1 + v_2)}{2ML} = \frac{3(1.1 \text{ kg})(5.0 \text{ m/s} + 2.0 \text{ m/s})}{2(4.5 \text{ kg})(1.5 \text{ m})} = 1.71 \text{ rad/s.}$$

(b) Linear momentum is not conserved; there is an external force exerted by the pivot. But the force on the pivot has zero torque. There is no external torque and angular momentum is conserved.

EVALUATE: $K_1 = \frac{1}{2}(1.1 \text{ kg})(5.0 \text{ m/s})^2 = 13.8 \text{ J}.$

$K_2 = \frac{1}{2}(1.1 \text{ kg})(2.0 \text{ m/s})^2 + \frac{1}{2}(\frac{1}{3}[4.5 \text{ kg}][1.5 \text{ m/s}]^2)(1.71 \text{ rad/s})^2 = 7.1 \text{ J}.$ This is an inelastic collision and

$K_2 < K_1.$

10.53.   **IDENTIFY:**   The precession angular velocity is $\Omega = \dfrac{wr}{I\omega}$, where ω is in rad/s. Also apply $\Sigma\vec{F} = m\vec{a}$ to the

gyroscope.

SET UP: The total mass of the gyroscope is $m_r + m_f = 0.140 \text{ kg} + 0.0250 \text{ kg} = 0.165 \text{ kg}.$

$\Omega = \dfrac{2\pi \text{ rad}}{T} = \dfrac{2\pi \text{ rad}}{2.20 \text{ s}} = 2.856 \text{ rad/s}.$

EXECUTE: **(a)** $F_p = w_{\text{tot}} = (0.165 \text{ kg})(9.80 \text{ m/s}^2) = 1.62 \text{ N}$

(b) $\omega = \dfrac{wr}{I\Omega} = \dfrac{(0.165 \text{ kg})(9.80 \text{ m/s}^2)(0.0400 \text{ m})}{(1.20\times10^{-4} \text{ kg} \cdot \text{m}^2)(2.856 \text{ rad/s})} = 189 \text{ rad/s} = 1.80\times10^3 \text{ rev/min}$

(c) If the figure in the problem is viewed from above, $\vec{\tau}$ is in the direction of the precession and $\vec{L}$ is
along the axis of the rotor, away from the pivot.

EVALUATE: There is no vertical component of acceleration associated with the motion, so the force from
the pivot equals the weight of the gyroscope. The larger ω is, the slower the rate of precession.

10.57.   **IDENTIFY:**   Apply $\Sigma\tau_z = I\alpha_z$ and constant acceleration equations to the motion of the grindstone.

SET UP: Let the direction of rotation of the grindstone be positive. The friction force is $f = \mu_k n$ and

produces torque fR. $\omega = (120\text{rev/min})\left(\dfrac{2\pi \text{ rad}}{1 \text{ rev}}\right)\left(\dfrac{1 \text{ min}}{60 \text{ s}}\right) = 4\pi \text{ rad/s}.$ $I = \frac{1}{2}MR^2 = 1.69 \text{ kg} \cdot \text{m}^2.$

EXECUTE: **(a)** The net torque must be

$$\tau = I\alpha = I\dfrac{\omega_z - \omega_{0z}}{t} = (1.69 \text{ kg} \cdot \text{m}^2)\dfrac{4\pi \text{ rad/s}}{9.00 \text{ s}} = 2.36 \text{ N} \cdot \text{m}.$$

This torque must be the sum of the applied force FR and the opposing frictional torques τ_f at the axle and

$fR = \mu_k nR$ due to the knife. $F = \dfrac{1}{R}(\tau + \tau_f + \mu_k nR).$

$F = \dfrac{1}{0.500 \text{ m}}((2.36 \text{ N} \cdot \text{m}) + (6.50 \text{ N} \cdot \text{m}) + (0.60)(160 \text{ N})(0.260 \text{ m})) = 67.6 \text{ N}.$

(b) To maintain a constant angular velocity, the net torque τ is zero, and the force F' is

$F' = \dfrac{1}{0.500 \text{ m}}(6.50 \text{ N} \cdot \text{m} + 24.96 \text{ N} \cdot \text{m}) = 62.9 \text{ N}.$

(c) The time t needed to come to a stop is found by taking the magnitudes in Eq. (10.29), with $\tau = \tau_f$

constant; $t = \dfrac{L}{\tau_f} = \dfrac{\omega I}{\tau_f} = \dfrac{(4\pi \text{ rad/s})(1.69 \text{ kg} \cdot \text{m}^2)}{6.50 \text{ N} \cdot \text{m}} = 3.27 \text{ s}.$

EVALUATE: The time for a given change in ω is proportional to α, which is in turn proportional to the

net torque, so the time in part (c) can also be found as $t = (9.00 \text{ s})\dfrac{2.36 \text{ N} \cdot \text{m}}{6.50 \text{ N} \cdot \text{m}}.$

10.59.   **IDENTIFY:**   Use the kinematic information to solve for the angular acceleration of the grindstone. Assume
that the grindstone is rotating counterclockwise and let that be the positive sense of rotation. Then apply
Eq. (10.7) to calculate the friction force and use $f_k = \mu_k n$ to calculate μ_k.

SET UP: $\omega_{0z} = 850 \text{ rev/min}(2\pi \text{ rad/1 rev})(1 \text{ min/60 s}) = 89.0 \text{ rad/s}$

$t = 7.50 \text{ s}$; $\omega_z = 0$ (comes to rest); $\alpha_z = ?$

EXECUTE: $\omega_z = \omega_{0z} + \alpha_z t$

$\alpha_z = \dfrac{0 - 89.0 \text{ rad/s}}{7.50 \text{ s}} = -11.9 \text{ rad/s}^2$

SET UP: Apply $\sum \tau_z = I\alpha_z$ to the grindstone. The free-body diagram is given in Figure 10.59.

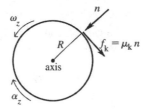

Figure 10.59

The normal force has zero moment arm for rotation about an axis at the center of the grindstone, and therefore zero torque. The only torque on the grindstone is that due to the friction force f_k exerted by the ax; for this force the moment arm is $l = R$ and the torque is negative.

EXECUTE: $\sum \tau_z = -f_k R = -\mu_k nR$

$I = \frac{1}{2}MR^2$ (solid disk, axis through center)

Thus $\sum \tau_z = I\alpha_z$ gives $-\mu_k nR = (\frac{1}{2}MR^2)\alpha_z$

$$\mu_k = -\frac{MR\alpha_z}{2n} = -\frac{(50.0 \text{ kg})(0.260 \text{ m})(-11.9 \text{ rad/s}^2)}{2(160 \text{ N})} = 0.483$$

EVALUATE: The friction torque is clockwise and slows down the counterclockwise rotation of the grindstone.

10.63. **IDENTIFY:** Use $\sum \tau_z = I\alpha_z$ to find the angular acceleration just after the ball falls off and use conservation of energy to find the angular velocity of the bar as it swings through the vertical position.

SET UP: The axis of rotation is at the axle. For this axis the bar has $I = \frac{1}{12}m_{bar}L^2$, where $m_{bar} = 3.80$ kg and $L = 0.800$ m. Energy conservation gives $K_1 + U_1 = K_2 + U_2$. The gravitational potential energy of the bar doesn't change. Let $y_1 = 0$, so $y_2 = -L/2$.

EXECUTE: **(a)** $\tau_z = m_{ball}g(L/2)$ and $I = I_{ball} + I_{bar} = \frac{1}{12}m_{bar}L^2 + m_{ball}(L/2)^2$. $\sum \tau_z = I\alpha_z$ gives

$$\alpha_z = \frac{m_{ball}g(L/2)}{\frac{1}{12}m_{bar}L^2 + m_{ball}(L/2)^2} = \frac{2g}{L}\left(\frac{m_{ball}}{m_{ball} + m_{bar}/3}\right) \text{ and}$$

$$\alpha_z = \frac{2(9.80 \text{ m/s}^2)}{0.800 \text{ m}}\left(\frac{2.50 \text{ kg}}{2.50 \text{ kg} + [3.80 \text{ kg}]/3}\right) = 16.3 \text{ rad/s}^2.$$

(b) As the bar rotates, the moment arm for the weight of the ball decreases and the angular acceleration of the bar decreases.

(c) $K_1 + U_1 = K_2 + U_2$. $0 = K_2 + U_2$. $\frac{1}{2}(I_{bar} + I_{ball})\omega^2 = -m_{ball}g(-L/2)$.

$$\omega = \sqrt{\frac{m_{ball}gL}{m_{ball}L^2/4 + m_{bar}L^2/12}} = \sqrt{\frac{g}{L}\left(\frac{4m_{ball}}{m_{ball} + m_{bar}/3}\right)} = \sqrt{\frac{9.80 \text{ m/s}^2}{0.800 \text{ m}}\left(\frac{4[2.50 \text{ kg}]}{2.50 \text{ kg} + [3.80 \text{ kg}]/3}\right)}$$

$\omega = 5.70$ rad/s.

EVALUATE: As the bar swings through the vertical, the linear speed of the ball that is still attached to the bar is $v = (0.400 \text{ m})(5.70 \text{ rad/s}) = 2.28$ m/s. A point mass in free-fall acquires a speed of 2.80 m/s after falling 0.400 m; the ball on the bar acquires a speed less than this.

10.67. **IDENTIFY:** Blocks A and B have linear acceleration and therefore obey the linear form of Newton's second law $\sum F_y = ma_y$. The wheel C has angular acceleration, so it obeys the rotational form of Newton's second law $\sum \tau_z = I\alpha_z$.

SET UP: *A* accelerates downward, *B* accelerates upward and the wheel turns clockwise. Apply $\sum F_y = ma_y$ to blocks *A* and *B*. Let +*y* be downward for *A* and +*y* be upward for *B*. Apply $\sum \tau_z = I\alpha_z$ to the wheel, with the clockwise sense of rotation positive. Each block has the same magnitude of acceleration, *a*, and $a = R\alpha$. Call the T_A the tension in the cord between *C* and *A* and T_B the tension between *C* and *B*.

EXECUTE: For *A*, $\sum F_y = ma_y$ gives $m_A g - T_A = m_A a$. For *B*, $\sum F_y = ma_y$ gives $T_B - m_B g = m_B a$. For

the wheel, $\sum \tau_z = I\alpha_z$ gives $T_A R - T_B R = I\alpha = I(a/R)w$ and $T_A - T_B = \left(\dfrac{I}{R^2}\right)a$. Adding these three

equations gives $(m_A - m_B)g = \left(m_A + m_B + \dfrac{I}{R^2}\right)a$. Solving for *a*, we have

$$a = \left(\frac{m_A - m_B}{m_A + m_B + I/R^2}\right)g = \left(\frac{4.00 \text{ kg} - 2.00 \text{ kg}}{4.00 \text{ kg} + 2.00 \text{ kg} + (0.300 \text{ kg}\cdot\text{m}^2)/(0.120 \text{ m})^2}\right)(9.80 \text{ m/s}^2) = 0.730 \text{ m/s}^2.$$

$$\alpha = \frac{a}{R} = \frac{0.730 \text{ m/s}^2}{0.120 \text{ m}} = 6.08 \text{ rad/s}^2.$$

$$T_A = m_A(g - a) = (4.00 \text{ kg})(9.80 \text{ m/s}^2 - 0.730 \text{ m/s}^2) = 36.3 \text{ N}.$$

$$T_B = m_B(g + a) = (2.00 \text{ kg})(9.80 \text{ m/s}^2 + 0.730 \text{ m/s}^2) = 21.1 \text{ N}.$$

EVALUATE: The tensions must be different in order to produce a torque that accelerates the wheel when the blocks accelerate.

10.69.   **IDENTIFY:**   Apply $\sum \vec{F}_{\text{ext}} = m\vec{a}_{\text{cm}}$ and $\sum \tau_z = I_{\text{cm}}\alpha_z$ to the roll.

SET UP: At the point of contact, the wall exerts a friction force *f* directed downward and a normal force *n* directed to the right. This is a situation where the net force on the roll is zero, but the net torque is *not* zero.

EXECUTE: **(a)** Balancing vertical forces, $F_{\text{rod}}\cos\theta = f + w + F$, and balancing horizontal forces

$F_{\text{rod}}\sin\theta = n$. With $f = \mu_k n$, these equations become $F_{\text{rod}}\cos\theta = \mu_k n + F + w$, $F_{\text{rod}}\sin\theta = n$. Eliminating

n and solving for F_{rod} gives $F_{\text{rod}} = \dfrac{w + F}{\cos\theta - \mu_k\sin\theta} = \dfrac{(16.0 \text{ kg})(9.80 \text{ m/s}^2) + (60.0 \text{ N})}{\cos 30° - (0.25)\sin 30°} = 293 \text{ N}.$

(b) With respect to the center of the roll, the rod and the normal force exert zero torque. The magnitude of the net torque is $(F - f)R$, and $f = \mu_k n$ may be found by insertion of the value found for F_{rod} into either of the above relations; i.e., $f = \mu_k F_{\text{rod}}\sin\theta = 36.57 \text{ N}.$ Then,

$$\alpha = \frac{\tau}{I} = \frac{(60.0 \text{ N} - 36.57 \text{ N})(18.0\times10^{-2} \text{ m})}{(0.260 \text{ kg}\cdot\text{m}^2)} = 16.2 \text{ rad/s}^2.$$

EVALUATE: If the applied force *F* is increased, F_{rod} increases and this causes *n* and *f* to increase. The angle θ changes as the amount of paper unrolls and this affects α for a given *F*.

10.73.   **IDENTIFY:**   Apply $\sum \vec{F} = m\vec{a}$ to each object and apply $\sum \tau_z = I\alpha_z$ to the pulley.

SET UP: Call the 75.0 N weight *A* and the 125 N weight *B*. Let T_A and T_B be the tensions in the cord to the left and to the right of the pulley. For the pulley, $I = \frac{1}{2}MR^2$, where $Mg = 80.0 \text{ N}$ and $R = 0.300 \text{ m}$.

The 125 N weight accelerates downward with acceleration *a*, the 75.0 N weight accelerates upward with acceleration *a* and the pulley rotates clockwise with angular acceleration α, where $a = R\alpha$.

EXECUTE: $\sum \vec{F} = m\vec{a}$ applied to the 75.0 N weight gives $T_A - w_A = m_A a$. $\sum \vec{F} = m\vec{a}$ applied to the 125.0 N

weight gives $w_B - T_B = m_B a$. $\sum \tau_z = I\alpha_z$ applied to the pulley gives $(T_B - T_A)R = (\frac{1}{2}MR^2)\alpha_z$ and

$T_B - T_A = \frac{1}{2}Ma$. Combining these three equations gives $w_B - w_A = (m_A + m_B + M/2)a$ and

$$a = \left(\frac{w_B - w_A}{w_A + w_B + w_{\text{pulley}}/2}\right)g = \left(\frac{125 \text{ N} - 75.0 \text{ N}}{75.0 \text{ N} + 125 \text{ N} + 40.0 \text{ N}}\right)g = 0.2083g.$$

$T_A = w_A(1 + a/g) = 1.2083w_A = 90.62 \text{ N}.$ $T_B = w_B(1 - a/g) = 0.792w_B = 98.96 \text{ N}.$ $\sum \vec{F} = m\vec{a}$ applied to the pulley gives that the force F applied by the hook to the pulley is $F = T_A + T_B + w_{\text{pulley}} = 270$ N. The force the ceiling applies to the hook is 270 N.

EVALUATE: The force the hook exerts on the pulley is less than the total weight of the system, since the net effect of the motion of the system is a downward acceleration of mass.

10.75.   **IDENTIFY:** Apply $\sum \vec{F}_{\text{ext}} = m\vec{a}_{\text{cm}}$ to the motion of the center of mass and apply $\sum \tau_z = I_{\text{cm}} \alpha_z$ to the rotation about the center of mass.

SET UP: $I = 2\left(\frac{1}{2}mR^2\right) = mR^2$. The moment arm for T is b.

EXECUTE: The tension is related to the acceleration of the yo-yo by $(2m)g - T = (2m)a$, and to the

angular acceleration by $Tb = I\alpha = I\dfrac{a}{b}$. Dividing the second equation by b and adding to the first to

eliminate T yields $a = g\dfrac{2m}{(2m + I/b^2)} = g\dfrac{2}{2 + (R/b)^2}$, $\alpha = g\dfrac{2}{2b + R^2/b}$. The tension is found by

substitution into either of the two equations:

$$T = (2m)(g - a) = (2mg)\left(1 - \frac{2}{2 + (R/b)^2}\right) = 2mg\frac{(R/b)^2}{2 + (R/b)^2} = \frac{2mg}{(2(b/R)^2 + 1)}.$$

EVALUATE: $a \to 0$ when $b \to 0$. As $b \to R$, $a \to 2g/3$.

10.79.   **IDENTIFY:** As it rolls down the rough slope, the basketball gains rotational kinetic energy as well as translational kinetic energy. But as it moves up the smooth slope, its rotational kinetic energy does not change since there is no friction.

SET UP: $I_{\text{cm}} = \frac{2}{3}mR^2$. When it rolls without slipping, $v_{\text{cm}} = R\omega$. When there is no friction the angular speed of rotation is constant. Take $+y$ upward and let $y = 0$ in the valley.

EXECUTE: **(a)** Find the speed v_{cm} in the level valley: $K_1 + U_1 = K_2 + U_2$. $y_1 = H_0$, $y_2 = 0$. $K_1 = 0$,

$U_2 = 0$. Therefore, $U_1 = K_2$. $mgH_0 = \frac{1}{2}mv_{\text{cm}}^2 + \frac{1}{2}I_{\text{cm}}\omega^2$. $\frac{1}{2}I_{\text{cm}}\omega^2 = \frac{1}{2}\left(\frac{2}{3}mR^2\right)\left(\dfrac{v_{\text{cm}}}{R}\right)^2 = \frac{1}{3}mv_{\text{cm}}^2$, so

$mgH_0 = \frac{5}{6}mv_{\text{cm}}^2$ and $v_{\text{cm}}^2 = \dfrac{6gH_0}{5}$. Find the height H it goes up the other side. Its rotational kinetic energy

stays constant as it rolls on the frictionless surface. $\frac{1}{2}mv_{\text{cm}}^2 + \frac{1}{2}I_{\text{cm}}\omega^2 = \frac{1}{2}I_{\text{cm}}\omega^2 + mgH$.

$H = \dfrac{v_{\text{cm}}^2}{2g} = \frac{3}{5}H_0$.

(b) Some of the initial potential energy has been converted into rotational kinetic energy so there is less potential energy at the second height H than at the first height H_0.

EVALUATE: Mechanical energy is conserved throughout this motion. But the initial gravitational potential energy on the rough slope is not all transformed into potential energy on the smooth slope because some of that energy remains as rotational kinetic energy at the highest point on the smooth slope.

10.81.   **IDENTIFY:** Apply conservation of energy to the motion of the boulder.

SET UP: $K = \frac{1}{2}mv^2 + \frac{1}{2}I\omega^2$ and $v = R\omega$ when there is rolling without slipping. $I = \frac{2}{5}mR^2$.

EXECUTE: Break into two parts, the rough and smooth sections.

Rough: $mgh_1 = \frac{1}{2}mv^2 + \frac{1}{2}I\omega^2$. $mgh_1 = \frac{1}{2}mv^2 + \frac{1}{2}\left(\frac{2}{5}mR^2\right)\left(\dfrac{v}{R}\right)^2$. $v^2 = \dfrac{10}{7}gh_1$.

Smooth: Rotational kinetic energy does not change. $mgh_2 + \dfrac{1}{2}mv^2 + K_{\text{rot}} = \dfrac{1}{2}mv_{\text{Bottom}}^2 + K_{\text{rot}}$.

$gh_2 + \dfrac{1}{2}\left(\dfrac{10}{7}gh_1\right) = \dfrac{1}{2}v_{\text{Bottom}}^2$. $v_{\text{Bottom}} = \sqrt{\dfrac{10}{7}gh_1 + 2gh_2} = \sqrt{\dfrac{10}{7}(9.80 \text{ m/s}^2)(25 \text{ m}) + 2(9.80 \text{ m/s}^2)(25 \text{ m})} = 29.0 \text{ m/s}.$

EVALUATE: If all the hill was rough enough to cause rolling without slipping,

$v_{\text{Bottom}} = \sqrt{\dfrac{10}{7}g(50 \text{ m})} = 26.5$ m/s. A smaller fraction of the initial gravitational potential energy goes into

translational kinetic energy of the center of mass than if part of the hill is smooth. If the entire hill is

smooth and the boulder slides without slipping, $v_{\text{Bottom}} = \sqrt{2g(50 \text{ m})} = 31.3$ m/s. In this case all the initial

gravitational potential energy goes into the kinetic energy of the translational motion.

10.87. **IDENTIFY:** Use conservation of energy to relate the speed of the block to the distance it has descended.
Then use a constant acceleration equation to relate these quantities to the acceleration.

SET UP: For the cylinder, $I = \frac{1}{2}M(2R)^2$, and for the pulley, $I = \frac{1}{2}MR^2$.

EXECUTE: Doing this problem using kinematics involves four unknowns (six, counting the two angular
accelerations), while using energy considerations simplifies the calculations greatly. If the block and the
cylinder both have speed v, the pulley has angular velocity v/R and the cylinder has angular velocity
$v/2R$, the total kinetic energy is

$$K = \frac{1}{2}\left[Mv^2 + \frac{M(2R)^2}{2}(v/2R)^2 + \frac{MR^2}{2}(v/R)^2 + Mv^2 \right] = \frac{3}{2}Mv^2.$$

This kinetic energy must be the work done by gravity; if the hanging mass descends a distance y,
$K = Mgy$, or $v^2 = (2/3)gy$. For constant acceleration, $v^2 = 2ay$, and comparison of the two expressions
gives $a = g/3$.

EVALUATE: If the pulley were massless and the cylinder slid without rolling, $Mg = 2Ma$ and $a = g/2$.
The rotation of the objects reduces the acceleration of the block.

10.89. **IDENTIFY:** Apply conservation of energy to the motion of the first ball before the collision and to the
motion of the second ball after the collision. Apply conservation of angular momentum to the collision
between the first ball and the bar.

SET UP: The speed of the ball just before it hits the bar is $v = \sqrt{2gy} = 15.34$ m/s. Use conservation of
angular momentum to find the angular velocity ω of the bar just after the collision. Take the axis at the
center of the bar.

EXECUTE: $L_1 = mvr = (5.00 \text{ kg})(15.34 \text{ m/s})(2.00 \text{ m}) = 153.4 \text{ kg} \cdot \text{m}^2$

Immediately after the collision the bar and both balls are rotating together.

$L_2 = I_{\text{tot}}\omega$

$I_{\text{tot}} = \frac{1}{12}Ml^2 + 2mr^2 = \frac{1}{12}(8.00 \text{ kg})(4.00 \text{ m})^2 + 2(5.00 \text{ kg})(2.00 \text{ m})^2 = 50.67 \text{ kg} \cdot \text{m}^2$

$L_2 = L_1 = 153.4 \text{ kg} \cdot \text{m}^2$

$\omega = L_2/I_{\text{tot}} = 3.027$ rad/s

Just after the collision the second ball has linear speed $v = r\omega = (2.00 \text{ m})(3.027 \text{ rad/s}) = 6.055$ m/s and is

moving upward. $\frac{1}{2}mv^2 = mgy$ gives $y = 1.87$ m for the height the second ball goes.

EVALUATE: Mechanical energy is lost in the inelastic collision and some of the final energy is in the
rotation of the bar with the first ball stuck to it. As a result, the second ball does not reach the height from
which the first ball was dropped.

10.95. **IDENTIFY:** Apply conservation of angular momentum to the collision between the bird and the bar and
apply conservation of energy to the motion of the bar after the collision.

SET UP: For conservation of angular momentum take the axis at the hinge. For this axis the initial angular
momentum of the bird is $m_{\text{bird}}(0.500 \text{ m})v$, where $m_{\text{bird}} = 0.500$ kg and $v = 2.25$ m/s . For this axis the

moment of inertia is $I = \frac{1}{3}m_{\text{bar}}L^2 = \frac{1}{3}(1.50 \text{ kg})(0.750 \text{ m})^2 = 0.281 \text{ kg} \cdot \text{m}^2$. For conservation of energy, the

gravitational potential energy of the bar is $U = m_{\text{bar}}gy_{\text{cm}}$, where y_{cm} is the height of the center of the bar.

Take $y_{\text{cm,1}} = 0$, so $y_{\text{cm,2}} = -0.375$ m.

EXECUTE: **(a)** $L_1 = L_2$ gives $m_{bird}(0.500 \text{ m})v = (\frac{1}{3}m_{bar}L^2)\omega$.

$$\omega = \frac{3m_{bird}(0.500 \text{ m})v}{m_{bar}L^2} = \frac{3(0.500 \text{ kg})(0.500 \text{ m})(2.25 \text{ m/s})}{(1.50 \text{ kg})(0.750 \text{ m})^2} = 2.00 \text{ rad/s}.$$

(b) $U_1 + K_1 = U_2 + K_2$ applied to the motion of the bar after the collision gives

$$\tfrac{1}{2}I\omega_1^2 = m_{bar}g(-0.375 \text{ m}) + \tfrac{1}{2}I\omega_2^2. \quad \omega_2 = \sqrt{\omega_1^2 + \frac{2}{I}m_{bar}g(0.375 \text{ m})}.$$

$$\omega_2 = \sqrt{(2.00 \text{ rad/s})^2 + \frac{2}{0.281 \text{ kg} \cdot \text{m}^2}(1.50 \text{ kg})(9.80 \text{ m/s}^2)(0.375 \text{ m})} = 6.58 \text{ rad/s}$$

EVALUATE: Mechanical energy is not conserved in the collision. The kinetic energy of the bar just after the collision is less than the kinetic energy of the bird just before the collision.

EQUILIBRIUM AND ELASTICITY

11.3. **IDENTIFY:** Treat the rod and clamp as point masses. The center of gravity of the rod is at its midpoint, and we know the location of the center of gravity of the rod-clamp system.

SET UP: $x_{cm} = \dfrac{m_1 x_1 + m_2 x_2}{m_1 + m_2}$.

EXECUTE: $1.20 \text{ m} = \dfrac{(1.80 \text{ kg})(1.00 \text{ m}) + (2.40 \text{ kg}) x_2}{1.80 \text{ kg} + 2.40 \text{ kg}}$.

$x_2 = \dfrac{(1.20 \text{ m})(1.80 \text{ kg} + 2.40 \text{ kg}) - (1.80 \text{ kg})(1.00 \text{ m})}{2.40 \text{ kg}} = 1.35 \text{ m}$

EVALUATE: The clamp is to the right of the center of gravity of the system, so the center of gravity of the system lies between that of the rod and the clamp, which is reasonable.

11.5. **IDENTIFY:** Apply $\sum \tau_z = 0$ to the ladder.

SET UP: Take the axis to be at point A. The free-body diagram for the ladder is given in Figure 11.5. The torque due to F must balance the torque due to the weight of the ladder.

EXECUTE: $F(8.0 \text{ m}) \sin 40° = (2800 \text{ N})(10.0 \text{ m})$, so $F = 5.45$ kN.

EVALUATE: The force required is greater than the weight of the ladder, because the moment arm for F is less than the moment arm for w.

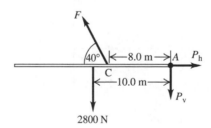

Figure 11.5

11.7. **IDENTIFY:** Apply $\sum F_y = 0$ and $\sum \tau_z = 0$ to the board.

SET UP: Let $+y$ be upward. Let x be the distance of the center of gravity of the motor from the end of the board where the 400 N force is applied.

EXECUTE: **(a)** If the board is taken to be massless, the weight of the motor is the sum of the applied forces, 1000 N. The motor is a distance $\dfrac{(2.00 \text{ m})(600 \text{ N})}{(1000 \text{ N})} = 1.20 \text{ m}$ from the end where the 400 N force is applied, and so is 0.800 m from the end where the 600 N force is applied.

(b) The weight of the motor is $400 \text{ N} + 600 \text{ N} - 200 \text{ N} = 800 \text{ N}$. Applying $\sum \tau_z = 0$ with the axis at the end of the board where the 400 N acts gives $(600 \text{ N})(2.00 \text{ m}) = (200 \text{ N})(1.00 \text{ m}) + (800 \text{ N})x$ and $x = 1.25 \text{ m}$.

The center of gravity of the motor is 0.75 m from the end of the board where the 600 N force is applied.

EVALUATE: The motor is closest to the end of the board where the larger force is applied.

11.9.   **IDENTIFY:** Apply the conditions for equilibrium to the bar. Set each tension equal to its maximum value.
SET UP: Let cable A be at the left-hand end. Take the axis to be at the left-hand end of the bar and x be the distance of the weight w from this end. The free-body diagram for the bar is given in Figure 11.9.
EXECUTE: **(a)** $\sum F_y = 0$ gives $T_A + T_B - w - w_{bar} = 0$ and

$w = T_A + T_B - w_{bar} = 500.0 \text{ N} + 400.0 \text{ N} - 350.0 \text{ N} = 550 \text{ N}$.

(b) $\sum \tau_z = 0$ gives $T_B(1.50 \text{ m}) - wx - w_{bar}(0.750 \text{ m}) = 0$.

$x = \dfrac{T_B(1.50 \text{ m}) - w_{bar}(0.750 \text{ m})}{w} = \dfrac{(400.0 \text{ N})(1.50 \text{ m}) - (350 \text{ N})(0.750 \text{ m})}{550 \text{ N}} = 0.614 \text{ m}$. The weight should

be placed 0.614 m from the left-hand end of the bar (cable A).
EVALUATE: If the weight is moved to the left, T_A exceeds 500.0 N and if it is moved to the right
T_B exceeds 400.0 N.

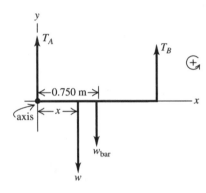

Figure 11.9

11.13.   **IDENTIFY:** Apply the first and second conditions of equilibrium to the strut.
(a) SET UP: The free-body diagram for the strut is given in Figure 11.13a. Take the origin of coordinates at the hinge (point A) and $+y$ upward. Let F_h and F_v be the horizontal and vertical components of the force $\vec{F}$ exerted on the strut by the pivot. The tension in the vertical cable is the weight w of the suspended object. The weight w of the strut can be taken to act at the center of the strut. Let L be the length of the strut.

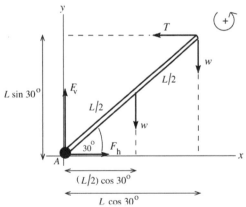

EXECUTE:
$\sum F_y = ma_y$
$F_v - w - w = 0$
$F_v = 2w$

Figure 11.13a

Sum torques about point A. The pivot force has zero moment arm for this axis and so doesn't enter into the torque equation.
$\tau_A = 0$
$TL\sin 30.0° - w((L/2)\cos 30.0°) - w(L\cos 30.0°) = 0$
$T\sin 30.0° - (3w/2)\cos 30.0° = 0$
$T = \dfrac{3w\cos 30.0°}{2\sin 30.0°} = 2.60w$

Then $\sum F_x = ma_x$ implies $T - F_h = 0$ and $F_h = 2.60w$.

We now have the components of $\vec{F}$ so can find its magnitude and direction (Figure 11.13b).

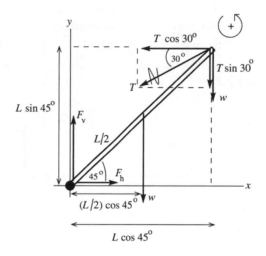

$$F = \sqrt{F_h^2 + F_v^2}$$

$$F = \sqrt{(2.60w)^2 + (2.00w)^2}$$

$$F = 3.28w$$

$$\tan\theta = \frac{F_v}{F_h} = \frac{2.00w}{2.60w}$$

$$\theta = 37.6°$$

Figure 11.13b

(b) SET UP: The free-body diagram for the strut is given in Figure 11.13c.

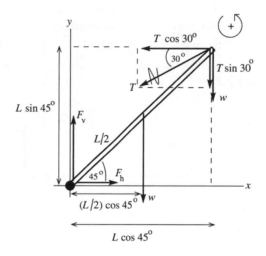

Figure 11.13c

The tension T has been replaced by its x and y components. The torque due to T equals the sum of the torques of its components, and the latter are easier to calculate.

EXECUTE: $\sum \tau_A = 0 + (T\cos 30.0°)(L\sin 45.0°) - (T\sin 30.0°)(L\cos 45.0°) -$

$w((L/2)\cos 45.0°) - w(L\cos 45.0°) = 0$

The length L divides out of the equation. The equation can also be simplified by noting that $\sin 45.0° = \cos 45.0°$.

Then $T(\cos 30.0° - \sin 30.0°) = 3w/2$.

$$T = \frac{3w}{2(\cos 30.0° - \sin 30.0°)} = 4.10w$$

$$\sum F_x = ma_x$$

$$F_h - T\cos 30.0° = 0$$

$$F_h = T\cos 30.0° = (4.10w)(\cos 30.0°) = 3.55w$$

$$\sum F_y = ma_y$$

$$F_v - w - w - T\sin 30.0° = 0$$

$$F_v = 2w + (4.10w)\sin 30.0° = 4.05w$$

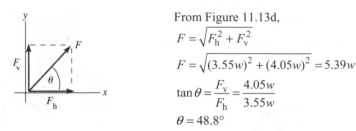

From Figure 11.13d,

$$F = \sqrt{F_h^2 + F_v^2}$$

$$F = \sqrt{(3.55w)^2 + (4.05w)^2} = 5.39w$$

$$\tan\theta = \frac{F_v}{F_h} = \frac{4.05w}{3.55w}$$

$$\theta = 48.8°$$

Figure 11.13d

EVALUATE: In each case the force exerted by the pivot does not act along the strut. Consider the net torque about the upper end of the strut. If the pivot force acted along the strut, it would have zero torque about this point. The two forces acting at this point also have zero torque and there would be one nonzero torque, due to the weight of the strut. The net torque about this point would then not be zero, violating the second condition of equilibrium.

11.15.   **IDENTIFY:** The athlete is in equilibrium, so the forces and torques on him must balance. The target variables are the forces on his hands and feet due to the floor.

SET UP: The free-body diagram is given in Figure 11.15. F_f is the force on each foot and F_h is the force on each hand. Use coordinates as shown. Take the pivot at his feet and let counterclockwise torques be positive. $\sum\tau_z = 0$ and $\sum F_y = 0$.

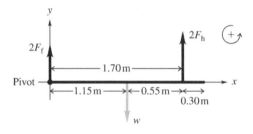

Figure 11.15

EXECUTE: $\sum\tau_z = 0$ gives $(2F_h)(1.70 \text{ m}) - w(1.15 \text{ m}) = 0$. Solving for F_h gives

$$F_h = w\frac{1.15 \text{ m}}{2(1.70 \text{ m})} = 0.338w = 272 \text{ N}.$$ Applying $\sum F_y = 0$, we get $2F_f + 2F_h - w = 0$ which gives

$$F_f = \tfrac{1}{2}w - F_h = 402 \text{ N} - 272 \text{ N} = 130 \text{ N}.$$

EVALUATE: His center of mass is closer to his hands than to his feet, so his hands exert a greater force.

11.19.   **IDENTIFY:** Apply the first and second conditions of equilibrium to the rod.

SET UP: The force diagram for the rod is given in Figure 11.19.

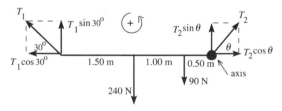

Figure 11.19

EXECUTE: $\sum\tau_z = 0$, axis at right end of rod, counterclockwise torque is positive

$$(240 \text{ N})(1.50 \text{ m}) + (90 \text{ N})(0.50 \text{ m}) - (T_1 \sin 30.0°)(3.00 \text{ m}) = 0$$

$$T_1 = \frac{360 \text{ N} \cdot \text{m} + 45 \text{ N} \cdot \text{m}}{1.50 \text{ m}} = 270 \text{ N}$$

$$\sum F_x = ma_x$$

$T_2 \cos\theta - T_1 \cos 30° = 0$ and $T_2 \cos\theta = 234$ N

$\Sigma F_y = ma_y$

$T_1 \sin 30° + T_2 \sin\theta - 240$ N $- 90$ N $= 0$

$T_2 \sin\theta = 330$ N $- (270$ N$)\sin 30° = 195$ N

Then $\dfrac{T_2 \sin\theta}{T_2 \cos\theta} = \dfrac{195 \text{ N}}{234 \text{ N}}$ gives $\tan\theta = 0.8333$ and $\theta = 40°$

And $T_2 = \dfrac{195 \text{ N}}{\sin 40°} = 303$ N.

EVALUATE: The monkey is closer to the right rope than to the left one, so the tension is larger in the right rope. The horizontal components of the tensions must be equal in magnitude and opposite in direction. Since $T_2 > T_1$, the rope on the right must be at a greater angle above the horizontal to have the same horizontal component as the tension in the other rope.

11.23. **IDENTIFY:** The student's head is at rest, so the torques on it must balance. The target variable is the tension in her neck muscles.

SET UP: Let the pivot be at point P and let counterclockwise torques be positive. $\Sigma \tau_z = 0$.

EXECUTE: **(a)** The free-body diagram is given in Figure 11.23.

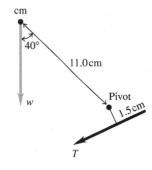

Figure 11.23

(b) $\Sigma \tau_z = 0$ gives $w(11.0 \text{ cm})(\sin 40.0°) - T(1.50 \text{ cm}) = 0$.

$T = \dfrac{(4.50 \text{ kg})(9.80 \text{ m/s}^2)(11.0 \text{ cm})\sin 40.0°}{1.50 \text{ cm}} = 208$ N.

EVALUATE: Her head weighs about 45 N but the tension in her neck muscles must be much larger because the tension has a small moment arm.

11.25. **IDENTIFY and SET UP:** Apply Eq. (11.10) and solve for A and then use $A = \pi r^2$ to get the radius and $d = 2r$ to calculate the diameter.

EXECUTE: $Y = \dfrac{l_0 F_\perp}{A \Delta l}$ so $A = \dfrac{l_0 F_\perp}{Y \Delta l}$ (A is the cross-section area of the wire)

For steel, $Y = 2.0 \times 10^{11}$ Pa (Table 11.1)

Thus $A = \dfrac{(2.00 \text{ m})(400 \text{ N})}{(2.0 \times 10^{11} \text{ Pa})(0.25 \times 10^{-2} \text{ m})} = 1.6 \times 10^{-6}$ m^2.

$A = \pi r^2$, so $r = \sqrt{A/\pi} = \sqrt{1.6 \times 10^{-6} \text{ m}^2/\pi} = 7.1 \times 10^{-4}$ m

$d = 2r = 1.4 \times 10^{-3}$ m $= 1.4$ mm

EVALUATE: Steel wire of this diameter doesn't stretch much; $\Delta l/l_0 = 0.12\%$.

11.27. **IDENTIFY:** $Y = \dfrac{l_0 F_\perp}{A \Delta l}$

SET UP: $A = 0.50$ cm$^2 = 0.50 \times 10^{-4}$ m^2 +

EXECUTE: $Y = \dfrac{(4.00 \text{ m})(5000 \text{ N})}{(0.50 \times 10^{-4} \text{ m}^2)(0.20 \times 10^{-2} \text{m})} = 2.0 \times 10^{11} \text{ Pa}$

EVALUATE: Our result is the same as that given for steel in Table 11.1.

11.31. **IDENTIFY:** The amount of compression depends on the bulk modulus of the bone.

SET UP: $\dfrac{\Delta V}{V_0} = -\dfrac{\Delta p}{B}$ and $1 \text{ atm} = 1.01 \times 10^5 \text{ Pa}$.

EXECUTE: **(a)** $\Delta p = -B \dfrac{\Delta V}{V_0} = -(15 \times 10^9 \text{ Pa})(-0.0010) = 1.5 \times 10^7 \text{ Pa} = 150 \text{ atm}$.

(b) The depth for a pressure increase of $1.5 \times 10^7 \text{ Pa}$ is 1.5 km.

EVALUATE: An extremely large pressure increase is needed for just a 0.10% bone compression, so pressure changes do not appreciably affect the bones. Unprotected dives do not approach a depth of 1.5 km, so bone compression is not a concern for divers.

11.33. **IDENTIFY:** Vigorous downhill hiking produces a shear force on the knee cartilage which could deform the cartilage. The target variable is the angle of deformation of the cartilage.

SET UP: $S = \dfrac{F_\parallel}{A\phi}$, where $\phi = x/h$. $F_\parallel = F \sin 12°$. ϕ is in radians. $F = 8mg$, with $m = 10$ kg. 1 rad $= 180°$.

EXECUTE: $\phi = \dfrac{F_\parallel}{AS} = \dfrac{8mg \sin 12°}{(10 \times 10^{-4} \text{ m}^2)(12 \times 10^6 \text{ Pa})} = 0.1494 \text{ rad} = 8.6°$.

EVALUATE: The shear modulus of cartilage is much less than the values for metals given in Table 11.1 in the text.

11.39. **IDENTIFY and SET UP:** Use Eq. (11.8).

EXECUTE: Tensile stress $= \dfrac{F_\perp}{A} = \dfrac{F_\perp}{\pi r^2} = \dfrac{90.8 \text{ N}}{\pi (0.92 \times 10^{-3} \text{ m})^2} = 3.41 \times 10^7 \text{ Pa}$

EVALUATE: A modest force produces a very large stress because the cross-sectional area is small.

11.41. **IDENTIFY:** The elastic limit is a value of the stress, $F_\perp / A$. Apply $\sum \vec{F} = m\vec{a}$ to the elevator in order to find the tension in the cable.

SET UP: $\dfrac{F_\perp}{A} = \tfrac{1}{3}(2.40 \times 10^8 \text{ Pa}) = 0.80 \times 10^8 \text{ Pa}$. The free-body diagram for the elevator is given in Figure 11.41. $F_\perp$ is the tension in the cable.

EXECUTE: $F_\perp = A(0.80 \times 10^8 \text{ Pa}) = (3.00 \times 10^{-4} \text{ m}^2)(0.80 \times 10^8 \text{ Pa}) = 2.40 \times 10^4 \text{ N}$. $\sum F_y = ma_y$ applied to

the elevator gives $F_\perp - mg = ma$ and $a = \dfrac{F_\perp}{m} - g = \dfrac{2.40 \times 10^4 \text{ N}}{1200 \text{ kg}} - 9.80 \text{ m/s}^2 = 10.2 \text{ m/s}^2$

EVALUATE: The tension in the cable is about twice the weight of the elevator.

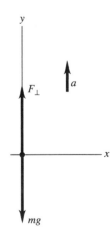

Figure 11.41

11.43. **IDENTIFY:** The center of gravity of the combined object must be at the fulcrum. Use Eq. (11.3) to calculate x_{cm}.

SET UP: The center of gravity of the sand is at the middle of the box. Use coordinates with the origin at the fulcrum and $+x$ to the right. Let $m_1 = 25.0$ kg, so $x_1 = 0.500$ m. Let $m_2 = m_{sand}$, so $x_2 = -0.625$ m. $x_{cm} = 0$.

EXECUTE: $x_{cm} = \dfrac{m_1 x_1 + m_2 x_2}{m_1 + m_2} = 0$ and $m_2 = -m_1 \dfrac{x_1}{x_2} = -(25.0 \text{ kg})\left(\dfrac{0.500 \text{ m}}{-0.625 \text{ m}}\right) = 20.0$ kg.

EVALUATE: The mass of sand required is less than the mass of the plank since the center of the box is farther from the fulcrum than the center of gravity of the plank is.

11.45. **IDENTIFY:** Apply the conditions of equilibrium to the climber. For the minimum coefficient of friction the static friction force has the value $f_s = \mu_s n$.

SET UP: The free-body diagram for the climber is given in Figure 11.45. f_s and n are the vertical and horizontal components of the force exerted by the cliff face on the climber. The moment arm for the force T is $(1.4 \text{ m})\cos 10°$.

EXECUTE: **(a)** $\sum \tau_z = 0$ gives $T(1.4 \text{ m})\cos 10° - w(1.1 \text{ m})\cos 35.0° = 0$.

$T = \dfrac{(1.1 \text{ m})\cos 35.0°}{(1.4 \text{ m})\cos 10°}(82.0 \text{ kg})(9.80 \text{ m/s}^2) = 525$ N

(b) $\sum F_x = 0$ gives $n = T \sin 25.0° = 222$ N. $\sum F_y = 0$ gives $f_s + T\cos 25° - w = 0$ and $f_s = (82.0 \text{ kg})(9.80 \text{ m/s}^2) - (525 \text{ N})\cos 25° = 328$ N.

(c) $\mu_s = \dfrac{f_s}{n} = \dfrac{328 \text{ N}}{222 \text{ N}} = 1.48$

EVALUATE: To achieve this large value of μ_s the climber must wear special rough-soled shoes.

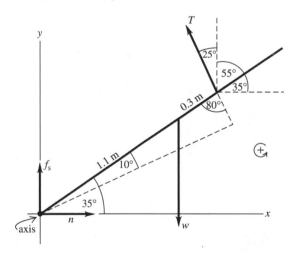

Figure 11.45

11.51. **IDENTIFY:** Apply the conditions of equilibrium to the horizontal beam. Since the two wires are symmetrically placed on either side of the middle of the sign, their tensions are equal and are each equal to $T_w = mg/2 = 137$ N.

SET UP: The free-body diagram for the beam is given in Figure 11.51. F_v and F_h are the horizontal and vertical forces exerted by the hinge on the sign. Since the cable is 2.00 m long and the beam is 1.50 m long, $\cos\theta = \dfrac{1.50 \text{ m}}{2.00 \text{ m}}$ and $\theta = 41.4°$. The tension T_c in the cable has been replaced by its horizontal and vertical components.

EXECUTE: **(a)** $\sum \tau_z = 0$ gives $T_c(\sin 41.4°)(1.50 \text{ m}) - w_{\text{beam}}(0.750 \text{ m}) - T_w(1.50 \text{ m}) - T_w(0.60 \text{ m}) = 0.$

$$T_c = \frac{(12.0 \text{ kg})(9.80 \text{ m/s}^2)(0.750 \text{ m}) + (137 \text{ N})(1.50 \text{ m} + 0.60 \text{ m})}{(1.50 \text{ m})(\sin 41.4°)} = 379 \text{ N}.$$

(b) $\sum F_y = 0$ gives $F_v + T_c \sin 41.4° - w_{\text{beam}} - 2T_w = 0$ and

$F_v = 2T_w + w_{\text{beam}} - T_c \sin 41.4° = 2(137 \text{ N}) + (12.0 \text{ kg})(9.80 \text{ m/s}^2) - (379 \text{ N})(\sin 41.4°) = 141 \text{ N}.$ The hinge must be able to supply a vertical force of 141 N.

EVALUATE: The force from the two wires could be replaced by the weight of the sign acting at a point 0.60 m to the left of the right-hand edge of the sign.

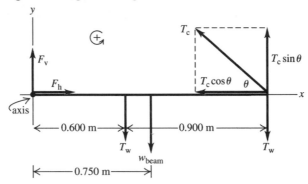

Figure 11.51

11.55.   **IDENTIFY:**   We want to locate the center of mass of the leg-cast system. We can treat each segment of the leg and cast as a point-mass located at its center of mass.
SET UP: The force diagram for the leg is given in Figure 11.55. The weight of each piece acts at the center of mass of that piece. The mass of the upper leg is $m_{\text{ul}} = (0.215)(37 \text{ kg}) = 7.955 \text{ kg}$. The mass of the lower leg is $m_{\text{ll}} = (0.140)(37 \text{ kg}) = 5.18 \text{ kg}$. Use the coordinates shown, with the origin at the hip and the x-axis along the leg, and use $x_{\text{cm}} = \dfrac{x_{\text{ul}}m_{\text{ul}} + x_{\text{ll}}m_{\text{ll}} + x_{\text{cast}}m_{\text{cast}}}{m_{\text{ul}} + m_{\text{ll}} + m_{\text{cast}}}$.

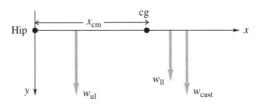

Figure 11.55

EXECUTE: Using $x_{\text{cm}} = \dfrac{x_{\text{ul}}m_{\text{ul}} + x_{\text{ll}}m_{\text{ll}} + x_{\text{cast}}m_{\text{cast}}}{m_{\text{ul}} + m_{\text{ll}} + m_{\text{cast}}}$, we have

$$x_{\text{cm}} = \frac{(18.0 \text{ cm})(7.955 \text{ kg}) + (69.0 \text{ cm})(5.18 \text{ kg}) + (78.0 \text{ cm})(5.50 \text{ kg})}{7.955 \text{ kg} + 5.18 \text{ kg} + 5.50 \text{ kg}} = 49.9 \text{ cm}$$

EVALUATE: The strap is attached to the left of the center of mass of the cast, but it is still supported by the rigid cast since the cast extends beyond its center of mass.

11.57.   **IDENTIFY:**   The leg is not rotating, so the external torques on it must balance.
SET UP: The free-body diagram for the leg is given in Figure 11.57. Take the pivot at the hip joint and let counterclockwise torque be positive. There are also forces on the leg exerted by the hip joint but these forces produce no torque and aren't shown. $\sum \tau_z = 0$ for no rotation.
EXECUTE: **(a)** $\sum \tau_z = 0$ gives $T(10 \text{ cm})(\sin \theta) - w(44 \text{ cm})(\cos \theta) = 0.$

$$T = \frac{4.4w\cos\theta}{\sin\theta} = \frac{4.4w}{\tan\theta} \text{ and for } \theta = 60°, \ T = \frac{4.4(15 \text{ kg})(9.80 \text{ m/s}^2)}{\tan 60°} = 370 \text{ N}.$$

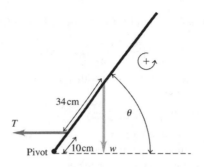

Figure 11.57

(b) For $\theta = 5°$, $T = 7400$ N. The tension is much greater when he just starts to raise his leg off the ground.

(c) $T \to \infty$ as $\theta \to 0$. The person could not raise his leg. If the leg is horizontal so θ is zero, the moment arm for T is zero and T produces no torque to rotate the leg against the torque due to its weight.

EVALUATE: Most of the exercise benefit of leg-raises occurs when the person just starts to raise his legs off the ground.

11.61. **IDENTIFY:** Apply $\sum \tau_z = 0$ to the beam.

SET UP: The free-body diagram for the beam is given in Figure 11.61.

EXECUTE: $\sum \tau_z = 0$, axis at hinge, gives $T(6.0 \text{ m})(\sin 40°) - w(3.75 \text{ m})(\cos 30°) = 0$ and $T = 4900$ N.

EVALUATE: The tension in the cable is less than the weight of the beam. $T \sin 40°$ is the component of T that is perpendicular to the beam.

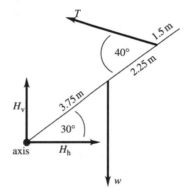

Figure 11.61

11.63. **IDENTIFY:** The amount the tendon stretches depends on Young's modulus for the tendon material. The foot is in rotational equilibrium, so the torques on it balance.

SET UP: $Y = \dfrac{F_T/A}{\Delta l / l_0}$. The foot is in rotational equilibrium, so $\sum \tau_z = 0$.

EXECUTE: **(a)** The free-body diagram for the foot is given in Figure 11.63. T is the tension in the tendon and A is the force exerted on the foot by the ankle. $n = (75 \text{ kg})g$, the weight of the person.

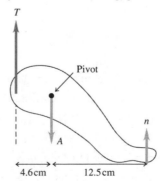

Figure 11.63

(b) Apply $\sum \tau_z = 0$, letting counterclockwise torques be positive and with the pivot at the ankle:

$$T(4.6 \text{ cm}) - n(12.5 \text{ cm}) = 0. \quad T = \left(\frac{12.5 \text{ cm}}{4.6 \text{ cm}}\right)(75 \text{ kg})(9.80 \text{ m/s}^2) = 2000 \text{ N, which is 2.72 times his weight.}$$

(c) The foot pulls downward on the tendon with a force of 2000 N.

$$\Delta l = \left(\frac{F_T}{YA}\right)l_0 = \frac{2000 \text{ N}}{(1470 \times 10^6 \text{ Pa})(78 \times 10^{-6} \text{ m}^2)}(25 \text{ cm}) = 4.4 \text{ mm.}$$

EVALUATE: The tension is quite large, but the Achilles tendon stretches about 4.4 mm, which is only about 1/6 of an inch, so it must be a strong tendon.

11.67. **IDENTIFY:** The torques must balance since the person is not rotating.
SET UP: Figure 11.67a shows the distances and angles. $\theta + \phi = 90°$. $\theta = 56.3°$ and $\phi = 33.7°$. The distances x_1 and x_2 are $x_1 = (90 \text{ cm})\cos\theta = 50.0 \text{ cm}$ and $x_2 = (135 \text{ cm})\cos\phi = 112 \text{ cm}$. The free-body diagram for the person is given in Figure 11.67b. $w_l = 277 \text{ N}$ is the weight of his feet and legs, and $w_t = 473 \text{ N}$ is the weight of his trunk. n_f and f_f are the total normal and friction forces exerted on his feet and n_h and f_h are those forces on his hands. The free-body diagram for his legs is given in Figure 11.67c. F is the force exerted on his legs by his hip joints. For balance, $\sum \tau_z = 0$.

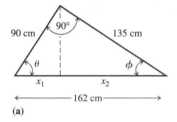

(a)

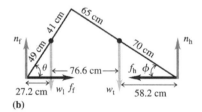

(b)

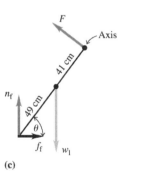

(c)

Figure 11.67

EXECUTE: **(a)** Consider the force diagram of Figure 11.67b. $\sum \tau_z = 0$ with the pivot at his feet and counterclockwise torques positive gives $n_h(162 \text{ cm}) - (277 \text{ N})(27.2 \text{ cm}) - (473 \text{ N})(103.8 \text{ cm}) = 0$.
$n_h = 350 \text{ N}$, so there is a normal force of 175 N at each hand. $n_f + n_h - w_l - w_t = 0$ so
$n_f = w_l + w_t - n_h = 750 \text{ N} - 350 \text{ N} = 400 \text{ N}$, so there is a normal force of 200 N at each foot.
(b) Consider the force diagram of Figure 11.67c. $\sum \tau_z = 0$ with the pivot at his hips and counterclockwise torques positive gives $f_f(74.9 \text{ cm}) + w_l(22.8 \text{ cm}) - n_f(50.0 \text{ cm}) = 0$.

$$f_f = \frac{(400 \text{ N})(50.0 \text{ cm}) - (277 \text{ N})(22.8 \text{ cm})}{74.9 \text{ cm}} = 182.7 \text{ N}. \text{ There is a friction force of 91 N at each foot.}$$

$\sum F_x = 0$ in Figure 11.67b gives $f_h = f_f$, so there is a friction force of 91 N at each hand.

EVALUATE: In this position the normal forces at his feet and at his hands don't differ very much.

11.69. **IDENTIFY:** Apply the equilibrium conditions to the crate. When the crate is on the verge of tipping it touches the floor only at its lower left-hand corner and the normal force acts at this point. The minimum coefficient of static friction is given by the equation $f_s = \mu_s n$.

SET UP: The free-body diagram for the crate when it is ready to tip is given in Figure 11.69.

EXECUTE: **(a)** $\sum \tau_z = 0$ gives $P(1.50 \text{ m})\sin 53.0° - w(1.10 \text{ m}) = 0$.

$$P = w\left(\frac{1.10 \text{ m}}{[1.50 \text{ m}][\sin 53.0°]}\right) = 1.15 \times 10^3 \text{ N}$$

(b) $\sum F_y = 0$ gives $n - w - P\cos 53.0° = 0$.

$n = w + P\cos 53.0° = 1250 \text{ N} + (1.15 \times 10^3 \text{ N})\cos 53° = 1.94 \times 10^3 \text{ N}$

(c) $\sum F_y = 0$ gives $f_s = P\sin 53.0° = (1.15 \times 10^3 \text{ N})\sin 53.0° = 918 \text{ N}.$

(d) $\mu_s = \dfrac{f_s}{n} = \dfrac{918 \text{ N}}{1.94 \times 10^3 \text{ N}} = 0.473$

EVALUATE: The normal force is greater than the weight because P has a downward component.

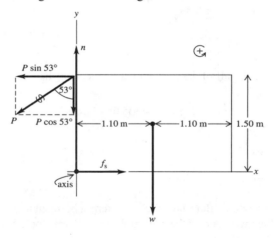

Figure 11.69

11.71. **IDENTIFY:** Apply the first and second conditions of equilibrium to the crate.
SET UP: The free-body diagram for the crate is given in Figure 11.71.

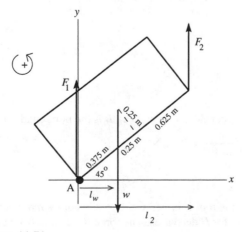

Figure 11.71

$l_w = (0.375 \text{ m})\cos 45°$

$l_2 = (1.25 \text{ m})\cos 45°$

Let $\vec{F}_1$ and $\vec{F}_2$ be the vertical forces exerted by you and your friend. Take the origin at the lower left-hand corner of the crate (point A).

EXECUTE: $\sum F_y = ma_y$ gives $F_1 + F_2 - w = 0$

$F_1 + F_2 = w = (200 \text{ kg})(9.80 \text{ m/s}^2) = 1960 \text{ N}$

$\sum \tau_A = 0$ gives $F_2 l_2 - w l_w = 0$

$F_2 = w\left(\dfrac{l_w}{l_2}\right) = 1960 \text{ N}\left(\dfrac{0.375 \text{ m} \cos 45°}{1.25 \text{ m} \cos 45°}\right) = 590 \text{ N}$

Then $F_1 = w - F_2 = 1960 \text{ N} - 590 \text{ N} = 1370 \text{ N}$.

EVALUATE: The person below (you) applies a force of 1370 N. The person above (your friend) applies a force of 590 N. It is better to be the person above. As the sketch shows, the moment arm for $\vec{F}_1$ is less than for $\vec{F}_2$, so must have $F_1 > F_2$ to compensate.

11.75.   **IDENTIFY:**   Apply $\sum \tau_z = 0$ first to the roof and then to one wall.

(a) SET UP: Consider the forces on the roof; see Figure 11.75a.

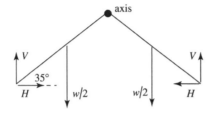

V and H are the vertical and horizontal forces each wall exerts on the roof.
$w = 20{,}000 \text{ N}$ is the total weight of the roof.
$2V = w$ so $V = w/2$

Figure 11.75a

Apply $\sum \tau_z = 0$ to one half of the roof, with the axis along the line where the two halves join. Let each half have length L.

EXECUTE: $(w/2)(L/2)(\cos 35.0°) + HL \sin 35.0° - VL \cos 35° = 0$

L divides out, and use $V = w/2$

$H \sin 35.0° = \frac{1}{4} w \cos 35.0°$

$H = \dfrac{w}{4 \tan 35.0°} = 7140 \text{ N}$

EVALUATE: By Newton's third law, the roof exerts a horizontal, outward force on the wall. For torque about an axis at the lower end of the wall, at the ground, this force has a larger moment arm and hence larger torque the taller the walls.

(b) SET UP: The force diagram for one wall is given in Figure 11.75b.

Consider the torques on this wall.

Figure 11.75b

H is the horizontal force exerted by the roof, as considered in part (a). B is the horizontal force exerted by the buttress. Now the angle is $40°$, so $H = \dfrac{w}{4 \tan 40°} = 5959 \text{ N}$.

EXECUTE: $\sum \tau_z = 0$, axis at the ground

$H(40 \text{ m}) - B(30 \text{ m}) = 0$ and $B = 7900 \text{ N}$.

EVALUATE: The horizontal force exerted by the roof is larger as the roof becomes more horizontal, since for torques applied to the roof the moment arm for H decreases. The force B required from the buttress is less the higher up on the wall this force is applied.

11.79.   **IDENTIFY:**   Apply the first and second conditions of equilibrium, first to both marbles considered as a composite object and then to the bottom marble.
(a) SET UP: The forces on each marble are shown in Figure 11.79.

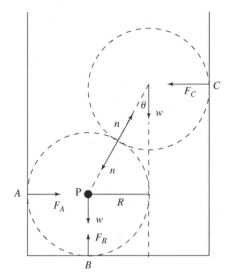

EXECUTE:
$F_B = 2w = 1.47$ N
$\sin\theta = R/2R$ so $\theta = 30°$
$\sum \tau_z = 0,$ axis at P
$F_C(2R\cos\theta) - wR = 0$

$$F_C = \frac{mg}{2\cos 30°} = 0.424 \text{ N}$$

$F_A = F_C = 0.424$ N

Figure 11.79

(b) Consider the forces on the bottom marble. The horizontal forces must sum to zero, so $F_A = n\sin\theta$.

$$n = \frac{F_A}{\sin 30°} = 0.848 \text{ N}$$

Could use instead that the vertical forces sum to zero

$$F_B - mg - n\cos\theta = 0$$

$$n = \frac{F_B - mg}{\cos 30°} = 0.848 \text{ N, which checks.}$$

EVALUATE: If we consider each marble separately, the line of action of every force passes through the center of the marble so there is clearly no torque about that point for each marble. We can use the results we obtained to show that $\sum F_x = 0$ and $\sum F_y = 0$ for the top marble.

11.81.   **IDENTIFY:**   Apply the first and second conditions of equilibrium to the bale.
(a) SET UP: Find the angle where the bale starts to tip. When it starts to tip only the lower left-hand corner of the bale makes contact with the conveyor belt. Therefore the line of action of the normal force n passes through the left-hand edge of the bale. Consider $\sum \tau_z = 0$ with point A at the lower left-hand corner. Then $\tau_n = 0$ and $\tau_f = 0$, so it must be that $\tau_{mg} = 0$ also. This means that the line of action of the gravity must pass through point A. Thus the free-body diagram must be as shown in Figure 11.81a.

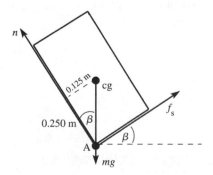

EXECUTE: $\tan\beta = \dfrac{0.125 \text{ m}}{0.250 \text{ m}}$

$\beta = 27°$, angle where tips

Figure 11.81a

SET UP: At the angle where the bale is ready to slip down the incline f_s has its maximum possible value, $f_s = \mu_s n$. The free-body diagram for the bale, with the origin of coordinates at the cg is given in Figure 11.81b.

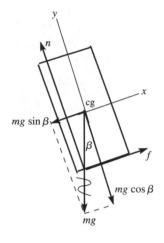

EXECUTE:

$\sum F_y = ma_y$

$n - mg \cos \beta = 0$

$n = mg \cos \beta$

$f_s = \mu_s mg \cos \beta$

(f_s has maximum value when bale ready to slip)

$\sum F_x = ma_x$

$f_s - mg \sin \beta = 0$

$\mu_s mg \cos \beta - mg \sin \beta = 0$

$\tan \beta = \mu_s$

$\mu_s = 0.60$ gives that $\beta = 31°$

Figure 11.81b

$\beta = 27°$ to tip; $\beta = 31°$ to slip, so tips first

(b) The magnitude of the friction force didn't enter into the calculation of the tipping angle; still tips at $\beta = 27°$. For $\mu_s = 0.40$ slips at $\beta = \arctan(0.40) = 22°$.

Now the bale will start to slide down the incline before it tips.

EVALUATE: With a smaller μ_s the slope angle β where the bale slips is smaller.

11.83.   **IDENTIFY:** Apply the first and second conditions of equilibrium to the door.

(a) SET UP: The free-body diagram for the door is given in Figure 11.83.

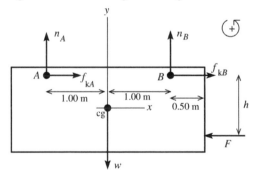

Figure 11.83

Take the origin of coordinates at the center of the door (at the cg). Let n_A, f_{kA}, n_B and f_{kB} be the normal and friction forces exerted on the door at each wheel.

EXECUTE: $\sum F_y = ma_y$

$n_A + n_B - w = 0$

$n_A + n_B = w = 950 \text{ N}$

$\sum F_x = ma_x$

$f_{kA} + f_{kB} - F = 0$

$F = f_{kA} + f_{kB}$

$f_{kA} = \mu_k n_A$, $f_{kB} = \mu_k n_B$, so $F = \mu_k(n_A + n_B) = \mu_k w = (0.52)(950 \text{ N}) = 494 \text{ N}$

$\sum \tau_B = 0$

n_B, f_{kA} and f_{kB} all have zero moment arms and hence zero torque about this point.

Thus $+w(1.00 \text{ m}) - n_A(2.00 \text{ m}) - F(h) = 0$

$$n_A = \frac{w(1.00 \text{ m}) - F(h)}{2.00 \text{ m}} = \frac{(950 \text{ N})(1.00 \text{ m}) - (494 \text{ N})(1.60 \text{ m})}{2.00 \text{ m}} = 80 \text{ N}$$

And then $n_B = 950 \text{ N} - n_A = 950 \text{ N} - 80 \text{ N} = 870 \text{ N}$.

(b) SET UP: If h is too large the torque of F will cause wheel A to leave the track. When wheel A just starts to lift off the track n_A and f_{kA} both go to zero.

EXECUTE: The equations in part (a) still apply.

$n_A + n_B - w = 0$ gives $n_B = w = 950 \text{ N}$

Then $f_{kB} = \mu_k n_B = 0.52(950 \text{ N}) = 494 \text{ N}$

$F = f_{kA} + f_{kB} = 494 \text{ N}$

$+w(1.00 \text{ m}) - n_A(2.00 \text{ m}) - F(h) = 0$

$$h = \frac{w(1.00 \text{ m})}{F} = \frac{(950 \text{ N})(1.00 \text{ m})}{494 \text{ N}} = 1.92 \text{ m}$$

EVALUATE: The result in part (b) is larger than the value of h in part (a). Increasing h increases the clockwise torque about B due to F and therefore decreases the clockwise torque that n_A must apply.

11.85.   **IDENTIFY:** Apply the first and second conditions of equilibrium to the pole.

(a) SET UP: The free-body diagram for the pole is given in Figure 11.85.

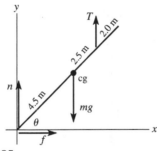

n and f are the vertical and horizontal components of the force the ground exerts on the pole.

$\sum F_x = ma_x$

$f = 0$

The force exerted by the ground has no horizontal component.

Figure 11.85

EXECUTE: $\sum \tau_A = 0$

$+T(7.0 \text{ m})\cos\theta - mg(4.5 \text{ m})\cos\theta = 0$

$T = mg(4.5 \text{ m}/7.0 \text{ m}) = (4.5/7.0)(5700 \text{ N}) = 3700 \text{ N}$

$\sum F_y = 0$

$n + T - mg = 0$

$n = mg - T = 5700 \text{ N} - 3700 \text{ N} = 2000 \text{ N}$

The force exerted by the ground is vertical (upward) and has magnitude 2000 N.

EVALUATE: We can verify that $\sum \tau_z = 0$ for an axis at the cg of the pole. $T > n$ since T acts at a point closer to the cg and therefore has a smaller moment arm for this axis than n does.

(b) In the $\sum \tau_A = 0$ equation the angle θ divided out. All forces on the pole are vertical and their moment arms are all proportional to $\cos\theta$.

11.89.   **IDENTIFY:** Apply Newton's second law to the mass to find the tension in the wire. Then apply Eq. (11.10) to the wire to find the elongation this tensile force produces.

(a) SET UP: Calculate the tension in the wire as the mass passes through the lowest point. The free-body diagram for the mass is given in Figure 11.89a.

The mass moves in an arc of a
circle with radius $R = 0.50$ m.
It has acceleration $\vec{a}_{\text{rad}}$ directed
in toward the center of the circle,
so at this point $\vec{a}_{\text{rad}}$ is upward.

Figure 11.89 a

EXECUTE: $\sum F_y = ma_y$

$T - mg = mR\omega^2$ so that $T = m(g + R\omega^2)$.

But ω must be in rad/s:

$\omega = (120 \text{ rev/min})(2\pi \text{ rad/1 rev})(1 \text{ min/60 s}) = 12.57 \text{ rad/s}.$

Then $T = (12.0 \text{ kg})(9.80 \text{ m/s}^2 + (0.50 \text{ m})(12.57 \text{ rad/s})^2) = 1066 \text{ N}.$

Now calculate the elongation Δl of the wire that this tensile force produces:

$Y = \dfrac{F_\perp l_0}{A\Delta l}$ so $\Delta l = \dfrac{F_\perp l_0}{YA} = \dfrac{(1066 \text{ N})(0.50 \text{ m})}{(7.0\times10^{10} \text{ Pa})(0.014\times10^{-4} \text{ m}^2)} = 0.54 \text{ cm}.$

(b) SET UP: The acceleration $\vec{a}_{\text{rad}}$ is directed in toward the center of the circular path, and at this point in
the motion this direction is downward. The free-body diagram is given in Figure 11.89b.

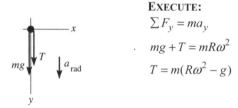

EXECUTE:

$\sum F_y = ma_y$

$mg + T = mR\omega^2$

$T = m(R\omega^2 - g)$

Figure 11.89 b

$T = (12.0 \text{ kg})((0.50 \text{ m})(12.57 \text{ rad/s})^2 - 9.80 \text{ m/s}^2) = 830 \text{ N}$

$\Delta l = \dfrac{F_\perp l_0}{YA} = \dfrac{(830 \text{ N})(0.50 \text{ m})}{(7.0\times10^{10} \text{ Pa})(0.014\times10^{-4} \text{ m}^2)} = 0.42 \text{ cm}.$

EVALUATE: At the lowest point T and w are in opposite directions and at the highest point they are in the
same direction, so T is greater at the lowest point and the elongation is greatest there. The elongation is at
most 1% of the length.

11.91.   **IDENTIFY:**   Use the second condition of equilibrium to relate the tension in the two wires to the distance w
is from the left end. Use Eqs. (11.8) and (11.10) to relate the tension in each wire to its stress and strain.
(a) SET UP: stress $= F_\perp/A$, so equal stress implies T/A same for each wire.

$T_A/2.00 \text{ mm}^2 = T_B/4.00 \text{ mm}^2$ so $T_B = 2.00T_A$

The question is where along the rod to hang the weight in order to produce this relation between the
tensions in the two wires. Let the weight be suspended at point C, a distance x to the right of wire A. The
free-body diagram for the rod is given in Figure 11.91.

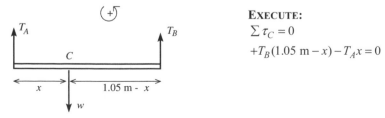

EXECUTE:

$\sum \tau_C = 0$

$+T_B(1.05 \text{ m} - x) - T_A x = 0$

Figure 11.91

But $T_B = 2.00 T_A$ so $2.00 T_A (1.05 \text{ m} - x) - T_A x = 0$

$2.10 \text{ m} - 2.00x = x$ and $x = 2.10 \text{ m}/3.00 = 0.70 \text{ m}$ (measured from A).

(b) SET UP: $Y = \text{stress/strain}$ gives that $\text{strain} = \text{stress}/Y = F_\perp / AY$.

EXECUTE: Equal strain thus implies

$$\frac{T_A}{(2.00 \text{ mm}^2)(1.80 \times 10^{11} \text{ Pa})} = \frac{T_B}{(4.00 \text{ mm}^2)(1.20 \times 10^{11} \text{ Pa})}$$

$$T_B = \left(\frac{4.00}{2.00}\right)\left(\frac{1.20}{1.80}\right) T_A = 1.333 T_A.$$

The $\sum \tau_C = 0$ equation still gives $T_B(1.05 \text{ m} - x) - T_A x = 0$.

But now $T_B = 1.333 T_A$ so $(1.333 T_A)(1.05 \text{ m} - x) - T_A x = 0$.

$1.40 \text{ m} = 2.33x$ and $x = 1.40 \text{ m}/2.33 = 0.60 \text{ m}$ (measured from A).

EVALUATE: Wire B has twice the diameter so it takes twice the tension to produce the same stress. For equal stress the moment arm for T_B (0.35 m) is half that for T_A (0.70 m), since the torques must be equal.

The smaller Y for B partially compensates for the larger area in determining the strain and for equal strain the moment arms are closer to being equal.

11.95.   **IDENTIFY:** Apply Eq. (11.13) and calculate ΔV.

SET UP: The pressure increase is w/A, where w is the weight of the bricks and A is the area πr^2 of the piston.

EXECUTE: $\Delta p = \dfrac{(1420 \text{ kg})(9.80 \text{ m/s}^2)}{\pi(0.150 \text{ m})^2} = 1.97 \times 10^5 \text{ Pa}$

$\Delta p = -B \dfrac{\Delta V}{V_0}$ gives $\Delta V = -\dfrac{(\Delta p)V_0}{B} = -\dfrac{(1.97 \times 10^5 \text{ Pa})(250 \text{ L})}{9.09 \times 10^8 \text{ Pa}} = -0.0542 \text{ L}$

EVALUATE: The fractional change in volume is only 0.022%, so this attempt is not worth the effort.

12

FLUID MECHANICS

12.5. **IDENTIFY:** Apply $\rho = m/V$ to relate the densities and volumes for the two spheres.

SET UP: For a sphere, $V = \frac{4}{3}\pi r^3$. For lead, $\rho_1 = 11.3 \times 10^3$ kg/m^3 and for aluminum, $\rho_a = 2.7 \times 10^3$ kg/m^3.

EXECUTE: $m = \rho V = \frac{4}{3}\pi r^3 \rho$. Same mass means $r_a^3 \rho_a = r_1^3 \rho_1$. $\dfrac{r_a}{r_1} = \left(\dfrac{\rho_1}{\rho_a}\right)^{1/3} = \left(\dfrac{11.3 \times 10^3}{2.7 \times 10^3}\right)^{1/3} = 1.6$.

EVALUATE: The aluminum sphere is larger, since its density is less.

12.7. **IDENTIFY:** $w = mg$ and $m = \rho V$. Find the volume V of the pipe.

SET UP: For a hollow cylinder with inner radius R_1, outer radius R_2, and length L the volume is $V = \pi(R_2^2 - R_1^2)L$. $R_1 = 1.25 \times 10^{-2}$ m and $R_2 = 1.75 \times 10^{-2}$ m.

EXECUTE: $V = \pi([0.0175 \text{ m}]^2 - [0.0125 \text{ m}]^2)(1.50 \text{ m}) = 7.07 \times 10^{-4}$ m^3.

$m = \rho V = (8.9 \times 10^3 \text{ kg/m}^3)(7.07 \times 10^{-4} \text{ m}^3) = 6.29$ kg. $w = mg = 61.6$ N.

EVALUATE: The pipe weights about 14 pounds.

12.9. **IDENTIFY:** The gauge pressure $p - p_0$ at depth h is $p - p_0 = \rho g h$.

SET UP: Freshwater has density 1.00×10^3 kg/m^3 and seawater has density 1.03×10^3 kg/m^3.

EXECUTE: **(a)** $p - p_0 = (1.00 \times 10^3 \text{ kg/m}^3)(3.71 \text{ m/s}^2)(500 \text{ m}) = 1.86 \times 10^6$ Pa.

(b) $h = \dfrac{p - p_0}{\rho g} = \dfrac{1.86 \times 10^6 \text{ Pa}}{(1.03 \times 10^3 \text{ kg/m}^3)(9.80 \text{ m/s}^2)} = 184$ m

EVALUATE: The pressure at a given depth is greater on earth because a cylinder of water of that height weighs more on earth than on Mars.

12.11. **IDENTIFY:** Apply $p = p_0 + \rho g h$.

SET UP: Gauge pressure is $p - p_{\text{air}}$.

EXECUTE: The pressure difference between the top and bottom of the tube must be at least 5980 Pa in order to force fluid into the vein: $\rho g h = 5980$ Pa and

$$h = \frac{5980 \text{ Pa}}{\rho h} = \frac{5980 \text{ N/m}^2}{(1050 \text{ kg/m}^3)(9.80 \text{ m/s}^2)} = 0.581 \text{ m}.$$

EVALUATE: The bag of fluid is typically hung from a vertical pole to achieve this height above the patient's arm.

12.15. **IDENTIFY:** The external pressure on the eardrum increases with depth in the ocean. This increased pressure could damage the eardrum.

SET UP: The density of seawater is 1.03×10^3 kg/m^3. The area of the eardrum is $A = \pi r^2$, with $r = 4.1$ mm. The pressure increase with depth is $\Delta p = \rho g h$ and $F = pA$.

EXECUTE: $\Delta F = (\Delta p)A = \rho g h A$. Solving for h gives

$$h = \frac{\Delta F}{\rho g A} = \frac{1.5 \text{ N}}{(1.03 \times 10^3 \text{ kg/m}^3)(9.80 \text{ m/s}^2)\pi(4.1 \times 10^{-3} \text{ m})^2} = 2.8 \text{ m}.$$

EVALUATE: 2.8 m is less than 10 ft, so it is probably a good idea to wear ear plugs if you scuba dive.

12.17. IDENTIFY: Apply $p = p_0 + \rho g h$.

SET UP: For water, $\rho = 1.00 \times 10^3 \text{ kg/m}^3$.

EXECUTE: $p - p_{\text{air}} = \rho g h = (1.00 \times 10^3 \text{ kg/m}^3)(9.80 \text{ m/s}^2)(6.1 \text{ m}) = 6.0 \times 10^4 \text{ Pa}$.

EVALUATE: The pressure difference increases linearly with depth.

12.19. IDENTIFY: $p = p_0 + \rho g h$. $F = pA$.

SET UP: For seawater, $\rho = 1.03 \times 10^3 \text{ kg/m}^3$

EXECUTE: The force F that must be applied is the difference between the upward force of the water and the downward forces of the air and the weight of the hatch. The difference between the pressure inside and out is the gauge pressure, so

$$F = (\rho g h) A - w = (1.03 \times 10^3 \text{ kg/m}^3)(9.80 \text{ m/s}^2)(30 \text{ m})(0.75 \text{ m}^2) - 300 \text{ N} = 2.27 \times 10^5 \text{ N}.$$

EVALUATE: The force due to the gauge pressure of the water is much larger than the weight of the hatch and would be impossible for the crew to apply it just by pushing.

12.21. IDENTIFY: The gauge pressure at the top of the oil column must produce a force on the disk that is equal to its weight.

SET UP: The area of the bottom of the disk is $A = \pi r^2 = \pi(0.150 \text{ m})^2 = 0.0707 \text{ m}^2$.

EXECUTE: **(a)** $p - p_0 = \dfrac{w}{A} = \dfrac{45.0 \text{ N}}{0.0707 \text{ m}^2} = 636 \text{ Pa}$.

(b) The increase in pressure produces a force on the disk equal to the increase in weight. By Pascal's law the increase in pressure is transmitted to all points in the oil.

(i) $\Delta p = \dfrac{83.0 \text{ N}}{0.0707 \text{ m}^2} = 1170 \text{ Pa}$. (ii) 1170 Pa

EVALUATE: The absolute pressure at the top of the oil produces an upward force on the disk but this force is partially balanced by the force due to the air pressure at the top of the disk.

12.23. IDENTIFY: $F_2 = \dfrac{A_2}{A_1} F_1$. F_2 must equal the weight $w = mg$ of the car.

SET UP: $A = \pi D^2/4$. D_1 is the diameter of the vessel at the piston where F_1 is applied and D_2 of the diameter at the car.

EXECUTE: $mg = \dfrac{\pi D_2^2/4}{\pi D_1^2/4} F_1$. $\dfrac{D_2}{D_1} = \sqrt{\dfrac{mg}{F_1}} = \sqrt{\dfrac{(1520 \text{ kg})(9.80 \text{ m/s}^2)}{125 \text{ N}}} = 10.9$

EVALUATE: The diameter is smaller where the force is smaller, so the pressure will be the same at both pistons.

12.27. IDENTIFY: Apply $\Sigma F_y = ma_y$ to the sample, with $+y$ upward. $B = \rho_{\text{water}} V_{\text{obj}} g$.

SET UP: $w = mg = 17.50 \text{ N}$ and $m = 1.79 \text{ kg}$.

EXECUTE: $T + B - mg = 0$. $B = mg - T = 17.50 \text{ N} - 11.20 \text{ N} = 6.30 \text{ N}$.

$$V_{\text{obj}} = \frac{B}{\rho_{\text{water}} g} = \frac{6.30 \text{ N}}{(1.00 \times 10^3 \text{ kg/m}^3)(9.80 \text{ m/s}^2)} = 6.43 \times 10^{-4} \text{ m}^3.$$

$$\rho = \frac{m}{V} = \frac{1.79 \text{ kg}}{6.43 \times 10^{-4} \text{ m}^3} = 2.78 \times 10^3 \text{ kg/m}^3.$$

EVALUATE: The density of the sample is greater than that of water and it doesn't float.

12.29. IDENTIFY: For a floating object, the weight of the object equals the upward buoyancy force, B, exerted by the fluid.

SET UP: $B = \rho_{\text{fluid}} V_{\text{submerged}} g$. The weight of the object can be written as $w = \rho_{\text{object}} V_{\text{object}} g$. For seawater, $\rho = 1.03 \times 10^3 \text{ kg/m}^3$.

EXECUTE: **(a)** The displaced fluid has less volume than the object but must weigh the same, so $\rho < \rho_{\text{fluid}}$.

(b) If the ship does not leak, much of the water will be displaced by air or cargo, and the average density of the floating ship is less than that of water.

(c) Let the portion submerged have volume V, and the total volume be V_0. Then $\rho V_0 = \rho_{\text{fluid}} V$, so $\frac{V}{V_0} = \frac{\rho}{\rho_{\text{fluid}}}$. The fraction above the fluid surface is then $1 - \frac{\rho}{\rho_{\text{fluid}}}$. If $\rho \to 0$, the entire object floats, and if $\rho \to \rho_{\text{fluid}}$, none of the object is above the surface.

(d) Using the result of part (c), $1 - \frac{\rho}{\rho_{\text{fluid}}} = 1 - \frac{(0.042 \text{ kg})/([5.0][4.0][3.0] \times 10^{-6} \text{m}^3)}{1030 \text{kg/m}^3} = 0.32 = 32\%$.

EVALUATE: For a given object, the fraction of the object above the surface increases when the density of the fluid in which it floats increases.

12.31. **IDENTIFY** and **SET UP:** Use Eq. (12.8) to calculate the gauge pressure at the two depths.
(a) The distances are shown in Figure 12.31a.

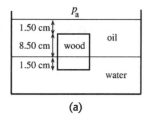

EXECUTE: $p - p_0 = \rho g h$

The upper face is 1.50 cm below the top of the oil, so

$p - p_0 = (790 \text{ kg/m}^3)(9.80 \text{ m/s}^2)(0.0150 \text{ m})$

$p - p_0 = 116 \text{ Pa}$

Figure 12.31a

(b) The pressure at the interface is $p_{\text{interface}} = p_a + \rho_{\text{oil}} g (0.100 \text{ m})$. The lower face of the block is 1.50 cm below the interface, so the pressure there is $p = p_{\text{interface}} + \rho_{\text{water}} g (0.0150 \text{ m})$. Combining these two equations gives

$p - p_a = \rho_{\text{oil}} g (0.100 \text{ m}) + \rho_{\text{water}} g (0.0150 \text{ m})$

$p - p_a = [(790 \text{ kg/m}^3)(0.100 \text{ m}) + (1000 \text{ kg/m}^3)(0.0150 \text{ m})](9.80 \text{ m/s}^2)$

$p - p_a = 921 \text{ Pa}$

(c) **IDENTIFY** and **SET UP:** Consider the forces on the block. The area of each face of the block is $A = (0.100 \text{ m})^2 = 0.0100 \text{ m}^2$. Let the absolute pressure at the top face be p_t and the pressure at the bottom face be p_b. In Eq. (12.3) use these pressures to calculate the force exerted by the fluids at the top and bottom of the block. The free-body diagram for the block is given in Figure 12.31b.

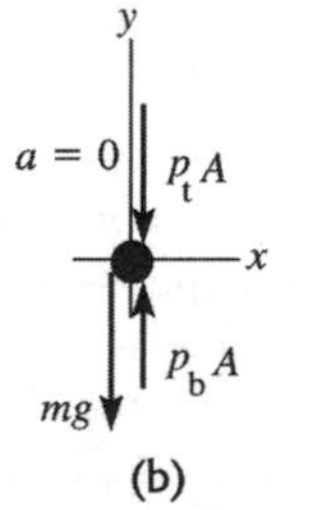

EXECUTE: $\Sigma F_y = m a_y$

$p_b A - p_t A - mg = 0$

$(p_b - p_t) A = mg$

Figure 12.31b

Note that $(p_b - p_t) = (p_b - p_a) - (p_t - p_a) = 921 \text{ Pa} - 116 \text{ Pa} = 805 \text{ Pa}$; the difference in absolute pressures equals the difference in gauge pressures.

$$m = \frac{(p_b - p_t) A}{g} = \frac{(805 \text{ Pa})(0.0100 \text{ m}^2)}{9.80 \text{ m/s}^2} = 0.821 \text{ kg}.$$

And then $\rho = m/V = 0.821 \text{ kg}/(0.100 \text{ m})^3 = 821 \text{ kg/m}^3$.

EVALUATE: We can calculate the buoyant force as $B = (\rho_{oil}V_{oil} + \rho_{water}V_{water})g$ where

$V_{oil} = (0.0100 \text{ m}^2)(0.0850 \text{ m}) = 8.50 \times 10^{-4} \text{ m}^3$ is the volume of oil displaced by the block and

$V_{water} = (0.0100 \text{ m}^2)(0.0150 \text{ m}) = 1.50 \times 10^{-4} \text{ m}^3$ is the volume of water displaced by the block. This gives

$B = (0.821 \text{ kg})g$. The mass of water displaced equals the mass of the block.

12.33. **IDENTIFY:** The vertical forces on the rock sum to zero. The buoyant force equals the weight of liquid displaced by the rock. $V = \frac{4}{3}\pi R^3$.

SET UP: The density of water is $1.00 \times 10^3 \text{ kg/m}^3$.

EXECUTE: The rock displaces a volume of water whose weight is $39.2 \text{ N} - 28.4 \text{ N} = 10.8 \text{ N}$. The mass of this much water is thus $10.8 \text{ N}/(9.80 \text{ m/s}^2) = 1.102 \text{ kg}$ and its volume, equal to the rock's volume, is

$\dfrac{1.102 \text{ kg}}{1.00 \times 10^3 \text{ kg/m}^3} = 1.102 \times 10^{-3} \text{ m}^3$. The weight of unknown liquid displaced is $39.2 \text{ N} - 18.6 \text{ N} = 20.6 \text{ N}$,

and its mass is $20.6 \text{ N}/(9.80 \text{ m/s}^2) = 2.102 \text{ kg}$. The liquid's density is thus

$2.102 \text{ kg}/(1.102 \times 10^{-3} \text{ m}^3) = 1.91 \times 10^3 \text{ kg/m}^3$.

EVALUATE: The density of the unknown liquid is roughly twice the density of water.

12.35. **IDENTIFY:** Apply the equation of continuity, $v_1 A_1 = v_2 A_2$.

SET UP: $A = \pi r^2$

EXECUTE: $v_2 = v_1(A_1/A_2)$. $A_1 = \pi(0.80 \text{ cm})^2$, $A_2 = 20\pi(0.10 \text{ cm})^2$. $v_2 = (3.0 \text{ m/s})\dfrac{\pi(0.80)^2}{20\pi(0.10)^2} = 9.6 \text{ m/s}$.

EVALUATE: The total area of the shower head openings is less than the cross-sectional area of the pipe, and the speed of the water in the shower head opening is greater than its speed in the pipe.

12.37. **IDENTIFY** and **SET UP:** Apply Eq. (12.10). In part (a) the target variable is V. In part (b) solve for A and then from that get the radius of the pipe.

EXECUTE: **(a)** $vA = 1.20 \text{ m}^3/\text{s}$

$v = \dfrac{1.20 \text{ m}^3/\text{s}}{A} = \dfrac{1.20 \text{ m}^3/\text{s}}{\pi r^2} = \dfrac{1.20 \text{ m}^3/\text{s}}{\pi(0.150 \text{ m})^2} = 17.0 \text{ m/s}$

(b) $vA = 1.20 \text{ m}^3/\text{s}$

$v\pi r^2 = 1.20 \text{ m}^3/\text{s}$

$r = \sqrt{\dfrac{1.20 \text{ m}^3/\text{s}}{v\pi}} = \sqrt{\dfrac{1.20 \text{ m}^3/\text{s}}{(3.80 \text{ m/s})\pi}} = 0.317 \text{ m}$

EVALUATE: The speed is greater where the area and radius are smaller.

12.41. **IDENTIFY** and **SET UP:**

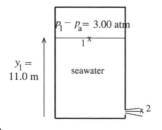

Apply Bernoulli's equation with points 1 and 2 chosen as shown in Figure 12.41. Let $y = 0$ at the bottom of the tank so $y_1 = 11.0 \text{ m}$ and $y_2 = 0$. The target variable is v_2.

Figure 12.41

$p_1 + \rho g y_1 + \frac{1}{2}\rho v_1^2 = p_2 + \rho g y_2 + \frac{1}{2}\rho v_2^2$

$A_1 v_1 = A_2 v_2$, so $v_1 = (A_2/A_1)v_2$. But the cross-sectional area of the tank (A_1) is much larger than the cross-sectional area of the hole (A_2), so $v_1 \ll v_2$ and the $\frac{1}{2}\rho v_1^2$ term can be neglected.

EXECUTE: This gives $\frac{1}{2}\rho v_2^2 = (p_1 - p_2) + \rho g y_1$.

Use $p_2 = p_a$ and solve for v_2:

$$v_2 = \sqrt{2(p_1 - p_a)/\rho + 2gy_1} = \sqrt{\frac{2(3.039\times10^5 \text{ Pa})}{1030 \text{ kg/m}^3} + 2(9.80 \text{ m/s}^2)(11.0 \text{ m})}$$

$v_2 = 28.4$ m/s

EVALUATE: If the pressure at the top surface of the water were air pressure, then Toricelli's theorem (Example: 12.8) gives $v_2 = \sqrt{2g(y_1 - y_2)} = 14.7$ m/s. The actual afflux speed is much larger than this due to the excess pressure at the top of the tank.

12.43.   **IDENTIFY** and **SET UP:**

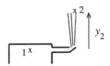

Apply Bernoulli's equation to points 1 and 2 as shown in Figure 12.43. Point 1 is in the mains and point 2 is at the maximum height reached by the stream, so $v_2 = 0$.

Figure 12.43

Solve for p_1 and then convert this absolute pressure to gauge pressure.

EXECUTE: $p_1 + \rho g y_1 + \frac{1}{2}\rho v_1^2 = p_2 + \rho g y_2 + \frac{1}{2}\rho v_2^2$

Let $y_1 = 0$, $y_2 = 15.0$ m. The mains have large diameter, so $v_1 \approx 0$.

Thus $p_1 = p_2 + \rho g y_2$.

But $p_2 = p_a$, so $p_1 - p_a = \rho g y_2 = (1000 \text{ kg/m}^3)(9.80 \text{ m/s}^2)(15.0 \text{ m}) = 1.47\times10^5$ Pa.

EVALUATE: This is the gauge pressure at the bottom of a column of water 15.0 m high.

12.47.   **IDENTIFY** and **SET UP:** Let point 1 be where $r_1 = 4.00$ cm and point 2 be where $r_2 = 2.00$ cm. The volume flow rate vA has the value 7200 cm^3/s at all points in the pipe. Apply Eq. (12.10) to find the fluid speed at points 1 and 2 and then use Bernoulli's equation for these two points to find p_2.

EXECUTE: $v_1 A_1 = v_1 \pi r_1^2 = 7200$ cm^3, so $v_1 = 1.43$ m/s

$v_2 A_2 = v_2 \pi r_2^2 = 7200$ cm^3/s, so $v_2 = 5.73$ m/s

$p_1 + \rho g y_1 + \frac{1}{2}\rho v_1^2 = p_2 + \rho g y_2 + \frac{1}{2}\rho v_2^2$

$y_1 = y_2$ and $p_1 = 2.40\times10^5$ Pa, so $p_2 = p_1 + \frac{1}{2}\rho(v_1^2 - v_2^2) = 2.25\times10^5$ Pa

EVALUATE: Where the area decreases the speed increases and the pressure decreases.

12.49.   **IDENTIFY:** Increasing the cross-sectional area of the artery will increase the amount of blood that flows through it per second.

SET UP: The flow rate, $\frac{\Delta V}{\Delta t}$, is related to the radius R or diameter D of the artery by Poiseuille's law:

$\frac{\Delta V}{\Delta t} = \frac{\pi R^4}{8\eta}\left(\frac{p_1 - p_2}{L}\right) = \frac{\pi D^4}{128\eta}\left(\frac{p_1 - p_2}{L}\right)$. Assume the pressure gradient $(p_1 - p_2)/L$ in the artery remains the same.

EXECUTE: $(\Delta V/\Delta t)/D^4 = \frac{\pi}{128\eta}\left(\frac{p_1 - p_2}{L}\right) = $ constant, so $(\Delta V/\Delta t)_{\text{old}}/D_{\text{old}}^4 = (\Delta V/\Delta t)_{\text{new}}/D_{\text{new}}^4$.

$(\Delta V/\Delta t)_{\text{new}} = 2(\Delta V/\Delta t)_{\text{old}}$ and $D_{\text{old}} = D$. This gives $D_{\text{new}} = D_{\text{old}}\left[\frac{(\Delta V/\Delta t)_{\text{new}}}{(\Delta V/\Delta t)_{\text{old}}}\right]^{1/4} = 2^{1/4}D = 1.19D$.

EVALUATE: Since the flow rate is proportional to D^4, a 19% increase in D doubles the flow rate.

12.53. **IDENTIFY:** In part (a), the force is the weight of the water. In part (b), the pressure due to the water at a depth h is $\rho g h$. $F = pA$ and $m = \rho V$.

SET UP: The density of water is 1.00×10^3 kg/m^3.

EXECUTE: **(a)** The weight of the water is

$\rho g V = (1.00 \times 10^3$ kg/m^3)(9.80 m/s^2)((5.00 m)(4.0 m)(3.0 m)) = 5.9×10^5 N.

(b) Integration gives the expected result that the force is what it would be if the pressure were uniform and equal to the pressure at the midpoint. If d is the depth of the pool and A is the area of one end of the pool,

then $F = \rho g A \dfrac{d}{2} = (1.00 \times 10^3$ kg/m^3)(9.80 m/s^2)((4.0 m)(3.0 m))(1.50 m) = 1.76×10^5 N.

EVALUATE: The answer to part (a) can be obtained as $F = pA$, where $p = \rho g d$ is the gauge pressure at the bottom of the pool and $A = (5.0$ m)(4.0 m) is the area of the bottom of the pool.

12.57. **IDENTIFY:** The buoyant force on an object in a liquid is equal to the weight of the liquid it displaces.

SET UP: $V = \dfrac{m}{\rho}$.

EXECUTE: When it is floating, the ice displaces an amount of glycerin equal to its weight. From Table 12.1, the density of glycerin is 1260 kg/m^3. The volume of this amount of glycerin is

$V = \dfrac{m}{\rho} = \dfrac{0.180 \text{ kg}}{1260 \text{ kg/m}^3} = 1.429 \times 10^{-4}$ m^3. The ice cube produces 0.180 kg of water. The volume of this

mass of water is $V = \dfrac{m}{\rho} = \dfrac{0.180 \text{ kg}}{1000 \text{ kg/m}^3} = 1.80 \times 10^{-4}$ m^3. The volume of water from the melted ice is greater

than the volume of glycerin displaced by the floating cube and the level of liquid in the cylinder rises. The

distance the level rises is $\dfrac{1.80 \times 10^{-4} \text{ m}^3 - 1.429 \times 10^{-4} \text{ m}^3}{\pi (0.0350 \text{ m})^2} = 9.64 \times 10^{-3}$ m = 0.964 cm.

EVALUATE: The melted ice has the same mass as the solid ice, but a different density.

12.61. **IDENTIFY:** Apply Newton's second law to the barge plus its contents. Apply Archimedes's principle to express the buoyancy force B in terms of the volume of the barge.

SET UP: The free-body diagram for the barge plus coal is given in Figure 12.61.

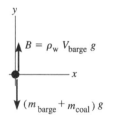

EXECUTE: $\sum F_y = ma_y$

$B - (m_{\text{barge}} + m_{\text{coal}})g = 0$

$\rho_w V_{\text{barge}} g = (m_{\text{barge}} + m_{\text{coal}})g$

$m_{\text{coal}} = \rho_w V_{\text{barge}} - m_{\text{barge}}$

Figure 12.61

$V_{\text{barge}} = (22$ m)(12 m)(40 m) = 1.056×10^4 m^3

The mass of the barge is $m_{\text{barge}} = \rho_s V_s$, where s refers to steel.

From Table 12.1, $\rho_s = 7800$ kg/m^3. The volume V_s is 0.040 m times the total area of the five pieces of steel that make up the barge

$V_s = (0.040$ m)$[2(22$ m)(12 m) $+ 2(40$ m)(12 m) $+ (22$ m)(40 m)] = 94.7 m^3.

Therefore, $m_{\text{barge}} = \rho_s V_s = (7800$ kg/m^3)(94.7 m^3) = 7.39×10^5 kg.

Then $m_{\text{coal}} = \rho_w V_{\text{barge}} - m_{\text{barge}} = (1000$ kg/m^3)(1.056 \times 10^4$ m^3) $- 7.39 \times 10^5$ kg = 9.8×10^6 kg.

The volume of this mass of coal is $V_{\text{coal}} = m_{\text{coal}}/\rho_{\text{coal}} = 9.8 \times 10^6$ kg/1500 kg/m^3 = 6500 m^3; this is less than V_{barge} so it will fit into the barge.

EVALUATE: The buoyancy force B must support both the weight of the coal and also the weight of the barge. The weight of the coal is about 13 times the weight of the barge. The buoyancy force increases when more of the barge is submerged, so when it holds the maximum mass of coal the barge is fully submerged.

12.65. **IDENTIFY:** For a floating object the buoyant force equals the weight of the object. The buoyant force when the wood sinks is $B = \rho_{\text{water}} V_{\text{tot}} g$, where V_{tot} is the volume of the wood plus the volume of the lead. $\rho = m/V$.

SET UP: The density of lead is 11.3×10^3 kg/m^3.

EXECUTE: $V_{\text{wood}} = (0.600 \text{ m})(0.250 \text{ m})(0.080 \text{ m}) = 0.0120 \text{ m}^3$.

$m_{\text{wood}} = \rho_{\text{wood}} V_{\text{wood}} = (700 \text{ kg/m}^3)(0.0120 \text{ m}^3) = 8.40 \text{ kg}$.

$B = (m_{\text{wood}} + m_{\text{lead}})g$. Using $B = \rho_{\text{water}} V_{\text{tot}} g$ and $V_{\text{tot}} = V_{\text{wood}} + V_{\text{lead}}$ gives

$\rho_{\text{water}}(V_{\text{wood}} + V_{\text{lead}})g = (m_{\text{wood}} + m_{\text{lead}})g$. $m_{\text{lead}} = \rho_{\text{lead}} V_{\text{lead}}$ then gives

$\rho_{\text{water}} V_{\text{wood}} + \rho_{\text{water}} V_{\text{lead}} = m_{\text{wood}} + \rho_{\text{lead}} V_{\text{lead}}$.

$V_{\text{lead}} = \dfrac{\rho_{\text{water}} V_{\text{wood}} - m_{\text{wood}}}{\rho_{\text{lead}} - \rho_{\text{water}}} = \dfrac{(1000 \text{ kg/m}^3)(0.0120 \text{ m}^3) - 8.40 \text{ kg}}{11.3 \times 10^3 \text{ kg/m}^3 - 1000 \text{ kg/m}^3} = 3.50 \times 10^{-4} \text{ m}^3$.

$m_{\text{lead}} = \rho_{\text{lead}} V_{\text{lead}} = 3.95 \text{ kg}$.

EVALUATE: The volume of the lead is only 2.9% of the volume of the wood. If the contribution of the volume of the lead to F_B is neglected, the calculation is simplified: $\rho_{\text{water}} V_{\text{wood}} g = (m_{\text{wood}} + m_{\text{lead}})g$ and $m_{\text{lead}} = 3.6$ kg. The result of this calculation is in error by about 9%.

12.67. **(a) IDENTIFY:** Apply Newton's second law to the airship. The buoyancy force is given by Archimedes's principle; the fluid that exerts this force is the air.

SET UP: The free-body diagram for the dirigible is given in Figure 12.67. The lift corresponds to a mass $m_{\text{lift}} = (90 \times 10^3 \text{ N})/(9.80 \text{ m/s}^2) = 9.184 \times 10^3$ kg. The mass m_{tot} is 9.184×10^3 kg plus the mass m_{gas} of the gas that fills the dirigible. B is the buoyant force exerted by the air.

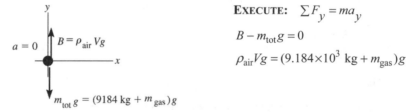

EXECUTE: $\sum F_y = ma_y$

$B - m_{\text{tot}} g = 0$

$\rho_{\text{air}} V g = (9.184 \times 10^3 \text{ kg} + m_{\text{gas}})g$

Figure 12.67

Write m_{gas} in terms of V: $m_{\text{gas}} = \rho_{\text{gas}} V$ and let g divide out; the equation becomes

$\rho_{\text{air}} V = 9.184 \times 10^3 \text{ kg} + \rho_{\text{gas}} V$.

$$V = \frac{9.184 \times 10^3 \text{ kg}}{1.20 \text{ kg/m}^3 - 0.0899 \text{ kg/m}^3} = 8.27 \times 10^3 \text{ m}^3$$

EVALUATE: The density of the airship is less than the density of air and the airship is totally submerged in the air, so the buoyancy force exceeds the weight of the airship.

(b) SET UP: Let m_{lift} be the mass that could be lifted.

EXECUTE: From part (a), $m_{\text{lift}} = (\rho_{\text{air}} - \rho_{\text{gas}})V = (1.20 \text{ kg/m}^3 - 0.166 \text{ kg/m}^3)(8.27 \times 10^3 \text{ m}^3) = 8550$ kg.

The lift force is $m_{\text{lift}} = (8550 \text{ kg})(9.80 \text{ m/s}^2) = 83.8$ kN.

EVALUATE: The density of helium is less than that of air but greater than that of hydrogen. Helium provides lift, but less lift than hydrogen. Hydrogen is not used because it is highly explosive in air.

12.69. **IDENTIFY:** Bernoulli's principle will give us the speed with which the acid leaves the hole in the tank, and two-dimensional projectile motion will give us how far the acid travels horizontally after it leaves the tank.

SET UP: Apply Bernoulli's principle, $p_1 + \rho g y_1 + \frac{1}{2}\rho v_1^2 = p_2 + \rho g y_2 + \frac{1}{2}\rho v_2^2$, with point 1 at the surface of the acid in the tank and point 2 in the stream as it emerges from the hole. $p_1 = p_2 = p_{air}$. Since the hole is small the level in the tank drops slowly and $v_1 \approx 0$. After a drop of acid exits the hole the only force on it is gravity and it moves in projectile motion. For the projectile motion take $+y$ downward, so $a_x = 0$ and $a_y = +9.80 \text{ m/s}^2$.

EXECUTE: Bernoulli's equation with $p_1 = p_2$ and $v_1 = 0$ gives $v_2 = \sqrt{2g(y_1 - y_2)} = \sqrt{2(9.80 \text{ m/s}^2)(0.75 \text{ m})} = 3.83 \text{ m/s}$. Now apply projectile motion. Use the vertical motion to find the time in the air. Combining $v_{0y} = 0$, $a_y = +9.80 \text{ m/s}^2$, $y - y_0 = 1.4 \text{ m}$.

$y - y_0 = v_{0y}t + \frac{1}{2}a_y t^2$ gives $t = \sqrt{\dfrac{2(y - y_0)}{a_y}} = \sqrt{\dfrac{2(1.4 \text{ m})}{9.80 \text{ m/s}^2}} = 0.535 \text{ s}$. The horizontal distance a drop travels in this time is $x - x_0 = v_{0x}t + \frac{1}{2}a_x t^2 = (3.83 \text{ m/s})(0.535 \text{ s}) = 2.05 \text{ m}$.

EVALUATE: If the depth of acid in the tank is increased, then the velocity of the stream as it emerges from the hole increases and the horizontal range of the stream increases.

12.73.   **IDENTIFY:**   Apply the second condition of equilibrium to the balance arm and apply the first condition of equilibrium to the block and to the brass mass. The buoyancy force on the wood is given by Archimedes's principle and the buoyancy force on the brass mass is ignored.

SET UP: The objects and forces are sketched in Figure 12.73a.

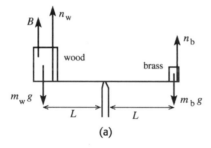

The buoyant force on the brass is neglected, but we include the buoyant force B on the block of wood. n_w and n_b are the normal forces exerted by the balance arm on which the objects sit.

Figure 12.73a

The free-body diagram for the balance arm is given in Figure 12.73b.

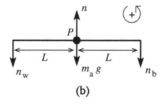

EXECUTE: $\tau_P = 0$

$n_w L - n_b L = 0$

$n_w = n_b$

Figure 12.73b

SET UP: The free-body diagram for the brass mass is given in Figure 12.73c.

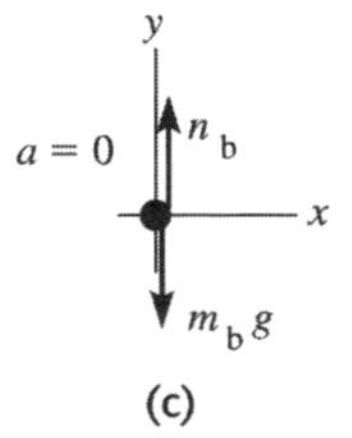

EXECUTE: $\sum F_y = ma_y$

$n_b - m_b g = 0$

$n_b = m_b g$

Figure 12.73c

The free-body diagram for the block of wood is given in Figure 12.73d.

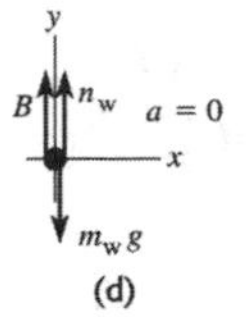

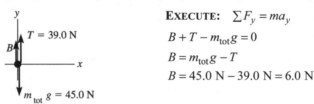

$$\sum F_y = ma_y$$
$$n_w + B - m_w g = 0$$
$$n_w = m_w g - B$$

(d)

Figure 12.73d

But $n_b = n_w$ implies $m_b g = m_w g - B$.

And $B = \rho_{air} V_w g = \rho_{air}(m_w/\rho_w)g$, so $m_b g = m_w g - \rho_{air}(m_w/\rho_w)g$.

$$m_w = \frac{m_b}{1 - \rho_{air}/\rho_w} = \frac{0.115 \text{ kg}}{1 - ((1.20 \text{ kg/m}^3)/(150 \text{ kg/m}^3))} = 0.116 \text{ kg}.$$

EVALUATE: The mass of the wood is greater than the mass of the brass; the wood is partially supported by the buoyancy force exerted by the air. The buoyancy in air of the brass can be neglected because the density of brass is much more than the density of air; the buoyancy force exerted on the brass by the air is much less than the weight of the brass. The density of the balsa wood is much less than the density of the brass, so the buoyancy force on the balsa wood is not such a small fraction of its weight.

12.75. **IDENTIFY:** Apply Newton's second law to the ingot. Use the expression for the buoyancy force given by Archimedes's principle to solve for the volume of the ingot. Then use the facts that the total mass is the mass of the gold plus the mass of the aluminum and that the volume of the ingot is the volume of the gold plus the volume of the aluminum.

SET UP: The free-body diagram for the piece of alloy is given in Figure 12.75.

$T = 39.0 \text{ N}$

B

x

$m_{tot}\, g = 45.0 \text{ N}$

EXECUTE: $\sum F_y = ma_y$
$$B + T - m_{tot} g = 0$$
$$B = m_{tot} g - T$$
$$B = 45.0 \text{ N} - 39.0 \text{ N} = 6.0 \text{ N}$$

Figure 12.75

Also, $m_{tot} g = 45.0 \text{ N}$ so $m_{tot} = 45.0 \text{ N}/(9.80 \text{ m/s}^2) = 4.59 \text{ kg}$.

We can use the known value of the buoyant force to calculate the volume of the object:
$B = \rho_w V_{obj} g = 6.0 \text{ N}$

$$V_{obj} = \frac{6.0 \text{ N}}{\rho_w g} = \frac{6.0 \text{ N}}{(1000 \text{ kg/m}^3)(9.80 \text{ m/s}^2)} = 6.122 \times 10^{-4} \text{ m}^3$$

We know two things:

(1) The mass m_g of the gold plus the mass m_a of the aluminum must add to m_{tot}: $m_g + m_a = m_{tot}$

We write this in terms of the volumes V_g and V_a of the gold and aluminum: $\rho_g V_g + \rho_a V_a = m_{tot}$

(2) The volumes V_a and V_g must add to give V_{obj}: $V_a + V_g = V_{obj}$ so that $V_a = V_{obj} - V_g$

Use this in the equation in (1) to eliminate V_a: $\rho_g V_g + \rho_a(V_{obj} - V_g) = m_{tot}$

$$V_g = \frac{m_{tot} - \rho_a V_{obj}}{\rho_g - \rho_a} = \frac{4.59 \text{ kg} - (2.7 \times 10^3 \text{ kg/m}^3)(6.122 \times 10^{-4} \text{ m}^3)}{19.3 \times 10^3 \text{ kg/m}^3 - 2.7 \times 10^3 \text{ kg/m}^3} = 1.769 \times 10^{-4} \text{ m}^3.$$

Then $m_g = \rho_g V_g = (19.3 \times 10^3 \text{ kg/m}^3)(1.769 \times 10^{-4} \text{ m}^3) = 3.41 \text{ kg}$ and the weight of gold is

$w_g = m_g g = 33.4 \text{ N}$.

EVALUATE: The gold is 29% of the volume but 74% of the mass, since the density of gold is much greater than the density of aluminum.

12.79. **IDENTIFY** and **SET UP:** Use Archimedes's principle for B.

(a) $B = \rho_{water} V_{tot} g$, where V_{tot} is the total volume of the object.

$V_{tot} = V_m + V_0$, where V_m is the volume of the metal.

EXECUTE: $V_m = w/g\rho_m$ so $V_{tot} = w/g\rho_m + V_0$

This gives $B = \rho_{water} g (w/g\rho_m + V_0)$.

Solving for V_0 gives $V_0 = B/(\rho_{water} g) - w/(\rho_m g)$, as was to be shown.

(b) The expression derived in part (a) gives

$$V_0 = \frac{20\ \text{N}}{(1000\ \text{kg/m}^3)(9.80\ \text{m/s}^2)} - \frac{156\ \text{N}}{(8.9 \times 10^3\ \text{kg/m}^3)(9.80\ \text{m/s}^2)} = 2.52 \times 10^{-4}\ \text{m}^3$$

$$V_{tot} = \frac{B}{\rho_{water} g} = \frac{20\ \text{N}}{(1000\ \text{kg/m}^3)(9.80\ \text{m/s}^2)} = 2.04 \times 10^{-3}\ \text{m}^3 \text{ and}$$

$V_0/V_{tot} = (2.52 \times 10^{-4}\ \text{m}^3)/(2.04 \times 10^{-3}\ \text{m}^3) = 0.124$.

EVALUATE: When $V_0 \to 0$, the object is solid and $V_{obj} = V_m = w/(\rho_m g)$. For $V_0 = 0$, the result in part (a) gives $B = (w/\rho_m)\rho_{water} = V_m \rho_{water} g = V_{obj} \rho_{water} g$, which agrees with Archimedes's principle. As V_0 increases with the weight kept fixed, the total volume of the object increases and there is an increase in B.

12.85. **IDENTIFY:** Consider the fluid in the horizontal part of the tube. This fluid, with mass $\rho A l$, is subject to a net force due to the pressure difference between the ends of the tube.

SET UP: The difference between the gauge pressures at the bottoms of the ends of the tubes is $\rho g(y_L - y_R)$.

EXECUTE: The net force on the horizontal part of the fluid is $\rho g(y_L - y_R)A = \rho A l a$, or, $(y_L - y_R) = \frac{a}{g}l$.

(b) Again consider the fluid in the horizontal part of the tube. As in part (a), the fluid is accelerating; the center of mass has a radial acceleration of magnitude $a_{rad} = \omega^2 l/2$, and so the difference in heights between the columns is $(\omega^2 l/2)(l/g) = \omega^2 l^2/2g$. An equivalent way to do part (b) is to break the fluid in the horizontal part of the tube into elements of thickness dr; the pressure difference between the sides of this piece is $dp = \rho(\omega^2 r)dr$ and integrating from $r = 0$ to $r = l$ gives $\Delta p = \rho \omega^2 l^2/2$, the same result.

EVALUATE: **(c)** The pressure at the bottom of each arm is proportional to ρ and the mass of fluid in the horizontal portion of the tube is proportional to ρ, so ρ divides out and the results are independent of the density of the fluid. The pressure at the bottom of a vertical arm is independent of the cross-sectional area of the arm. Newton's second law could be applied to a cross-sectional of fluid smaller than that of the tubes. Therefore, the results are independent and of the size and shape of all parts of the tube.

12.93. **IDENTIFY:** Apply Bernoulli's equation and the equation of continuity.

SET UP: Example 12.8 shows that the speed of efflux at point D is $\sqrt{2gh_1}$.

EXECUTE: Applying the equation of continuity to points at C and D gives that the fluid speed is $\sqrt{8gh_1}$ at C. Applying Bernoulli's equation to points A and C gives that the gauge pressure at C is $\rho g h_1 - 4\rho g h_1 = -3\rho g h_1$, and this is the gauge pressure at the surface of the fluid at E. The height of the fluid in the column is $h_2 = 3h_1$.

EVALUATE: The gauge pressure at C is less than the gauge pressure $\rho g h_1$ at the bottom of tank A because of the speed of the fluid at C.

GRAVITATION

13.5. **IDENTIFY:** Use Eq. (13.1) to find the force exerted by each large sphere. Add these forces as vectors to get the net force and then use Newton's 2nd law to calculate the acceleration.
SET UP: The forces are shown in Figure 13.5.

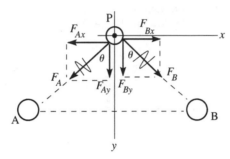

$\sin\theta = 0.80$
$\cos\theta = 0.60$
Take the origin of coordinate at point P.

Figure 13.5

EXECUTE: $F_A = G\dfrac{m_A m}{r^2} = G\dfrac{(0.26 \text{ kg})(0.010 \text{ kg})}{(0.100 \text{ m})^2} = 1.735\times10^{-11}$ N

$F_B = G\dfrac{m_B m}{r^2} = 1.735\times10^{-11}$ N

$F_{Ax} = -F_A\sin\theta = -(1.735\times10^{-11}\text{ N})(0.80) = -1.39\times10^{-11}$ N

$F_{Ay} = +F_A\cos\theta = +(1.735\times10^{-11}\text{ N})(0.60) = +1.04\times10^{-11}$ N

$F_{Bx} = +F_B\sin\theta = +1.39\times10^{-11}$ N

$F_{By} = +F_B\cos\theta = +1.04\times10^{-11}$ N

$\Sigma F_x = ma_x$ gives $F_{Ax} + F_{Bx} = ma_x$

$0 = ma_x$ so $a_x = 0$

$\Sigma F_y = ma_y$ gives $F_{Ay} + F_{By} = ma_y$

$2(1.04\times10^{-11}\text{ N}) = (0.010\text{ kg})a_y$

$a_y = 2.1\times10^{-9}$ m/s^2, directed downward midway between A and B

EVALUATE: For ordinary size objects the gravitational force is very small, so the initial acceleration is very small. By symmetry there is no x-component of net force and the y-component is in the direction of the two large spheres, since they attract the small sphere.

13.9. **IDENTIFY:** Use Eq. (13.2) to calculate the gravitational force each particle exerts on the third mass. The equilibrium is stable when for a displacement from equilibrium the net force is directed toward the equilibrium position and it is unstable when the net force is directed away from the equilibrium position.

SET UP: For the net force to be zero, the two forces on M must be in opposite directions. This is the case only when M is on the line connecting the two particles and between them. The free-body diagram for M is given in Figure 13.9. $m_1 = 3m$ and $m_2 = m$. If M is a distance x from m_1, it is a distance $1.00 \text{ m} - x$ from m_2.

EXECUTE: (a) $F_x = F_{1x} + F_{2x} = -G\dfrac{3mM}{x^2} + G\dfrac{mM}{(1.00 \text{ m} - x)^2} = 0. \ 3 \ (1.00 \text{ m} - x)^2 = x^2.$

$1.00 \text{ m} - x = \pm x/\sqrt{3}.$ Since M is between the two particles, x must be less than 1.00 m and

$x = \dfrac{1.00 \text{ m}}{1 + 1/\sqrt{3}} = 0.634$ m. M must be placed at a point that is 0.634 m from the particle of mass $3m$ and

0.366 m from the particle of mass m.

(b) (i) If M is displaced slightly to the right in Figure 13.9, the attractive force from m is larger than the force from $3m$ and the net force is to the right. If M is displaced slightly to the left in Figure 13.9, the attractive force from $3m$ is larger than the force from m and the net force is to the left. In each case the net force is away from equilibrium and the equilibrium is unstable.

(ii) If M is displaced a very small distance along the y axis in Figure 13.9, the net force is directed opposite to the direction of the displacement and therefore the equilibrium is stable.

EVALUATE: The point where the net force on M is zero is closer to the smaller mass.

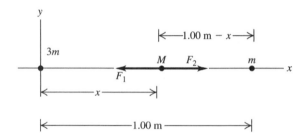

Figure 13.9

13.11.   IDENTIFY: $F_g = G\dfrac{mm_E}{r^2}$, so $a_g = G\dfrac{m_E}{r^2}$, where r is the distance of the object from the center of the earth.

SET UP: $r = h + R_E$, where h is the distance of the object above the surface of the earth and $R_E = 6.38 \times 10^6$ m is the radius of the earth.

EXECUTE: To decrease the acceleration due to gravity by one-tenth, the distance from the center of the earth must be increased by a factor of $\sqrt{10}$, and so the distance above the surface of the earth is $(\sqrt{10} - 1)R_E = 1.38 \times 10^7$ m.

EVALUATE: This height is about twice the radius of the earth.

13.15.   IDENTIFY: Apply Eq. (13.2) to the astronaut.

SET UP: $m_E = 5.97 \times 10^{24}$ kg and $R_E = 6.38 \times 10^6$ m.

EXECUTE: $F_g - G\dfrac{mm_E}{r^2}.$ $r = 600 \times 10^3 \text{ m} + R_E$ so $F_g = 610$ N. At the surface of the earth, $w = mg = 735$ N. The gravity force is not zero in orbit. The satellite and the astronaut have the same acceleration so the astronaut's apparent weight is zero.

EVALUATE: In Eq. (13.2), r is the distance of the object from the center of the earth.

13.19.   IDENTIFY: Apply Newton's second law to the motion of the satellite and obtain an equation that relates the orbital speed v to the orbital radius r.

SET UP: The distances are shown in Figure 13.19a.

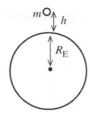

The radius of the orbit is $r = h + R_E$.

$r = 7.80 \times 10^5 \text{ m} + 6.38 \times 10^6 \text{ m} = 7.16 \times 10^6 \text{ m}.$

Figure 13.19a

The free-body diagram for the satellite is given in Figure 13.19b.

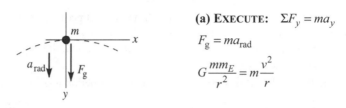

(a) EXECUTE: $\Sigma F_y = ma_y$

$F_g = ma_{\text{rad}}$

$G\dfrac{mm_E}{r^2} = m\dfrac{v^2}{r}$

Figure 13.19b

$v = \sqrt{\dfrac{Gm_E}{r}} = \sqrt{\dfrac{(6.673 \times 10^{-11} \text{ N} \cdot \text{m}^2/\text{kg}^2)(5.97 \times 10^{24} \text{ kg})}{7.16 \times 10^6 \text{ m}}} = 7.46 \times 10^3 \text{ m/s}$

(b) $T = \dfrac{2\pi r}{v} = \dfrac{2\pi(7.16 \times 10^6 \text{ m})}{7.46 \times 10^3 \text{ m/s}} = 6030 \text{ s} = 1.68 \text{ h}.$

EVALUATE: Note that $r = h + R_E$ is the radius of the orbit, measured from the center of the earth. For this satellite r is greater than for the satellite in Example 13.6, so its orbital speed is less.

13.21. **IDENTIFY:** We know orbital data (speed and orbital radius) for one satellite and want to use it to find the orbital speed of another satellite having a known orbital radius. Newton's second law and the law of universal gravitation apply to both satellites.

SET UP: For circular motion, $F_{\text{net}} = ma = mv^2/r$, which in this case is $G\dfrac{mm_p}{r^2} = m\dfrac{v^2}{r}$.

EXECUTE: Using $G\dfrac{mm_p}{r^2} = m\dfrac{v^2}{r}$, we get $Gm_p = rv^2 = \text{constant}$. $r_1v_1^2 = r_2v_2^2$.

$v_2 = v_1\sqrt{\dfrac{r_1}{r_2}} = (4800 \text{ m/s})\sqrt{\dfrac{5.00 \times 10^7 \text{ m}}{3.00 \times 10^7 \text{ m}}} = 6200 \text{ m/s}.$

EVALUATE: The more distant satellite moves slower than the closer satellite, which is reasonable since the planet's gravity decreases with distance. The masses of the satellites do not affect their orbits.

13.23. **IDENTIFY:** Apply $\Sigma \vec{F} = m\vec{a}$ to the motion of the baseball. $v = \dfrac{2\pi r}{T}$.

SET UP: $r_D = 6 \times 10^3 \text{ m}$.

EXECUTE: **(a)** $F_g = ma_{\text{rad}}$ gives $G\dfrac{m_D m}{r_D^2} = m\dfrac{v^2}{r_D}$.

$v = \sqrt{\dfrac{Gm_D}{r_D}} = \sqrt{\dfrac{(6.673 \times 10^{-11} \text{ N} \cdot \text{m}^2/\text{kg}^2)(2.0 \times 10^{15} \text{ kg})}{6 \times 10^3 \text{ m}}} = 4.7 \text{ m/s}$

$4.7 \text{ m/s} = 11 \text{ mph}$, which is easy to achieve.

(b) $T = \dfrac{2\pi r}{v} = \dfrac{2\pi(6 \times 10^3 \text{ m})}{4.7 \text{ m/s}} = 8020 \text{ s} = 134 \text{ min} = 2.23 \text{ h}.$ The game would last a long time.

EVALUATE: The speed v is relative to the center of Deimos. The baseball would already have some speed before we throw it, because of the rotational motion of Deimos.

13.25.   **IDENTIFY:** The orbital speed is given by $v = \sqrt{Gm/r}$, where m is the mass of the star. The orbital period is given by $T = \dfrac{2\pi r}{v}$.

SET UP: The sun has mass $m_S = 1.99 \times 10^{30}$ kg. The orbit radius of the earth is 1.50×10^{11} m.

EXECUTE: (a) $v = \sqrt{Gm/r}$.

$v = \sqrt{(6.673 \times 10^{-11} \text{ N} \cdot \text{m}^2/\text{kg}^2)(0.85 \times 1.99 \times 10^{30} \text{ kg})/((1.50 \times 10^{11} \text{ m})(0.11))} = 8.27 \times 10^4$ m/s.

(b) $2\pi r/v = 1.25 \times 10^6$ s $= 14.5$ days (about two weeks).

EVALUATE: The orbital period is less than the 88-day orbital period of Mercury; this planet is orbiting very close to its star, compared to the orbital radius of Mercury.

13.29.   **IDENTIFY:** Knowing the orbital radius and orbital period of a satellite, we can calculate the mass of the object about which it is revolving.

SET UP: The radius of the orbit is $r = 10.5 \times 10^9$ m and its period is $T = 6.3$ days $= 5.443 \times 10^5$ s. The mass of the sun is $m_S = 1.99 \times 10^{30}$ kg. The orbital period is given by $T = \dfrac{2\pi r^{3/2}}{\sqrt{Gm_{HD}}}$.

EXECUTE: Solving $T = \dfrac{2\pi r^{3/2}}{\sqrt{Gm_{HD}}}$ for the mass of the star gives

$m_{HD} = \dfrac{4\pi^2 r^3}{T^2 G} = \dfrac{4\pi^2 (10.5 \times 10^9 \text{ m})^3}{(5.443 \times 10^5 \text{ s})^2 (6.673 \times 10^{-11} \text{ N} \cdot \text{m}^2/\text{kg}^2)} = 2.3 \times 10^{30}$ kg, which is $m_{HD} = 1.2 m_S$.

EVALUATE: The mass of the star is only 20% greater than that of our sun, yet the orbital period of the planet is much shorter than that of the earth, so the planet must be much closer to the star than the earth is.

13.31.   **IDENTIFY:** Section 13.6 states that for a point mass outside a uniform sphere the gravitational force is the same as if all the mass of the sphere were concentrated at its center. It also states that for a point mass a distance r from the center of a uniform sphere, where r is less than the radius of the sphere, the gravitational force on the point mass is the same as though we removed all the mass at points farther than r from the center and concentrated all the remaining mass at the center.

SET UP: The density of the sphere is $\rho = \dfrac{M}{\frac{4}{3}\pi R^3}$, where M is the mass of the sphere and R is its radius.

The mass inside a volume of radius $r < R$ is $M_r = \rho V_r = \left(\dfrac{M}{\frac{4}{3}\pi R^3}\right)\left(\frac{4}{3}\pi r^3\right) = M\left(\dfrac{r}{R}\right)^3$. $r = 5.01$ m is outside the sphere and $r = 2.50$ m is inside the sphere.

EXECUTE: (a) (i) $F_g = \dfrac{GMm}{r^2} = (6.67 \times 10^{-11} \text{ N} \cdot \text{m}^2/\text{kg}^2)\dfrac{(1000.0 \text{ kg})(2.00 \text{ kg})}{(5.01 \text{ m})^2} = 5.31 \times 10^{-9}$ N.

(ii) $F_g = \dfrac{GM'm}{r^2}$. $M' = M\left(\dfrac{r}{R}\right)^3 = (1000.0 \text{ kg})\left(\dfrac{2.50 \text{ m}}{5.00 \text{ m}}\right)^3 = 125$ kg.

$F_g = (6.67 \times 10^{-11} \text{ N} \cdot \text{m}^2/\text{kg}^2)\dfrac{(125 \text{ kg})(2.00 \text{ kg})}{(2.50 \text{ m})^2} = 2.67 \times 10^{-9}$ N.

(b) $F_g = \dfrac{GM(r/R)^3 m}{r^2} = \left(\dfrac{GMm}{R^3}\right)r$ for $r < R$ and $F_g = \dfrac{GMm}{r^2}$ for $r > R$. The graph of F_g versus r is sketched in Figure 13.31.

EVALUATE: At points outside the sphere the force on a point mass is the same as for a shell of the same mass and radius. For $r < R$ the force is different in the two cases of uniform sphere versus hollow shell.

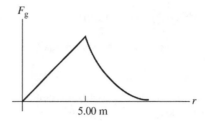

Figure 13.31

13.35. **IDENTIFY** and **SET UP:** At the north pole, $F_g = w_0 = mg_0$, where g_0 is given by Eq. (13.4) applied to Neptune. At the equator, the apparent weight is given by Eq. (13.28). The orbital speed v is obtained from the rotational period using Eq. (13.12).

EXECUTE: **(a)** $g_0 = Gm/R^2 = (6.673 \times 10^{-11} \text{ N} \cdot \text{m}^2/\text{kg}^2)(1.0 \times 10^{26} \text{ kg})/(2.5 \times 10^7 \text{ m})^2 = 10.7 \text{ m/s}^2$. This agrees with the value of g given in the problem.

$F = w_0 = mg_0 = (5.0 \text{ kg})(10.7 \text{ m/s}^2) = 53 \text{ N}$; this is the true weight of the object.

(b) From Eq. (13.28), $w = w_0 - mv^2/R$

$T = \dfrac{2\pi r}{v}$ gives $v = \dfrac{2\pi r}{T} = \dfrac{2\pi(2.5 \times 10^7 \text{ m})}{(16 \text{ h})(3600 \text{ s/1 h})} = 2.727 \times 10^3 \text{ m/s}$

$v^2/R = (2.727 \times 10^3 \text{ m/s})^2/2.5 \times 10^7 \text{ m} = 0.297 \text{ m/s}^2$

Then $w = 53 \text{ N} - (5.0 \text{ kg})(0.297 \text{ m/s}^2) = 52 \text{ N}$.

EVALUATE: The apparent weight is less than the true weight. This effect is larger on Neptune than on earth.

13.37. **IDENTIFY:** The orbital speed for an object a distance r from an object of mass M is $v = \sqrt{\dfrac{GM}{r}}$. The mass M of a black hole and its Schwarzschild radius R_S are related by Eq. (13.30).

SET UP: $c = 3.00 \times 10^8 \text{ m/s}$. $1 \text{ ly} = 9.461 \times 10^{15} \text{ m}$.

EXECUTE: **(a)**

$M = \dfrac{rv^2}{G} = \dfrac{(7.5 \text{ ly})(9.461 \times 10^{15} \text{ m/ly})(200 \times 10^3 \text{ m/s})^2}{(6.673 \times 10^{-11} \text{ N} \cdot \text{m}^2/\text{kg}^2)} = 4.3 \times 10^{37} \text{ kg} = 2.1 \times 10^7 \text{ M}_S$.

(b) No, the object has a mass very much greater than 50 solar masses.

(c) $R_S = \dfrac{2GM}{c^2} = \dfrac{2v^2 r}{c^2} = 6.32 \times 10^{10} \text{ m}$, which does fit.

EVALUATE: The Schwarzschild radius of a black hole is approximately the same as the radius of Mercury's orbit around the sun.

13.41. **IDENTIFY:** $g_n = G\dfrac{m_n}{R_n^2}$, where the subscript n refers to the neutron star. $w = mg$.

SET UP: $R_n = 10.0 \times 10^3 \text{ m}$. $m_n = 1.99 \times 10^{30} \text{ kg}$. Your mass is $m = \dfrac{w}{g} = \dfrac{675 \text{ N}}{9.80 \text{ m/s}^2} = 68.9 \text{ kg}$.

EXECUTE: $g_n = (6.673 \times 10^{-11} \text{ N} \cdot \text{m}^2/\text{kg}^2)\dfrac{1.99 \times 10^{30} \text{ kg}}{(10.0 \times 10^3 \text{ m})^2} = 1.33 \times 10^{12} \text{ m/s}^2$

Your weight on the neutron star would be $w_n = mg_n = (68.9 \text{ kg})(1.33 \times 10^{12} \text{ m/s}^2) = 9.16 \times 10^{13} \text{ N}$.

EVALUATE: Since R_n is much less than the radius of the sun, the gravitational force exerted by the neutron star on an object at its surface is immense.

13.45. **IDENTIFY:** The mass and radius of the moon determine the acceleration due to gravity at its surface. This in turn determines the normal force on the hip, which then determines the kinetic friction force while walking.

SET UP: $m_M = 7.35 \times 10^{22}$ kg, $R_M = 1.74 \times 10^6$ m. The mass supported by the hip is

$(0.65)(65 \text{ kg}) + 43 \text{ kg} = 85.25 \text{ kg}$. The acceleration due to gravity on the moon is $g_M = \dfrac{Gm_M}{R_M{}^2}$ and

$f_k = \mu_k n$.

EXECUTE: **(a)** The acceleration due to gravity on the moon is

$g_M = \dfrac{Gm_M}{R_M{}^2} = \dfrac{(6.673 \times 10^{-11} \text{ N} \cdot \text{m}^2/\text{kg}^2)(7.35 \times 10^{22} \text{ kg})}{(1.74 \times 10^6 \text{ m})^2} = 1.62 \text{ m/s}^2.$

(b) $n = (85.25 \text{ kg})g_M = 138 \text{ N}$ and $f_k = \mu_k n = (0.0050)(138 \text{ N}) = 0.69 \text{ N}.$

(c) $n = (85.25 \text{ kg})g_E = 835 \text{ N}$ and $f_k = \mu_k n = 4.2 \text{ N}.$

EVALUATE: Walking on the moon should produce much less wear on the hip joints than on the earth.

13.47. **IDENTIFY:** Knowing the density and radius of Toro, we can calculate its mass and then the acceleration due to gravity at its surface. We can then use orbital mechanics to determine its orbital speed knowing the radius of its orbit.

SET UP: Density is $\rho = m/V$, and the volume of a sphere is $\frac{4}{3}\pi R^3$. Use the assumption that the density

of Toro is the same as that of earth to calculate the mass of Toro. Then $g_T = G\dfrac{m_T}{R_T{}^2}$. Apply $\Sigma\vec{F} = m\vec{a}$ to

the object to find its speed when it is in a circular orbit around Toro.

EXECUTE: **(a)** Toro and the earth are assumed to have the same densities, so $\dfrac{m_E}{\frac{4}{3}\pi R_E{}^3} = \dfrac{m_T}{\frac{4}{3}\pi R_T{}^3}$ gives

$m_T = m_E \left(\dfrac{R_T}{R_E}\right)^3 = (5.97 \times 10^{24} \text{ kg})\left(\dfrac{5.0 \times 10^3 \text{ m}}{6.38 \times 10^6 \text{ m}}\right)^3 = 2.9 \times 10^{15} \text{ kg}.$

$g_T = G\dfrac{m_T}{R_T{}^2} = \dfrac{(6.673 \times 10^{-11} \text{ N} \cdot \text{m}^2/\text{kg}^2)(2.9 \times 10^{15} \text{ kg})}{(5.0 \times 10^3 \text{ m})^2} = 7.7 \times 10^{-3} \text{ m/s}^2.$

(b) The force of gravity on the object is mg_T. In a circular orbit just above the surface of Toro, its

acceleration is $\dfrac{v^2}{R_T}$. Then $\Sigma\vec{F} = m\vec{a}$ gives $mg_T = m\dfrac{v^2}{R_T}$ and

$v = \sqrt{g_T R_T} = \sqrt{(7.7 \times 10^{-3} \text{ m/s}^2)(5.0 \times 10^3 \text{ m})} = 6.2 \text{ m/s}.$

EVALUATE: A speed of 6.2 m/s corresponds to running 100 m in 16.1 s, which is barely possible for the average person, but a well-conditioned athlete might do it.

13.49. **IDENTIFY:** Apply conservation of energy and conservation of linear momentum to the motion of the two spheres.

SET UP: Denote the 50.0-kg sphere by a subscript 1 and the 100-kg sphere by a subscript 2.

EXECUTE: **(a)** Linear momentum is conserved because we are ignoring all other forces, that is, the net external force on the system is zero. Hence, $m_1 v_1 = m_2 v_2$.

(b) (i) From the work-energy theorem in the form $K_i + U_i = K_f + U_f$, with the initial kinetic energy

$K_i = 0$ and $U = -G\dfrac{m_1 m_2}{r}$, $Gm_1 m_2\left[\dfrac{1}{r_f} - \dfrac{1}{r_i}\right] = \dfrac{1}{2}(m_1 v_1^2 + m_2 v_2^2).$ Using the conservation of momentum

relation $m_1 v_1 = m_2 v_2$ to eliminate v_2 in favor of v_1 and simplifying yields $v_1^2 = \dfrac{2Gm_2^2}{m_1 + m_2}\left[\dfrac{1}{r_f} - \dfrac{1}{r_i}\right]$, with a

similar expression for v_2. Substitution of numerical values gives $v_1 = 1.49 \times 10^{-5}$ m/s, $v_2 = 7.46 \times 10^{-6}$ m/s.

(ii) The magnitude of the relative velocity is the sum of the speeds, 2.24×10^{-5} m/s.

(c) The distance the centers of the spheres travel (x_1 and x_2) is proportional to their acceleration, and

$\dfrac{x_1}{x_2} = \dfrac{a_1}{a_2} = \dfrac{m_2}{m_1}$, or $x_1 = 2x_2$. When the spheres finally make contact, their centers will be a distance of

$2r$ apart, or $x_1 + x_2 + 2r = 40$ m, or $2x_2 + x_2 + 2r = 40$ m. Thus, $x_2 = 40/3 \text{ m} - 2r/3$, and $x_1 = 80/3 \text{ m} - 4r/3$.
The point of contact of the surfaces is $80/3 \text{ m} - r/3 = 26.6 \text{ m}$ from the initial position of the center of the
50.0-kg sphere.

EVALUATE: The result $x_1/x_2 = 2$ can also be obtained from the conservation of momentum result that

$\dfrac{v_1}{v_2} = \dfrac{m_2}{m_1}$, at every point in the motion.

13.51.   **IDENTIFY** and **SET UP:** (a) To stay above the same point on the surface of the earth the orbital period of
the satellite must equal the orbital period of the earth:

$T = 1 \text{ d}(24 \text{ h/1 d})(3600 \text{ s/1 h}) = 8.64 \times 10^4 \text{ s}$

Eq. (13.14) gives the relation between the orbit radius and the period:

EXECUTE: $T = \dfrac{2\pi r^{3/2}}{\sqrt{Gm_E}}$ and $T^2 = \dfrac{4\pi^2 r^3}{Gm_E}$

$r = \left(\dfrac{T^2 Gm_E}{4\pi^2} \right)^{1/3} = \left(\dfrac{(8.64 \times 10^4 \text{ s})^2 (6.673 \times 10^{-11} \text{ N} \cdot \text{m}^2/\text{kg}^2)(5.97 \times 10^{24} \text{ kg})}{4\pi^2} \right)^{1/3} = 4.23 \times 10^7 \text{ m}$

This is the radius of the orbit; it is related to the height h above the earth's surface and the radius R_E of the

earth by $r = h + R_E$. Thus $h = r - R_E = 4.23 \times 10^7 \text{ m} - 6.38 \times 10^6 \text{ m} = 3.59 \times 10^7 \text{ m}$.

EVALUATE: The orbital speed of the geosynchronous satellite is $2\pi r/T = 3080$ m/s. The altitude is much
larger and the speed is much less than for the satellite in Example 13.6.
(b) Consider Figure 13.51.

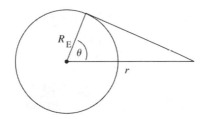

$\cos\theta = \dfrac{R_E}{r} = \dfrac{6.38 \times 10^6 \text{ m}}{4.23 \times 10^7 \text{ m}}$

$\theta = 81.3°$

Figure 13.51

A line from the satellite is tangent to a point on the earth that is at an angle of 81.3° above the equator.
The sketch shows that points at higher latitudes are blocked by the earth from viewing the satellite.

13.53.   **IDENTIFY:** From Example 13.5, the escape speed is $v = \sqrt{\dfrac{2GM}{R}}$. Use $\rho = M/V$ to write this expression

in terms of ρ.

SET UP: For a sphere $V = \frac{4}{3}\pi R^3$.

EXECUTE: In terms of the density ρ, the ratio M/R is $(4\pi/3)\rho R^2$, and so the escape speed is

$v = \sqrt{(8\pi/3)(6.673 \times 10^{-11} \text{ N} \cdot \text{m}^2/\text{kg}^2)(2500 \text{ kg/m}^3)(150 \times 10^3 \text{ m})^2} = 177 \text{ m/s}.$

EVALUATE: This is much less than the escape speed for the earth, 11,200 m/s.

13.55.   **IDENTIFY** and **SET UP:** The observed period allows you to calculate the angular velocity of the satellite
relative to you. You know your angular velocity as you rotate with the earth, so you can find the angular
velocity of the satellite in a space-fixed reference frame. $v = r\omega$ gives the orbital speed of the satellite and
Newton's second law relates this to the orbit radius of the satellite.

EXECUTE: **(a)** The satellite is revolving west to east, in the same direction the earth is rotating. If the angular speed of the satellite is ω_s and the angular speed of the earth is ω_E, the angular speed ω_{rel} of the satellite relative to you is $\omega_{rel} = \omega_s - \omega_E$.

$\omega_{rel} = (1 \text{ rev})/(12 \text{ h}) = \left(\frac{1}{12}\right) \text{ rev/h}$

$\omega_E = \left(\frac{1}{24}\right) \text{ rev/h}$

$\omega_s = \omega_{rel} + \omega_E = \left(\frac{1}{8}\right) \text{ rev/h} = 2.18 \times 10^{-4} \text{ rad/s}$

$\Sigma \vec{F} = m\vec{a}$ says $G\dfrac{mm_E}{r^2} = m\dfrac{v^2}{r}$

$v^2 = \dfrac{Gm_E}{r}$ and with $v = r\omega$ this gives $r^3 = \dfrac{Gm_E}{\omega^2}$; $r = 2.03 \times 10^7 \text{ m}$

This is the radius of the satellite's orbit. Its height h above the surface of the earth is

$h = r - R_E = 1.39 \times 10^7 \text{ m}$.

(b) Now the satellite is revolving opposite to the rotation of the earth. If west to east is positive, then

$\omega_{rel} = \left(-\frac{1}{12}\right) \text{ rev/h}$

$\omega_s = \omega_{rel} + \omega_E = \left(-\frac{1}{24}\right) \text{ rev/h} = -7.27 \times 10^{-5} \text{ rad/s}$

$r^3 = \dfrac{Gm_E}{\omega^2}$ gives $r = 4.22 \times 10^7 \text{ m}$ and $h = 3.59 \times 10^7 \text{ m}$

EVALUATE: In part (a) the satellite is revolving faster than the earth's rotation and in part (b) it is revolving slower. Slower v and ω means larger orbit radius r.

13.59. **IDENTIFY:** The free-fall time of the rock will give us the acceleration due to gravity at the surface of the planet. Applying Newton's second law and the law of universal gravitation will give us the mass of the planet since we know its radius.

SET UP: For constant acceleration, $y - y_0 = v_{0y}t + \frac{1}{2}a_y t^2$. At the surface of the planet, Newton's second law gives $m_{rock}g = \dfrac{Gm_{rock}m_p}{R_p^2}$.

EXECUTE: First find $a_y = g$. $y - y_0 = v_{0y}t + \frac{1}{2}a_y t^2$. $a_y = \dfrac{2(y - y_0)}{t^2} = \dfrac{2(1.90 \text{ m})}{(0.480 \text{ s})^2} = 16.49 \text{ m/s}^2 = g$.

$g = 16.49 \text{ m/s}^2$. $m_p = \dfrac{gR_p^2}{G} = \dfrac{(16.49 \text{ m/s})(8.60 \times 10^7 \text{ m})^2}{6.674 \times 10^{-11} \text{ N} \cdot \text{m}^2/\text{kg}^2} = 1.83 \times 10^{27} \text{ kg}$.

EVALUATE: The planet's mass is over 100 times that of the earth, which is reasonable since it is larger (in size) than the earth yet has a greater acceleration due to gravity at its surface.

13.63. **IDENTIFY and SET UP:** First use the radius of the orbit to find the initial orbital speed, from Eq. (13.10) applied to the moon.

EXECUTE: $v = \sqrt{Gm/r}$ and $r = R_M + h = 1.74 \times 10^6 \text{ m} + 50.0 \times 10^3 \text{ m} = 1.79 \times 10^6 \text{ m}$

Thus $v = \sqrt{\dfrac{(6.673 \times 10^{-11} \text{ N} \cdot \text{m}^2/\text{kg}^2)(7.35 \times 10^{22} \text{ kg})}{1.79 \times 10^6 \text{ m}}} = 1.655 \times 10^3 \text{ m/s}$

After the speed decreases by 20.0 m/s it becomes $1.655 \times 10^3 \text{ m/s} - 20.0 \text{ m/s} = 1.635 \times 10^3 \text{ m/s}$.

IDENTIFY and SET UP: Use conservation of energy to find the speed when the spacecraft reaches the lunar surface.

$K_1 + U_1 + W_{other} = K_2 + U_2$

Gravity is the only force that does work so $W_{other} = 0$ and $K_2 = K_1 + U_1 - U_2$

EXECUTE: $U_1 = -Gm_\text{m}m/r;\ \ U_2 = -Gm_\text{m}m/R_\text{m}$

$\frac{1}{2}mv_2^2 = \frac{1}{2}mv_1^2 + Gmm_\text{m}(1/R_\text{m} - 1/r)$

And the mass m divides out to give $v_2 = \sqrt{v_1^2 + 2Gm_\text{m}(1/R_\text{m} - 1/r)}$

$v_2 = 1.682 \times 10^3$ m/s(1 km/1000 m)(3600 s/1 h) $= 6060$ km/h

EVALUATE: After the thruster fires the spacecraft is moving too slowly to be in a stable orbit; the gravitational force is larger than what is needed to maintain a circular orbit. The spacecraft gains energy as it is accelerated toward the surface.

13.67.   **IDENTIFY:**   At the escape speed, $E = K + U = 0$.

SET UP: At the surface of the earth the satellite is a distance $R_\text{E} = 6.38 \times 10^6$ m from the center of the earth and a distance $R_\text{ES} = 1.50 \times 10^{11}$ m from the sun. The orbital speed of the earth is $\frac{2\pi R_\text{ES}}{T}$, where $T = 3.156 \times 10^7$ s is the orbital period. The speed of a point on the surface of the earth at an angle ϕ from the equator is $v = \frac{2\pi R_\text{E} \cos\phi}{T}$, where $T = 86{,}400$ s is the rotational period of the earth.

EXECUTE: **(a)** The escape speed will be $v = \sqrt{2G\left[\dfrac{m_\text{E}}{R_\text{E}} + \dfrac{m_\text{s}}{R_\text{ES}}\right]} = 4.35 \times 10^4$ m/s. Making the simplifying assumption that the direction of launch is the direction of the earth's motion in its orbit, the speed relative to the center of the earth is $v - \dfrac{2\pi R_\text{ES}}{T} = 4.35 \times 10^4$ m/s $- \dfrac{2\pi(1.50 \times 10^{11}\text{ m})}{(3.156 \times 10^7\text{s})} = 1.37 \times 10^4$ m/s.

(b) The rotational speed at Cape Canaveral is $\dfrac{2\pi(6.38 \times 10^6\text{ m}) \cos 28.5°}{86{,}400\text{ s}} = 4.09 \times 10^2$ m/s, so the speed relative to the surface of the earth is 1.33×10^4 m/s.

(c) In French Guiana, the rotational speed is 4.63×10^2 m/s, so the speed relative to the surface of the earth is 1.32×10^4 m/s.

EVALUATE: The orbital speed of the earth is a large fraction of the escape speed, but the rotational speed of a point on the surface of the earth is much less.

13.69.   **IDENTIFY:**   Eq. (13.12) relates orbital period and orbital radius for a circular orbit.

SET UP: The mass of the sun is $M = 1.99 \times 10^{30}$ kg.

EXECUTE: **(a)** The period of the asteroid is $T = \dfrac{2\pi a^{3/2}}{\sqrt{GM}}$. Inserting (i) 3×10^{11} m for a gives

2.84 y and (ii) 5×10^{11} m gives a period of 6.11 y.

(b) If the period is 5.93 y, then $a = 4.90 \times 10^{11}$ m.

(c) This happens because $0.4 = 2/5$, another ratio of integers. So once every 5 orbits of the asteroid and 2 orbits of Jupiter, the asteroid is at its perijove distance. Solving when $T = 4.74$ y, $a = 4.22 \times 10^{11}$ m.

EVALUATE: The orbit radius for Jupiter is 7.78×10^{11} m and for Mars it is 2.28×10^{11} m. The asteroid belt lies between Mars and Jupiter. The mass of Jupiter is about 3000 times that of Mars, so the effect of Jupiter on the asteroids is much larger.

13.73.   **IDENTIFY** and **SET UP:**   Use conservation of energy, $K_1 + U_1 + W_\text{other} = K_2 + U_2$. The gravity force exerted by the sun is the only force that does work on the comet, so $W_\text{other} = 0$.

EXECUTE: $K_1 = \frac{1}{2}mv_1^2,\ \ v_1 = 2.0 \times 10^4$ m/s

$U_1 = -Gm_\text{S}m/r_1,\ \ r_1 = 2.5 \times 10^{11}$ m

$K_2 = \frac{1}{2}mv_2^2$

$U_2 = -Gm_S m/r_2, \ r_2 = 5.0 \times 10^{10}$ m

$\frac{1}{2} m v_1^2 - Gm_S m/r_1 = \frac{1}{2} m v_2^2 - Gm_S m/r_2$

$v_2^2 = v_1^2 + 2Gm_S \left(\dfrac{1}{r_2} - \dfrac{1}{r_1} \right) = v_1^2 + 2Gm_S \left(\dfrac{r_1 - r_2}{r_1 r_2} \right)$

$v_2 = 6.8 \times 10^4$ m/s

EVALUATE: The comet has greater speed when it is closer to the sun.

13.77. **(a) IDENTIFY and SET UP:** Use Eq. (13.17), applied to the satellites orbiting the earth rather than the sun.
EXECUTE: Find the value of a for the elliptical orbit:

$2a = r_a + r_p = R_E + h_a + R_E + h_p$, where h_a and h_p are the heights at apogee and perigee, respectively.

$a = R_E + (h_a + h_p)/2$

$a = 6.38 \times 10^6$ m $+ (400 \times 10^3$ m $+ 4000 \times 10^3$ m$)/2 = 8.58 \times 10^6$ m

$T = \dfrac{2\pi a^{3/2}}{\sqrt{Gm_E}} = \dfrac{2\pi (8.58 \times 10^6 \text{ m})^{3/2}}{\sqrt{(6.673 \times 10^{-11} \text{ N} \cdot \text{m}^2/\text{kg}^2)(5.97 \times 10^{24} \text{ kg})}} = 7.91 \times 10^3$ s

(b) Conservation of angular momentum gives $r_a v_a = r_p v_p$

$\dfrac{v_p}{v_a} = \dfrac{r_a}{r_p} = \dfrac{6.38 \times 10^6 \text{ m} + 4.00 \times 10^6 \text{ m}}{6.38 \times 10^6 \text{ m} + 4.00 \times 10^5 \text{ m}} = 1.53$

(c) Conservation of energy applied to apogee and perigee gives $K_a + U_a = K_p + U_p$

$\frac{1}{2} m v_a^2 - Gm_E m/r_a = \frac{1}{2} m v_p^2 - Gm_E m/r_p$

$v_p^2 - v_a^2 = 2Gm_E (1/r_p - 1/r_a) = 2Gm_E (r_a - r_p)/r_a r_p$

But $v_p = 1.532 v_a$, so $1.347 v_a^2 = 2Gm_E (r_a - r_p)/r_a r_p$

$v_a = 5.51 \times 10^3$ m/s, $v_p = 8.43 \times 10^3$ m/s

(d) Need v so that $E = 0$, where $E = K + U$.

at perigee: $\frac{1}{2} m v_p^2 - Gm_E m/r_p = 0$

$v_p = \sqrt{2Gm_E/r_p} = \sqrt{2(6.673 \times 10^{-11} \text{ N} \cdot \text{m}^2/\text{kg}^2)(5.97 \times 10^{24} \text{ kg})/6.78 \times 10^6 \text{ m}} = 1.084 \times 10^4$ m/s

This means an increase of 1.084×10^4 m/s $- 8.43 \times 10^3$ m/s $= 2.41 \times 10^3$ m/s.

at apogee:

$v_a = \sqrt{2Gm_E/r_a} = \sqrt{2(6.673 \times 10^{-11} \text{ N} \cdot \text{m}^2/\text{kg}^2)(5.97 \times 10^{24} \text{ kg})/1.038 \times 10^7 \text{ m}} = 8.761 \times 10^3$ m/s

This means an increase of 8.761×10^3 m/s $- 5.51 \times 10^3$ m/s $= 3.25 \times 10^3$ m/s.

EVALUATE: Perigee is more efficient. At this point r is smaller so v is larger and the satellite has more kinetic energy and more total energy.

13.79. **IDENTIFY and SET UP:** Apply conservation of energy (Eq. (7.13)) and solve for W_{other}. Only $r = h + R_E$ is given, so use Eq. (13.10) to relate r and v.
EXECUTE: $K_1 + U_1 + W_{\text{other}} = K_2 + U_2$

$U_1 = -Gm_M m/r_1$, where m_M is the mass of Mars and $r_1 = R_M + h$, where R_M is the radius of Mars and $h = 2000 \times 10^3$ m.

$U_1 = -(6.673 \times 10^{-11} \text{ N} \cdot \text{m}^2/\text{kg}^2) \dfrac{(6.42 \times 10^{23} \text{ kg})(5000 \text{ kg})}{3.40 \times 10^6 \text{ m} + 2000 \times 10^3 \text{ m}} = -3.9667 \times 10^{10}$ J

$U_2 = -Gm_M m/r_2$, where r_2 is the new orbit radius.

$U_2 = -(6.673 \times 10^{-11} \text{ N} \cdot \text{m}^2/\text{kg}^2) \dfrac{(6.42 \times 10^{23} \text{ kg})(5000 \text{ kg})}{3.40 \times 10^6 \text{ m} + 4000 \times 10^3 \text{ m}} = -2.8950 \times 10^{10}$ J

For a circular orbit $v = \sqrt{Gm_M/r}$ (Eq. (13.10)), with the mass of Mars rather than the mass of the earth).

Using this gives $K = \frac{1}{2}mv^2 = \frac{1}{2}m(Gm_M/r) = \frac{1}{2}Gm_Mm/r$, so $K = -\frac{1}{2}U$.

$K_1 = -\frac{1}{2}U_1 = +1.9833 \times 10^{10}$ J and $K_2 = -\frac{1}{2}U_2 = +1.4475 \times 10^{10}$ J

Then $K_1 + U_1 + W_{other} = K_2 + U_2$ gives

$W_{other} = (K_2 - K_1) + (U_2 - U) = (1.4475 \times 10^{10}$ J $- 1.9833 \times 10^{10}$J$) + (+3.9667 \times 10^{10}$ J $- 2.8950 \times 10^{10}$ J$)$

$W_{other} = -5.3580 \times 10^9$ J $+ 1.0717 \times 10^{10}$ J $= 5.36 \times 10^9$ J.

EVALUATE: When the orbit radius increases the kinetic energy decreases and the gravitational potential energy increases. $K = -U/2$ so $E = K + U = -U/2$ and the total energy also increases (becomes less negative). Positive work must be done to increase the total energy of the satellite.

13.81. **IDENTIFY:** Integrate $dm = \rho dV$ to find the mass of the planet. Outside the planet, the planet behaves like a point mass, so at the surface $g = GM/R^2$.

SET UP: A thin spherical shell with thickness dr has volume $dV = 4\pi r^2 dr$. The earth has radius $R_E = 6.38 \times 10^6$ m.

EXECUTE: Get M: $M = \int dm = \int \rho dV = \int \rho 4\pi r^2 dr$. The density is $\rho = \rho_0 - br$, where

$\rho_0 = 15.0 \times 10^3$ kg/m^3 at the center and at the surface, $\rho_S = 2.0 \times 10^3$ kg/m^3, so $b = \dfrac{\rho_0 - \rho_s}{R}$.

$M = \int_0^R (\rho_0 - br) 4\pi r^2 dr = \dfrac{4\pi}{3}\rho_0 R^3 - \pi b R^4 = \dfrac{4}{3}\pi R^3 \rho_0 - \pi R^4 \left(\dfrac{\rho_0 - \rho_s}{R}\right) = \pi R^3 \left(\dfrac{1}{3}\rho_0 + \rho_s\right)$ and

$M = 5.71 \times 10^{24}$ kg. Then $g = \dfrac{GM}{R^2} = \dfrac{G\pi R^3\left(\frac{1}{3}\rho_0 + \rho_s\right)}{R^2} = \pi R G\left(\dfrac{1}{3}\rho_0 + \rho_s\right)$.

$g = \pi(6.38 \times 10^6 \text{m})(6.67 \times 10^{-11} \text{ N} \cdot \text{m}^2/\text{kg}^2)\left(\dfrac{15.0 \times 10^3 \text{ kg/m}^3}{3} + 2.0 \times 10^3 \text{ kg/m}^3\right)$.

$g = 9.36$ m/s^2.

EVALUATE: The average density of the planet is

$\rho_{av} = \dfrac{M}{V} = \dfrac{M}{\frac{4}{3}\pi R^3} = \dfrac{3(5.71 \times 10^{24} \text{ kg})}{4\pi(6.38 \times 10^6 \text{ m})^3} = 5.25 \times 10^3$ kg/m^3. Note that this is not $(\rho_0 + \rho_s)/2$.

13.83. **IDENTIFY:** The direct calculation of the force that the sphere exerts on the ring is slightly more involved than the calculation of the force that the ring exerts on the sphere. These forces are equal in magnitude but opposite in direction, so it will suffice to do the latter calculation. By symmetry, the force on the sphere will be along the axis of the ring in Figure E13.33 in the textbook, toward the ring.
SET UP: Divide the ring into infinitesimal elements with mass dM.

EXECUTE: Each mass element dM of the ring exerts a force of magnitude $\dfrac{(Gm)dM}{a^2 + x^2}$ on the sphere,

and the x-component of this force is $\dfrac{GmdM}{a^2 + x^2}\dfrac{x}{\sqrt{a^2 + x^2}} = \dfrac{GmdMx}{(a^2 + x^2)^{3/2}}$.

Therefore, the force on the sphere is $GmMx/(a^2 + x^2)^{3/2}$, in the $-x$-direction. The sphere attracts the ring with a force of the same magnitude.

EVALUATE: As $x \gg a$ the denominator approaches x^3 and $F \rightarrow \dfrac{GMm}{x^2}$, as expected.

PERIODIC MOTION

14.7. **IDENTIFY** and **SET UP:** Use Eq. (14.1) to calculate T, Eq. (14.2) to calculate ω and Eq. (14.10) for m.
EXECUTE: **(a)** $T = 1/f = 1/6.00 \text{ Hz} = 0.167 \text{ s}$

(b) $\omega = 2\pi f = 2\pi(6.00 \text{ Hz}) = 37.7 \text{ rad/s}$

(c) $\omega = \sqrt{k/m}$ implies $m = k/\omega^2 = (120 \text{ N/m})/(37.7 \text{ rad/s})^2 = 0.0844 \text{ kg}$

EVALUATE: We can verify that k/ω^2 has units of mass.

14.9. **IDENTIFY:** For SHM the motion is sinusoidal.
SET UP: $x(t) = A\cos(\omega t)$.

EXECUTE: $x(t) = A\cos(\omega t)$, where $A = 0.320 \text{ m}$ and $\omega = \dfrac{2\pi}{T} = \dfrac{2\pi}{0.900 \text{ s}} = 6.981 \text{ rad/s}$.

(a) $x = 0.320 \text{ m}$ at $t_1 = 0$. Let t_2 be the instant when $x = 0.160 \text{ m}$. Then we have

$0.160 \text{ m} = (0.320 \text{ m}) \cos(\omega t_2)$. $\cos(\omega t_2) = 0.500$. $\omega t_2 = 1.047 \text{ rad}$. $t_2 = \dfrac{1.047 \text{ rad}}{6.981 \text{ rad/s}} = 0.150 \text{ s}$. It takes

$t_2 - t_1 = 0.150 \text{ s}$.

(b) Let t_3 be when $x = 0$. Then we have $\cos(\omega t_3) = 0$ and $\omega t_3 = 1.571 \text{ rad}$. $t_3 = \dfrac{1.571 \text{ rad}}{6.981 \text{ rad/s}} = 0.225 \text{ s}$. It

takes $t_3 - t_2 = 0.225 \text{ s} - 0.150 \text{ s} = 0.0750 \text{ s}$.

EVALUATE: Note that it takes twice as long to go from $x = 0.320 \text{ m}$ to $x = 0.160 \text{ m}$ than to go from $x = 0.160 \text{ m}$ to $x = 0$, even though the two distances are the same, because the speeds are different over the two distances.

14.13. **IDENTIFY:** For SHM, $a_x = -\omega^2 x = -(2\pi f)^2 x$. Apply Eqs. (14.13), (14.15) and (14.16), with A and ϕ from Eqs. (14.18) and (14.19).
SET UP: $x = 1.1 \text{ cm}$, $v_{0x} = -15 \text{ cm/s}$. $\omega = 2\pi f$, with $f = 2.5 \text{ Hz}$.

EXECUTE: **(a)** $a_x = -(2\pi(2.5 \text{ Hz}))^2 (1.1 \times 10^{-2} \text{ m}) = -2.71 \text{ m/s}^2$.

(b) From Eq. (14.19) the amplitude is 1.46 cm, and from Eq. (14.18) the phase angle is 0.715 rad. The angular frequency is $2\pi f = 15.7 \text{ rad/s}$, so $x = (1.46 \text{ cm}) \cos ((15.7 \text{ rad/s})t + 0.715 \text{ rad})$,

$v_x = (-22.9 \text{ cm/s}) \sin ((15.7 \text{ rad/s})t + 0.715 \text{ rad})$ and $a_x = (-359 \text{ cm/s}^2) \cos ((15.7 \text{ rad/s})t + 0.715 \text{ rad})$.

EVALUATE: We can verify that our equations for x, v_x and a_x give the specified values at $t = 0$.

14.17. **IDENTIFY:** $T = 2\pi\sqrt{\dfrac{m}{k}}$. $a_x = -\dfrac{k}{m}x$ so $a_{max} = \dfrac{k}{m}A$. $F = -kx$.

SET UP: a_x is proportional to x so a_x goes through one cycle when the displacement goes through one cycle. From the graph, one cycle of a_x extends from $t = 0.10 \text{ s}$ to $t = 0.30 \text{ s}$, so the period is $T = 0.20 \text{ s}$. $k = 2.50 \text{ N/cm} = 250 \text{ N/m}$. From the graph the maximum acceleration is 12.0 m/s^2.

EXECUTE: (a) $T = 2\pi\sqrt{\dfrac{m}{k}}$ gives $m = k\left(\dfrac{T}{2\pi}\right)^2 = (250\ \text{N/m})\left(\dfrac{0.20\ \text{s}}{2\pi}\right)^2 = 0.253\ \text{kg}$

(b) $A = \dfrac{ma_{\max}}{k} = \dfrac{(0.253\ \text{kg})(12.0\ \text{m/s}^2)}{250\ \text{N/m}} = 0.0121\ \text{m} = 1.21\ \text{cm}$

(c) $F_{\max} = kA = (250\ \text{N/m})(0.0121\ \text{m}) = 3.03\ \text{N}$

EVALUATE: We can also calculate the maximum force from the maximum acceleration:

$F_{\max} = ma_{\max} = (0.253\ \text{kg})(12.0\ \text{m/s}^2) = 3.04\ \text{N},$ which agrees with our previous results.

14.19.   **IDENTIFY:** Compare the specific $x(t)$ given in the problem to the general form of Eq. (14.13).

SET UP: $A = 7.40$ cm, $\omega = 4.16$ rad/s, and $\phi = -2.42$ rad.

EXECUTE: (a) $T = \dfrac{2\pi}{\omega} = \dfrac{2\pi}{4.16\ \text{rad/s}} = 1.51\ \text{s}.$

(b) $\omega = \sqrt{\dfrac{k}{m}}$ so $k = m\omega^2 = (1.50\ \text{kg})(4.16\ \text{rad/s})^2 = 26.0\ \text{N/m}$

(c) $v_{\max} = \omega A = (4.16\ \text{rad/s})(7.40\ \text{cm}) = 30.8\ \text{cm/s}$

(d) $F_x = -kx$ so $F_{\max} = kA = (26.0\ \text{N/m})(0.0740\ \text{m}) = 1.92\ \text{N}.$

(e) $x(t)$ evaluated at $t = 1.00$ s gives $x = -0.0125$ m. $v_x = -\omega A \sin(\omega t + \phi) = 30.4$ cm/s.

$a_x = -kx/m = -\omega^2 x = +0.216\ \text{m/s}^2.$

(f) $F_x = -kx = -(26.0\ \text{N/m})(-0.0125\ \text{m}) = +0.325\ \text{N}$

EVALUATE: The maximum speed occurs when $x = 0$ and the maximum force is when $x = \pm A.$

14.21.   **IDENTIFY and SET UP:** Use Eqs. (14.13), (14.15) and (14.16).

EXECUTE: $f = 440$ Hz, $A = 3.0$ mm, $\phi = 0$

(a) $x = A\cos(\omega t + \phi)$

$\omega = 2\pi f = 2\pi(440\ \text{Hz}) = 2.76 \times 10^3\ \text{rad/s}$

$x = (3.0 \times 10^{-3}\ \text{m})\cos((2.76 \times 10^3\ \text{rad/s})t)$

(b) $v_x = -\omega A \sin(\omega t + \phi)$

$v_{\max} = \omega A = (2.76 \times 10^3\ \text{rad/s})(3.0 \times 10^{-3}\ \text{m}) = 8.3\ \text{m/s}$ (maximum magnitude of velocity)

$a_x = -\omega^2 A \cos(\omega t + \phi)$

$a_{\max} = \omega^2 A = (2.76 \times 10^3\ \text{rad/s})^2 (3.0 \times 10^{-3}\ \text{m}) = 2.3 \times 10^4\ \text{m/s}^2$ (maximum magnitude of acceleration)

(c) $a_x = -\omega^2 A \cos \omega t$

$da_x/dt = +\omega^3 A \sin \omega t = [2\pi(440\ \text{Hz})]^3 (3.0 \times 10^{-3}\ \text{m})\sin([2.76 \times 10^3\ \text{rad/s}]t) =$

$(6.3 \times 10^7\ \text{m/s}^3)\sin([2.76 \times 10^3\ \text{rad/s}]t)$

Maximum magnitude of the jerk is $\omega^3 A = 6.3 \times 10^7\ \text{m/s}^3$

EVALUATE: The period of the motion is small, so the maximum acceleration and jerk are large.

14.25.   **IDENTIFY:** $v_{\max} = \omega A = 2\pi fA.$ $K_{\max} = \frac{1}{2}mv_{\max}^2$

SET UP: The fly has the same speed as the tip of the tuning fork.

EXECUTE: (a) $v_{\max} = 2\pi fA = 2\pi(392\ \text{Hz})(0.600 \times 10^{-3}\ \text{m}) = 1.48\ \text{m/s}$

(b) $K_{\max} = \frac{1}{2}mv_{\max}^2 = \frac{1}{2}(0.0270 \times 10^{-3}\ \text{kg})(1.48\ \text{m/s})^2 = 2.96 \times 10^{-5}\ \text{J}$

EVALUATE: $v_{\max}$ is directly proportional to the frequency and to the amplitude of the motion.

14.27.   **IDENTIFY:** Velocity and position are related by $E = \frac{1}{2}kA^2 = \frac{1}{2}mv_x^2 + \frac{1}{2}kx^2.$ Acceleration and position are related by $-kx = ma_x.$

SET UP: The maximum speed is at $x = 0$ and the maximum magnitude of acceleration is at $x = \pm A.$

EXECUTE: **(a)** For $x = 0$, $\frac{1}{2}mv_{max}^2 = \frac{1}{2}kA^2$ and $v_{max} = A\sqrt{\frac{k}{m}} = (0.040 \text{ m})\sqrt{\frac{450 \text{ N/m}}{0.500 \text{ kg}}} = 1.20 \text{ m/s}$

(b) $v_x = \pm\sqrt{\frac{k}{m}}\sqrt{A^2 - x^2} = \pm\sqrt{\frac{450 \text{ N/m}}{0.500 \text{ kg}}}\sqrt{(0.040 \text{ m})^2 - (0.015 \text{ m})^2} = \pm 1.11 \text{ m/s}$.

The speed is $v = 1.11 \text{ m/s}$.

(c) For $x = \pm A$, $a_{max} = \frac{k}{m}A = \left(\frac{450 \text{ N/m}}{0.500 \text{ kg}}\right)(0.040 \text{ m}) = 36 \text{ m/s}^2$

(d) $a_x = -\frac{kx}{m} = -\frac{(450 \text{ N/m})(-0.015 \text{ m})}{0.500 \text{ kg}} = +13.5 \text{ m/s}^2$

(e) $E = \frac{1}{2}kA^2 = \frac{1}{2}(450 \text{ N/m})(0.040 \text{ m})^2 = 0.360 \text{ J}$

EVALUATE: The speed and acceleration at $x = -0.015 \text{ m}$ are less than their maximum values.

14.31.   **IDENTIFY:**   Conservation of energy says $\frac{1}{2}mv^2 + \frac{1}{2}kx^2 = \frac{1}{2}kA^2$ and Newton's second law says $-kx = ma_x$.

SET UP: Let $+x$ be to the right. Let the mass of the object be m.

EXECUTE: $k = -\frac{ma_x}{x} = -m\left(\frac{-8.40 \text{ m/s}^2}{0.600 \text{ m}}\right) = (14.0 \text{ s}^{-2})m$.

$A = \sqrt{x^2(m/k)v^2} = \sqrt{(0.600 \text{ m})^2 + \left(\frac{m}{[14.0 \text{ s}^{-2}]m}\right)(2.20 \text{ m/s})^2} = 0.840 \text{ m}$. The object will therefore

travel $0.840 \text{ m} - 0.600 \text{ m} = 0.240 \text{ m}$ to the right before stopping at its maximum amplitude.

EVALUATE: The acceleration is not constant and we cannot use the constant acceleration kinematic equations.

14.35.   **IDENTIFY:**   Work in an inertial frame moving with the vehicle after the engines have shut off. The acceleration before engine shut-off determines the amount the spring is initially stretched. The initial speed of the ball relative to the vehicle is zero.

SET UP: Before the engine shut-off the ball has acceleration $a = 5.00 \text{ m/s}^2$.

EXECUTE: **(a)** $F_x = -kx = ma_x$ gives $A = \frac{ma}{k} = \frac{(3.50 \text{ kg})(5.00 \text{ m/s}^2)}{225 \text{ N/m}} = 0.0778 \text{ m}$. This is the amplitude

of the subsequent motion.

(b) $f = \frac{1}{2\pi}\sqrt{\frac{k}{m}} = \frac{1}{2\pi}\sqrt{\frac{225 \text{ N/m}}{3.50 \text{ kg}}} = 1.28 \text{ Hz}$

(c) Energy conservation gives $\frac{1}{2}kA^2 = \frac{1}{2}mv_{max}^2$ and $v_{max} = \sqrt{\frac{k}{m}}A = \sqrt{\frac{225 \text{ N/m}}{3.50 \text{ kg}}}(0.0778 \text{ m}) = 0.624 \text{ m/s}$.

EVALUATE: During the simple harmonic motion of the ball its maximum acceleration, when $x = \pm A$, continues to have magnitude 5.00 m/s^2.

14.39.   **IDENTIFY:**   The location of the equilibrium position, the position where the downward gravity force is balanced by the upward spring force, changes when the mass of the suspended object changes.

SET UP: At the equilibrium position, the spring is stretched a distance d. The amplitude is the maximum distance of the object from the equilibrium position.

EXECUTE: **(a)** The force of the glue on the lower ball is the upward force that accelerates that ball upward. The upward acceleration of the two balls is greatest when they have the greatest downward displacement, so this is when the force of the glue must be greatest.

(b) With both balls, the distance d_1 that the spring is stretched at equilibrium is given by

$kd_1 = (1.50 \text{ kg} + 2.00 \text{ kg})g$ and $d_1 = 20.8 \text{ cm}$. At the lowest point the spring is stretched

$20.8 \text{ cm} + 15.0 \text{ cm} = 35.8 \text{ cm}$. After the 1.50 kg ball falls off the distance d_2 that the spring is stretched at equilibrium is given by $kd_2 = (2.00 \text{ kg})g$ and $d_2 = 11.9 \text{ cm}$. The new amplitude is

35.8 cm -11.9 cm $= 23.9$ cm. The new frequency is $f = \dfrac{1}{2\pi}\sqrt{\dfrac{k}{m}} = \dfrac{1}{2\pi}\sqrt{\dfrac{165 \text{ N/m}}{2.00 \text{ kg}}} = 1.45 \text{ Hz}.$

EVALUATE: The potential energy stored in the spring doesn't change when the lower ball comes loose.

14.41.   **IDENTIFY and SET UP:** The number of ticks per second tells us the period and therefore the frequency. We can use a formula from Table 9.2 to calculate I. Then Eq. (14.24) allows us to calculate the torsion constant κ.

       EXECUTE: Ticks four times each second implies 0.25 s per tick. Each tick is half a period, so $T = 0.50$ s and $f = 1/T = 1/0.50$ s $= 2.00$ Hz.

       (a) Thin rim implies $I = MR^2$ (from Table 9.2). $I = (0.900 \times 10^{-3} \text{ kg})(0.55 \times 10^{-2} \text{ m})^2 = 2.7 \times 10^{-8} \text{ kg} \cdot \text{m}^2$

       (b) $T = 2\pi\sqrt{I/\kappa}$ so $\kappa = I(2\pi/T)^2 = (2.7 \times 10^{-8} \text{ kg} \cdot \text{m}^2)(2\pi/0.50 \text{ s})^2 = 4.3 \times 10^{-6} \text{ N} \cdot \text{m/rad}$

       EVALUATE: Both I and κ are small numbers.

14.43.   **IDENTIFY:** $f = \dfrac{1}{2\pi}\sqrt{\dfrac{\kappa}{I}}.$

       SET UP: $f = 125/(265 \text{ s})$, the number of oscillations per second.

       EXECUTE: $I = \dfrac{\kappa}{(2\pi f)^2} = \dfrac{0.450 \text{ N} \cdot \text{m/rad}}{(2\pi(125)/(265 \text{ s}))^2} = 0.0512 \text{ kg} \cdot \text{m}^2.$

       EVALUATE: For a larger I, f is smaller.

14.49.   **IDENTIFY:** Apply $T = 2\pi\sqrt{L/g}$

       SET UP: The period of the pendulum is $T = (136 \text{ s})/100 = 1.36$ s.

       EXECUTE: $g = \dfrac{4\pi^2 L}{T^2} = \dfrac{4\pi^2(0.500 \text{ m})}{(1.36 \text{ s})^2} = 10.7 \text{ m/s}^2.$

       EVALUATE: The same pendulum on earth, where g is smaller, would have a larger period.

14.53.   **IDENTIFY:** $T = 2\pi\sqrt{I/mgd}.$

       SET UP: $d = 0.200$ m. $T = (120 \text{ s})/100.$

       EXECUTE: $I = mgd\left(\dfrac{T}{2\pi}\right)^2 = (1.80 \text{ kg})(9.80 \text{ m/s}^2)(0.200 \text{ m})\left(\dfrac{120 \text{ s}/100}{2\pi}\right)^2 = 0.129 \text{ kg.m}^2.$

       EVALUATE: If the rod were uniform, its center of gravity would be at its geometrical center and it would have length $l = 0.400$ m. For a uniform rod with an axis at one end, $I = \frac{1}{3}ml^2 = 0.096 \text{ kg} \cdot \text{m}^2$. The value of I for the actual rod is about 34% larger than this value.

14.57.   **IDENTIFY:** Pendulum A can be treated as a simple pendulum. Pendulum B is a physical pendulum. Use the parallel-axis theorem to find the moment of inertia of the ball in B for an axis at the top of the string.

       SET UP: For pendulum B the center of gravity is at the center of the ball, so $d = L$. For a solid sphere with an axis through its center, $I_{cm} = \frac{2}{5}MR^2$. $R = L/2$ and $I_{cm} = \frac{1}{10}ML^2$.

       EXECUTE: Pendulum A: $T_A = 2\pi\sqrt{\dfrac{L}{g}}.$

Pendulum B: The parallel-axis theorem says $I = I_{cm} + ML^2 = \frac{11}{10}ML^2.$

$$T = 2\pi\sqrt{\dfrac{I}{mgd}} = 2\pi\sqrt{\dfrac{11ML^2}{10MgL}} = \sqrt{\dfrac{11}{10}}\left(2\pi\sqrt{\dfrac{L}{g}}\right) = \sqrt{\dfrac{11}{10}}T_A = 1.05T_A.$$ It takes pendulum B longer to complete a swing.

       EVALUATE: The center of the ball is the same distance from the top of the string for both pendulums, but the mass is distributed differently and I is larger for pendulum B, even though the masses are the same.

14.59. **IDENTIFY** and **SET UP:** Use Eq. (14.43) to calculate ω', and then $f' = \omega'/2\pi$.

(a) EXECUTE: $\omega' = \sqrt{(k/m)-(b^2/4m^2)} = \sqrt{\dfrac{2.50 \text{ N/m}}{0.300 \text{ kg}} - \dfrac{(0.900 \text{ kg/s})^2}{4(0.300 \text{ kg})^2}} = 2.47 \text{ rad/s}$

$f' = \omega'/2\pi = (2.47 \text{ rad/s})/2\pi = 0.393 \text{ Hz}$

(b) IDENTIFY and **SET UP:** The condition for critical damping is $b = 2\sqrt{km}$ (Eq.14.44).

EXECUTE: $b = 2\sqrt{(2.50 \text{ N/m})(0.300 \text{ kg})} = 1.73 \text{ kg/s}$

EVALUATE: The value of b in part (a) is less than the critical damping value found in part (b). With no damping, the frequency is $f = 0.459$ Hz; the damping reduces the oscillation frequency.

14.61. **IDENTIFY:** $x(t)$ is given by Eq. (14.42). $v_x = dx/dt$ and $a_x = dv_x/dt$.

SET UP: $d(\cos \omega't)/dt = -\omega' \sin \omega't$. $d(\sin \omega't)/dt = \omega' \cos \omega't$. $d(e^{-\alpha t})/dt = -\alpha e^{-\alpha t}$.

EXECUTE: **(a)** With $\phi = 0$, $x(0) = A$.

(b) $v_x = \dfrac{dx}{dt} = Ae^{(b/2m)t}\left[-\dfrac{b}{2m}\cos \omega't - \omega' \sin \omega't\right]$, and at $t = 0, v_x = -Ab/2m$; the graph of x versus t near $t = 0$ slopes down.

(c) $a_x = \dfrac{dv_x}{dt} = Ae^{-(b/2m)t}\left[\left(\dfrac{b^2}{4m^2}-\omega'^2\right)\cos \omega't + \dfrac{\omega'b}{2m}\sin \omega't\right]$, and at $t = 0$,

$a_x = A\left(\dfrac{b^2}{4m^2}-\omega'^2\right) = A\left(\dfrac{b^2}{2m^2}-\dfrac{k}{m}\right)$. (Note that this is $(-bv_0 - kx_0)/m$.) This will be negative if $b < \sqrt{2km}$, zero if $b = \sqrt{2km}$ and positive if $b > \sqrt{2km}$. The graph in the three cases will be curved down, not curved, or curved up, respectively.

EVALUATE: $a_x(0) = 0$ corresponds to the situation of critical damping.

14.65. **IDENTIFY** and **SET UP:** Calculate x using Eq. (14.13). Use T to calculate ω and x_0 to calculate ϕ.

EXECUTE: $x = 0$ at $t = 0$ implies that $\phi = \pm \pi/2$ rad

Thus $x = A\cos(\omega t \pm \pi/2)$.

$T = 2\pi/\omega$ so $\omega = 2\pi/T = 2\pi/1.20 \text{ s} = 5.236 \text{ rad/s}$

$x = (0.600 \text{ m})\cos([5.236 \text{ rad/s}][0.480 \text{ s}] \pm \pi/2) = \mp 0.353 \text{ m}$.

The distance of the object from the equilibrium position is 0.353 m.

EVALUATE: The problem doesn't specify whether the object is moving in the $+x$ or $-x$-direction at $t = 0$.

14.67. **IDENTIFY:** $ma_x = -kx$ so $a_{max} = \dfrac{k}{m}A = \omega^2 A$ is the magnitude of the acceleration when $x = \pm A$.

$v_{max} = \sqrt{\dfrac{k}{m}}A = \omega A$. $P = \dfrac{W}{t} = \dfrac{\Delta K}{t}$.

SET UP: $A = 0.0500$ m. $\omega = 4500$ rpm $= 471.24$ rad/s.

EXECUTE: **(a)** $a_{max} = \omega^2 A = (471.24 \text{ rad/s})^2(0.0500 \text{ m}) = 1.11 \times 10^4 \text{ m/s}^2$.

(b) $F_{max} = ma_{max} = (0.450 \text{ kg})(1.11 \times 10^4 \text{ m/s}^2) = 5.00 \times 10^3 \text{ N}$.

(c) $v_{max} = \omega A = (471.24 \text{ rad/s})(0.0500 \text{ m}) = 23.6 \text{ m/s}$.

$K_{max} = \frac{1}{2}mv_{max}^2 = \frac{1}{2}(0.450 \text{ kg})(23.6 \text{ m/s})^2 = 125 \text{ J}$.

(d) $P = \dfrac{K_{max}}{t}$ and $t = \dfrac{T}{4} = \dfrac{2\pi}{4\omega} = \dfrac{\pi}{2\omega}$, so

$P = \dfrac{K_{max}}{t} = \dfrac{K_{max}}{\pi/2\omega} = \dfrac{2\omega K_{max}}{\pi} = \dfrac{2(471.24 \text{ rad/s})(125 \text{ J})}{\pi} = 3.75 \times 10^4 \text{ W}$.

(e) a_{max} is proportional to ω^2, so F_{max} increases by a factor of 4500/7000, to 1.21×10^4 N. v_{max} is proportional to ω, so v_{max} increases by a factor of 4500/7000, to 36.7 m/s, and K_{max} increases by a factor of $(7000/4500)^2$, to 302 J. In part (d), t decreases by a factor of 4500/7000 and K increases by a factor of $(7000/4500)^2$, so P_{max} increases by a factor of $(7000/4500)^3$ and becomes 1.41×10^5 W.

EVALUATE: For a given amplitude, the maximum acceleration and maximum velocity increase when the frequency of the motion increases and the period decreases.

14.71. **IDENTIFY:** The largest downward acceleration the ball can have is g whereas the downward acceleration of the tray depends on the spring force. When the downward acceleration of the tray is greater than g, then the ball leaves the tray. $y(t) = A\cos(\omega t + \phi)$.

SET UP: The downward force exerted by the spring is $F = kd$, where d is the distance of the object above the equilibrium point. The downward acceleration of the tray has magnitude $\dfrac{F}{m} = \dfrac{kd}{m}$, where m is the total mass of the ball and tray. $x = A$ at $t = 0$, so the phase angle ϕ is zero and $+x$ is downward.

EXECUTE: **(a)** $\dfrac{kd}{m} = g$ gives $d = \dfrac{mg}{k} = \dfrac{(1.775\text{ kg})(9.80\text{ m/s}^2)}{185\text{ N/m}} = 9.40$ cm. This point is 9.40 cm above the equilibrium point so is $9.40\text{ cm} + 15.0\text{ cm} = 24.4$ cm above point A.

(b) $\omega = \sqrt{\dfrac{k}{m}} = \sqrt{\dfrac{185\text{ N/m}}{1.775\text{ kg}}} = 10.2$ rad/s. The point in (a) is above the equilibrium point so $x = -9.40$ cm.

$x = A\cos(\omega t)$ gives $\omega t = \arccos\left(\dfrac{x}{A}\right) = \arccos\left(\dfrac{-9.40\text{ cm}}{15.0\text{ cm}}\right) = 2.25$ rad. $t = \dfrac{2.25\text{ rad}}{10.2\text{ rad/s}} = 0.221$ s.

(c) $\frac{1}{2}kx^2 + \frac{1}{2}mv^2 = \frac{1}{2}kA^2$ gives $v = \sqrt{\dfrac{k}{m}(A^2 - x^2)} = \sqrt{\dfrac{185\text{ N/m}}{1.775\text{ kg}}([0.150\text{ m}]^2 - [-0.0940\text{ m}]^2)} = 1.19$ m/s.

EVALUATE: The period is $T = 2\pi\sqrt{\dfrac{m}{k}} = 0.615$ s. To go from the lowest point to the highest point takes time $T/2 = 0.308$ s. The time in (b) is less than this, as it should be.

14.75. **IDENTIFY and SET UP:** The bounce frequency is given by Eq. (14.11) and the pendulum frequency by Eq. (14.33). Use the relation between these two frequencies that is specified in the problem to calculate the equilibrium length L of the spring, when the apple hangs at rest on the end of the spring.

EXECUTE: vertical SHM: $f_b = \dfrac{1}{2\pi}\sqrt{\dfrac{k}{m}}$

pendulum motion (small amplitude): $f_p = \dfrac{1}{2\pi}\sqrt{\dfrac{g}{L}}$

The problem specifies that $f_p = \frac{1}{2}f_b$.

$\dfrac{1}{2\pi}\sqrt{\dfrac{g}{L}} = \dfrac{1}{2}\dfrac{1}{2\pi}\sqrt{\dfrac{k}{m}}$

$g/L = k/4m$ so $L = 4gm/k = 4w/k = 4(1.00\text{ N})/1.50\text{ N/m} = 2.67$ m

EVALUATE: This is the *stretched* length of the spring, its length when the apple is hanging from it. (Note: Small angle of swing means v is small as the apple passes through the lowest point, so a_{rad} is small and the component of mg perpendicular to the spring is small. Thus the amount the spring is stretched changes very little as the apple swings back and forth.)

IDENTIFY: Use Newton's second law to calculate the distance the spring is stretched from its unstretched length when the apple hangs from it.

SET UP: The free-body diagram for the apple hanging at rest on the end of the spring is given in Figure 14.75.

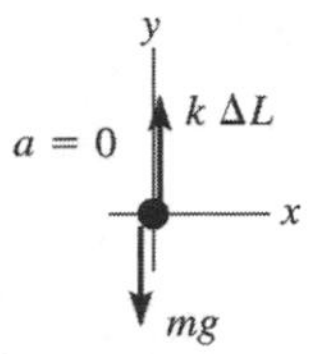

$$\text{EXECUTE:} \quad \sum F_y = ma_y$$

$a = 0$ $k \Delta L$

y

x

mg

$$k\Delta L - mg = 0$$
$$\Delta L = mg/k = w/k =$$
$$1.00 \text{ N}/1.50 \text{ N/m} = 0.667 \text{ m}$$

Figure 14.75

Thus the unstretched length of the spring is $2.67 \text{ m} - 0.67 \text{ m} = 2.00 \text{ m}$.

EVALUATE: The spring shortens to its unstretched length when the apple is removed.

14.79. **IDENTIFY:** The object oscillates as a physical pendulum, so $f = \frac{1}{2\pi}\sqrt{\frac{m_{\text{object}}gd}{I}}$. Use the parallel-axis

theorem, $I = I_{\text{cm}} + Md^2$, to find the moment of inertia of each stick about an axis at the hook.

SET UP: The center of mass of the square object is at its geometrical center, so its distance from the hook is $L\cos 45° = L/\sqrt{2}$. The center of mass of each stick is at its geometrical center. For each stick,

$I_{\text{cm}} = \frac{1}{12}mL^2$.

EXECUTE: The parallel-axis theorem gives I for each stick for an axis at the center of the square to be

$\frac{1}{12}mL^2 + m(L/2)^2 = \frac{1}{3}mL^2$ and the total I for this axis is $\frac{4}{3}mL^2$. For the entire object and an axis at the

hook, applying the parallel-axis theorem again to the object of mass $4m$ gives

$I = \frac{4}{3}mL^2 + 4m(L/\sqrt{2})^2 = \frac{10}{3}mL^2$.

$$f = \frac{1}{2\pi}\sqrt{\frac{m_{\text{object}}gd}{I}} = \frac{1}{2\pi}\sqrt{\frac{4m_{\text{object}}gL/\sqrt{2}}{\frac{10}{3}m_{\text{object}}L^2}} = \sqrt{\frac{6}{5\sqrt{2}}}\left(\frac{1}{2\pi}\sqrt{\frac{g}{L}}\right) = 0.921\left(\frac{1}{2\pi}\sqrt{\frac{g}{L}}\right).$$

EVALUATE: Just as for a simple pendulum, the frequency is independent of the mass. A simple pendulum

of length L has frequency $f = \frac{1}{2\pi}\sqrt{\frac{g}{L}}$ and this object has a frequency that is slightly less than this.

14.81. **IDENTIFY:** $T = 2\pi\sqrt{\frac{m}{k}}$ so the period changes because the mass changes.

SET UP: $\frac{dm}{dt} = -2.00 \times 10^{-3}$ kg/s. The rate of change of the period is $\frac{dT}{dt}$.

EXECUTE: **(a)** When the bucket is half full, $m = 7.00$ kg. $T = 2\pi\sqrt{\frac{7.00 \text{ kg}}{125 \text{ N/m}}} = 1.49$ s.

(b) $\frac{dT}{dt} = \frac{2\pi}{\sqrt{k}}\frac{d}{dt}(m^{1/2}) = \frac{2\pi}{\sqrt{k}}\frac{1}{2}m^{-1/2}\frac{dm}{dt} = \frac{\pi}{\sqrt{mk}}\frac{dm}{dt}$.

$\frac{dT}{dt} = \frac{\pi}{\sqrt{(7.00 \text{ kg})(125 \text{ N/m})}}(-2.00 \times 10^{-3} \text{ kg/s}) = -2.12 \times 10^{-4}$ s per s. $\frac{dT}{dt}$ is negative; the period is

getting shorter.

(c) The shortest period is when all the water has leaked out and $m = 2.00$ kg. Then $T = 0.795$ s.

EVALUATE: The rate at which the period changes is not constant but instead increases in time, even though the rate at which the water flows out is constant.

14.83. **IDENTIFY and SET UP:** Measure x from the equilibrium position of the object, where the gravity and spring forces balance. Let $+x$ be downward.

(a) Use conservation of energy (Eq.14.21) to relate v_x and x. Use Eq. (14.21) to relate T to k/m.

EXECUTE: $\frac{1}{2}mv_x^2 + \frac{1}{2}kx^2 = \frac{1}{2}kA^2$

For $x = 0, \frac{1}{2}mv_x^2 = \frac{1}{2}kA^2$ and $v = A\sqrt{k/m}$, just as for horizontal SHM. We can use the period to calculate

$\sqrt{k/m}$: $T = 2\pi\sqrt{m/k}$ implies $\sqrt{k/m} = 2\pi/T$. Thus $v = 2\pi A/T = 2\pi(0.100 \text{ m})/4.20 \text{ s} = 0.150$ m/s.

(b) IDENTIFY and **SET UP:**   Use Eq. (14.4) to relate a_x and x.

EXECUTE: $ma_x = -kx$ so $a_x = -(k/m)x$

$+x$-direction is downward, so here $x = -0.050$ m

$a_x = -(2\pi/T)^2(-0.050 \text{ m}) = +(2\pi/4.20 \text{ s})^2(0.050 \text{ m}) = 0.112 \text{ m/s}^2$ (positive, so direction is downward)

(c) IDENTIFY and **SET UP:**   Use Eq. (14.13) to relate x and t. The time asked for is twice the time it takes to go from $x = 0$ to $x = +0.050$ m.

EXECUTE: $x(t) = A\cos(\omega t + \phi)$

Let $\phi = -\pi/2$, so $x = 0$ at $t = 0$. Then $x = A\cos(\omega t - \pi/2) = A\sin \omega t = A\sin(2\pi t/T)$. Find the time t that gives $x = +0.050$ m: $0.050 \text{ m} = (0.100 \text{ m}) \sin(2\pi t/T)$

$2\pi t/T = \arcsin(0.50) = \pi/6$ and $t = T/12 = 4.20 \text{ s}/12 = 0.350$ s

The time asked for in the problem is twice this, 0.700 s.

(d) IDENTIFY: The problem is asking for the distance d that the spring stretches when the object hangs at rest from it. Apply Newton's second law to the object.

SET UP: The free-body diagram for the object is given in Figure 14.83.

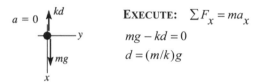

EXECUTE: $\sum F_x = ma_x$

$mg - kd = 0$

$d = (m/k)g$

Figure 14.83

But $\sqrt{k/m} = 2\pi/T$ (part (a)) and $m/k = (T/2\pi)^2$

$$d = \left(\frac{T}{2\pi}\right)^2 g = \left(\frac{4.20 \text{ s}}{2\pi}\right)^2 (9.80 \text{ m/s}^2) = 4.38 \text{ m}.$$

EVALUATE: When the displacement is upward (part (b)), the acceleration is downward. The mass of the partridge is never entered into the calculation. We used just the ratio k/m, that is determined from T.

14.85.   **IDENTIFY:**   Apply conservation of linear momentum to the collision between the steak and the pan. Then apply conservation of energy to the motion after the collision to find the amplitude of the subsequent SHM. Use Eq. (14.12) to calculate the period.

(a) SET UP: First find the speed of the steak just before it strikes the pan. Use a coordinate system with $+y$ downward.

$v_{0y} = 0$ (released from the rest); $y - y_0 = 0.40$ m; $a_y = +9.80 \text{ m/s}^2$; $v_y = ?$

$v_y^2 = v_{0_y}^2 + 2a_y(y - y_0)$

EXECUTE: $v_y = +\sqrt{2a_y(y - y_0)} = +\sqrt{2(9.80 \text{ m/s}^2)(0.40 \text{ m})} = +2.80 \text{ m/s}$

SET UP: Apply conservation of momentum to the collision between the steak and the pan. After the collision the steak and the pan are moving together with common velocity v_2. Let A be the steak and B be the pan. The system before and after the collision is shown in Figure 14.85.

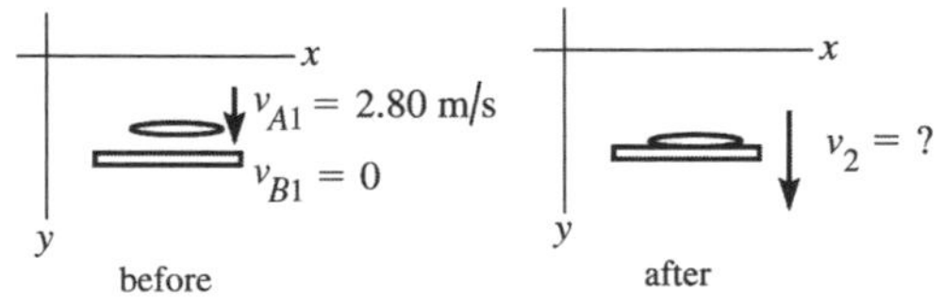

$v_{A1} = 2.80$ m/s	$v_2 = ?$
$v_{B1} = 0$	
before	after

Figure 14.85

EXECUTE: P_y conserved: $m_A v_{A1y} + m_B v_{B1y} = (m_A + m_B)v_{2y}$

$m_A v_{A1} = (m_A + m_B)v_2$

$v_2 = \left(\dfrac{m_A}{m_A + m_B}\right)v_{A1} = \left(\dfrac{2.2 \text{ kg}}{2.2 \text{ kg} + 0.20 \text{ kg}}\right)(2.80 \text{ m/s}) = 2.57 \text{ m/s}$

(b) SET UP: Conservation of energy applied to the SHM gives: $\frac{1}{2}mv_0^2 + \frac{1}{2}kx_0^2 = \frac{1}{2}kA^2$ where v_0 and x_0

are the initial speed and displacement of the object and where the displacement is measured from the
equilibrium position of the object.
EXECUTE: The weight of the steak will stretch the spring an additional distance d given by $kd = mg$ so

$d = \dfrac{mg}{k} = \dfrac{(2.2 \text{ kg})(9.80 \text{ m/s}^2)}{400 \text{ N/m}} = 0.0539 \text{ m}.$ So just after the steak hits the pan, before the pan has had time

to move, the steak plus pan is 0.0539 m above the equilibrium position of the combined object. Thus
$x_0 = 0.0539 \text{ m}.$ From part (a) $v_0 = 2.57 \text{ m/s},$ the speed of the combined object just after the collision.

Then $\frac{1}{2}mv_0^2 + \frac{1}{2}kx_0^2 = \frac{1}{2}kA^2$ gives

$$A = \sqrt{\dfrac{mv_0^2 + kx_0^2}{k}} = \sqrt{\dfrac{2.4 \text{ kg}(2.57 \text{ m/s})^2 + (400 \text{ N/m})(0.0539 \text{ m})^2}{400 \text{ N/m}}} = 0.21 \text{ m}$$

(c) $T = 2\pi\sqrt{m/k} = 2\pi\sqrt{\dfrac{2.4 \text{ kg}}{400 \text{ N/m}}} = 0.49 \text{ s}$

EVALUATE: The amplitude is less than the initial height of the steak above the pan because mechanical
energy is lost in the inelastic collision.

14.87.   **IDENTIFY** and **SET UP:**   Use Eq. (14.12) to calculate g and use Eq. (14.4) applied to Newtonia to relate g
to the mass of the planet.
EXECUTE: The pendulum swings through $\frac{1}{2}$ cycle in 1.42 s, so $T = 2.84$ s. $L = 1.85$ m. Use T to find g:

$T = 2\pi\sqrt{L/g}$ so $g = L(2\pi/T)^2 = 9.055 \text{ m/s}^2$

Use g to find the mass M_p of Newtonia: $g = GM_p/R_p^2$

$2\pi R_p = 5.14 \times 10^7$ m, so $R_p = 8.18 \times 10^6$ m

$m_p = \dfrac{gR_p^2}{G} = 9.08 \times 10^{24} \text{ kg}$

EVALUATE: g is similar to that at the surface of the earth. The radius of Newtonia is a little less than
earth's radius and its mass is a little more.

14.89.   **IDENTIFY:**   Use Eq. (14.13) to relate x and t. $T = 3.5$ s.
SET UP: The motion of the raft is sketched in Figure 14.89.

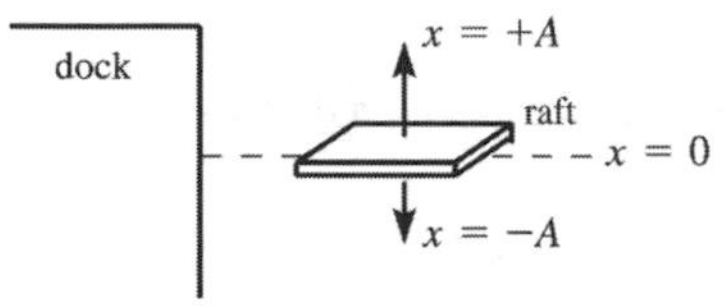

Let the raft be at $x = +A$ when $t = 0$.
Then $\phi = 0$ and $x(t) = A\cos\omega t$.

Figure 14.89

EXECUTE: Calculate the time it takes the raft to move from $x = +A = +0.200$ m to $x = A - 0.100$ m $= 0.100$ m.
Write the equation for $x(t)$ in terms of T rather than ω. $\omega = 2\pi/T$ gives that $x(t) = A\cos(2\pi t/T)$
$x = A$ at $t = 0$
$x = 0.100$ m implies 0.100 m $= (0.200 \text{ m})\cos(2\pi t/T)$
$\cos(2\pi t/T) = 0.500$ so $2\pi t/T = \arccos(0.500) = 1.047$ rad
$t = (T/2\pi)(1.047 \text{ rad}) = (3.5 \text{ s}/2\pi)(1.047 \text{ rad}) = 0.583$ s

This is the time for the raft to move down from $x = 0.200$ m to $x = 0.100$ m. But people can also get off while the raft is moving up from $x = 0.100$ m to $x = 0.200$ m, so during each period of the motion the time the people have to get off is $2t = 2(0.583 \text{ s}) = 1.17$ s.

EVALUATE: The time to go from $x = 0$ to $x = A$ and return is $T/2 = 1.75$ s. The time to go from $x = A/2$ to A and return is less than this.

14.91.   **IDENTIFY:** During the collision, linear momentum is conserved. After the collision, mechanical energy is conserved and the motion is SHM.

SET UP: The linear momentum is $p_x = mv_x$, the kinetic energy is $\frac{1}{2}mv^2$, and the potential energy is $\frac{1}{2}kx^2$. The period is $T = 2\pi\sqrt{\dfrac{m}{k}}$, which is the target variable.

EXECUTE: Apply conservation of linear momentum to the collision:

$(8.00 \times 10^{-3} \text{ kg})(280 \text{ m/s}) = (1.00 \text{ kg})v.$ $v = 2.24$ m/s. This is v_{max} for the SHM. $A = 0.180$ m (given).

So $\dfrac{1}{2}mv^2_{\text{max}} = \dfrac{1}{2}kA^2.$ $k = \left(\dfrac{v_{\text{max}}}{A}\right)^2 m = \left(\dfrac{2.24 \text{ m/s}}{0.180 \text{ m}}\right)^2 (1.00 \text{ kg}) = 154.9$ N/m.

$T = 2\pi\sqrt{\dfrac{m}{k}} = 2\pi\sqrt{\dfrac{1.00 \text{ kg}}{154.9 \text{ N/m}}} = 0.505$ s.

EVALUATE: This block would weigh about 2 pounds, which is rather heavy, but the spring constant is large enough to keep the period within an easily observable range.

14.95.   **IDENTIFY:** Apply conservation of energy to the motion before and after the collision. Apply conservation of linear momentum to the collision. After the collision the system moves as a simple pendulum. If the maximum angular displacement is small, $f = \dfrac{1}{2\pi}\sqrt{\dfrac{g}{L}}$.

SET UP: In the motion before and after the collision there is energy conversion between gravitational potential energy mgh, where h is the height above the lowest point in the motion, and kinetic energy.

EXECUTE: Energy conservation during downward swing: $m_2 gh_0 = \frac{1}{2}m_2 v^2$ and

$v = \sqrt{2gh_0} = \sqrt{2(9.8 \text{ m/s}^2)(0.100 \text{ m})} = 1.40$ m/s.

Momentum conservation during collision: $m_2 v = (m_2 + m_3)V$ and

$V = \dfrac{m_2 v}{m_2 + m_3} = \dfrac{(2.00 \text{ kg})(1.40 \text{ m/s})}{5.00 \text{ kg}} = 0.560$ m/s.

Energy conservation during upward swing: $Mgh_f = \dfrac{1}{2}MV^2$ and

$h_f = V^2/2g = \dfrac{(0.560 \text{ m/s})^2}{2(9.80 \text{ m/s}^2)} = 0.0160 \text{ m} = 1.60$ cm.

Figure 14.95 shows how the maximum angular displacement is calculated from h_f . $\cos\theta = \dfrac{48.4 \text{ cm}}{50.0 \text{ cm}}$ and

$\theta = 14.5°.$ $f = \dfrac{1}{2\pi}\sqrt{\dfrac{g}{l}} = \dfrac{1}{2\pi}\sqrt{\dfrac{9.80 \text{ m/s}^2}{0.500 \text{ m}}} = 0.705$ Hz.

EVALUATE: $14.5° = 0.253$ rad. $\sin(0.253 \text{ rad}) = 0.250.$ $\sin\theta \approx \theta$ and Eq. (14.34) is accurate.

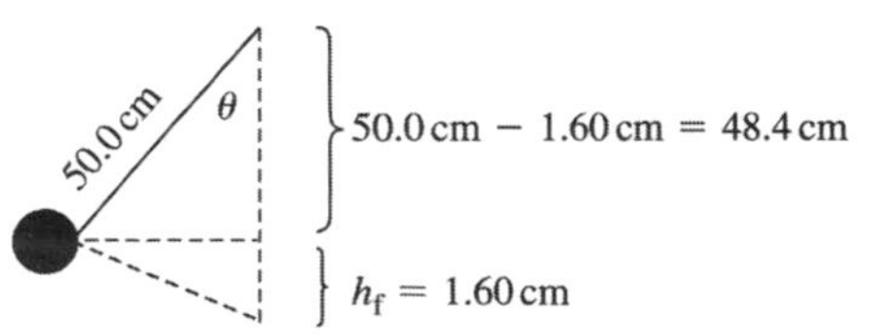

Figure 14.95

14.97.   **IDENTIFY:**   The motion is simple harmonic if the equation of motion for the angular oscillations is of the

form $\dfrac{d^2\theta}{dt^2} = -\dfrac{\kappa}{I}\theta,$ and in this case the period is $T = 2\pi\sqrt{I/\kappa}.$

SET UP: For a slender rod pivoted about its center, $I = \tfrac{1}{12}ML^2.$

EXECUTE: The torque on the rod about the pivot is $\tau = -\left(k\dfrac{L}{2}\theta\right)\dfrac{L}{2}.$ $\tau = I\alpha = I\dfrac{d^2\theta}{dt^2}$ gives

$\dfrac{d^2\theta}{dt^2} = -k\dfrac{L^2/4}{I}\theta = -\dfrac{3k}{M}\theta.$ $\dfrac{d^2\theta}{dt^2}$ is proportional to θ and the motion is angular SHM. $\dfrac{\kappa}{I} = \dfrac{3k}{M},$

$T = 2\pi\sqrt{\dfrac{M}{3k}}.$

EVALUATE: The expression we used for the torque, $\tau = -\left(k\dfrac{L}{2}\theta\right)\dfrac{L}{2},$ is valid only when θ is small

enough for $\sin\theta \approx \theta$ and $\cos\theta \approx 1.$

14.99.   **IDENTIFY:**   The object oscillates as a physical pendulum, with $f = \dfrac{1}{2\pi}\sqrt{\dfrac{mgd}{I}},$ where m is the total mass

of the object.

SET UP: The moment of inertia about the pivot is $2(1/3)ML^2 = (2/3)\,ML^2,$ and the center of gravity when

balanced is a distance $d = L/(2\sqrt{2})$ below the pivot.

EXECUTE: The frequency is $f = \dfrac{1}{T} = \dfrac{1}{2\pi}\sqrt{\dfrac{6g}{4\sqrt{2}L}} = \dfrac{1}{4\pi}\sqrt{\dfrac{6g}{\sqrt{2}L}}.$

EVALUATE: If $f_{sp} = \dfrac{1}{2\pi}\sqrt{\dfrac{g}{L}}$ is the frequency for a simple pendulum of length $L,$

$f = \dfrac{1}{2}\sqrt{\dfrac{6}{\sqrt{2}}}f_{sp} = 1.03f_{sp}.$

15

MECHANICAL WAVES

15.7. **IDENTIFY:** Use Eq. (15.1) to calculate v. $T = 1/f$ and k is defined by Eq. (15.5). The general form of the wave function is given by Eq. (15.8), which is the equation for the transverse displacement.

SET UP: $v = 8.00$ m/s, $A = 0.0700$ m, $\lambda = 0.320$ m

EXECUTE: **(a)** $v = f\lambda$ so $f = v/\lambda = (8.00 \text{ m/s})/(0.320 \text{ m}) = 25.0$ Hz

$T = 1/f = 1/25.0 \text{ Hz} = 0.0400$ s

$k = 2\pi/\lambda = 2\pi \text{ rad}/0.320 \text{ m} = 19.6$ rad/m

(b) For a wave traveling in the $-x$-direction,

$y(x, t) = A\cos 2\pi(x/\lambda + t/T)$ (Eq. (15.8).)

At $x = 0$, $y(0, t) = A\cos 2\pi(t/T)$, so $y = A$ at $t = 0$. This equation describes the wave specified in the problem.

Substitute in numerical values:

$y(x, t) = (0.0700 \text{ m})\cos(2\pi(x/0.320 \text{ m} + t/0.0400 \text{ s}))$.

Or, $y(x, t) = (0.0700 \text{ m})\cos((19.6 \text{ m}^{-1})x + (157 \text{ rad/s})t)$.

(c) From part (b), $y = (0.0700 \text{ m})\cos(2\pi(x/0.320 \text{ m} + t/0.0400 \text{ s}))$.

Plug in $x = 0.360$ m and $t = 0.150$ s:

$y = (0.0700 \text{ m})\cos(2\pi(0.360 \text{ m}/0.320 \text{ m} + 0.150 \text{ s}/0.0400 \text{ s}))$

$y = (0.0700 \text{ m})\cos[2\pi(4.875 \text{ rad})] = +0.0495 \text{ m} = +4.95$ cm

(d) In part (c) $t = 0.150$ s.

$y = A$ means $\cos(2\pi(x/\lambda + t/T)) = 1$

$\cos\theta = 1$ for $\theta = 0, 2\pi, 4\pi,\ldots = n(2\pi)$ or $n = 0, 1, 2,\ldots$

So $y = A$ when $2\pi(x/\lambda + t/T) = n(2\pi)$ or $x/\lambda + t/T = n$

$t = T(n - x/\lambda) = (0.0400 \text{ s})(n - 0.360 \text{ m}/0.320 \text{ m}) = (0.0400 \text{ s})(n - 1.125)$

For $n = 4$, $t = 0.1150$ s (before the instant in part (c))

For $n = 5$, $t = 0.1550$ s (the first occurrence of $y = A$ after the instant in part (c)). Thus the elapsed time is $0.1550 \text{ s} - 0.1500 \text{ s} = 0.0050$ s.

EVALUATE: Part (d) says $y = A$ at 0.115 s and next at 0.155 s; the difference between these two times is 0.040 s, which is the period. At $t = 0.150$ s the particle at $x = 0.360$ m is at $y = 4.95$ cm and traveling upward. It takes $T/4 = 0.0100$ s for it to travel from $y = 0$ to $y = A$, so our answer of 0.0050 s is reasonable.

15.11. **IDENTIFY and SET UP:** Read A and T from the graph. Apply Eq. (15.4) to determine λ and then use Eq. (15.1) to calculate v.

EXECUTE: **(a)** The maximum y is 4 mm (read from graph).

(b) For either x the time for one full cycle is 0.040 s; this is the period.

(c) Since $y = 0$ for $x = 0$ and $t = 0$ and since the wave is traveling in the $+x$-direction then

$y(x, t) = A\sin[2\pi(t/T - x/\lambda)]$. (The phase is different from the wave described by Eq. (15.4); for that wave

$y = A$ for $x = 0$, $t = 0$.) From the graph, if the wave is traveling in the $+x$-direction and if $x = 0$ and $x = 0.090$ m are within one wavelength the peak at $t = 0.01$ s for $x = 0$ moves so that it occurs at $t = 0.035$ s (read from graph so is approximate) for $x = 0.090$ m. The peak for $x = 0$ is the first peak past $t = 0$ so corresponds to the first maximum in $\sin[2\pi(t/T - x/\lambda)]$ and hence occurs at $2\pi(t/T - x/\lambda) = \pi/2$. If this same peak moves to $t_1 = 0.035$ s at $x_1 = 0.090$ m, then $2\pi(t/T - x/\lambda) = \pi/2$.

Solve for λ: $t_1/T - x_1/\lambda = 1/4$

$x_1/\lambda = t_1/T - 1/4 = 0.035$ s/0.040 s $- 0.25 = 0.625$

$\lambda = x_1/0.625 = 0.090$ m/$0.625 = 0.14$ m.

Then $v = f\lambda = \lambda/T = 0.14$ m/0.040 s $= 3.5$ m/s.

(d) If the wave is traveling in the $-x$-direction, then $y(x, t) = A\sin(2\pi(t/T + x/\lambda))$ and the peak at $t = 0.050$ s for $x = 0$ corresponds to the peak at $t_1 = 0.035$ s for $x_1 = 0.090$ m. This peak at $x = 0$ is the second peak past the origin so corresponds to $2\pi(t/T + x/\lambda) = 5\pi/2$. If this same peak moves to $t_1 = 0.035$ s for $x_1 = 0.090$ m, then $2\pi(t_1/T + x_1/\lambda) = 5\pi/2$.

$t_1/T + x_1/\lambda = 5/4$

$x_1/\lambda = 5/4 - t_1/T = 5/4 - 0.035$ s/0.040 s $= 0.375$

$\lambda = x_1/0.375 = 0.090$ m/$0.375 = 0.24$ m.

Then $v = f\lambda = \lambda/T = 0.24$ m/0.040 s $= 6.0$ m/s.

EVALUATE: (e) No. Wouldn't know which point in the wave at $x = 0$ moved to which point at $x = 0.090$ m.

15.17.   **IDENTIFY:**   The speed of the wave depends on the tension in the wire and its mass density. The target variable is the mass of the wire of known length.

SET UP: $v = \sqrt{\dfrac{F}{\mu}}$ and $\mu = m/L$.

EXECUTE: First find the speed of the wave: $v = \dfrac{3.80 \text{ m}}{0.0492 \text{ s}} = 77.24$ m/s. $v = \sqrt{\dfrac{F}{\mu}}$.

$\mu = \dfrac{F}{v^2} = \dfrac{(54.0 \text{ kg})(9.8 \text{ m/s}^2)}{(77.24 \text{ m/s})^2} = 0.08870$ kg/m. The mass of the wire is

$m = \mu L = (0.08870 \text{ kg/m})(3.80 \text{ m}) = 0.337$ kg.

EVALUATE: This mass is 337 g, which is a bit large for a wire 3.80 m long. It must be fairly thick.

15.19.   **IDENTIFY:**   For transverse waves on a string, $v = \sqrt{F/\mu}$. $v = f\lambda$.

SET UP: The wire has $\mu = m/L = (0.0165 \text{ kg})/(0.750 \text{ m}) = 0.0220$ kg/m.

EXECUTE: (a) $v = f\lambda = (875 \text{ Hz})(3.33 \times 10^{-2} \text{ m}) = 29.1$ m/s. The tension is

$F = \mu v^2 = (0.0220 \text{ kg/m})(29.1 \text{ m/s})^2 = 18.6$ N.

(b) $v = 29.1$ m/s

EVALUATE: If λ is kept fixed, the wave speed and the frequency increase when the tension is increased.

15.21.   **IDENTIFY:** $v = \sqrt{F/\mu}$. $v = f\lambda$. The general form for $y(x, t)$ is given in Eq. (15.4), where $T = 1/f$.

Eq. (15.10) says that the maximum transverse acceleration is $a_{\max} = \omega^2 A = (2\pi f)^2 A$.

SET UP: $\mu = 0.0500$ kg/m

EXECUTE: (a) $v = \sqrt{F/\mu} = \sqrt{(5.00 \text{ N})/(0.0500) \text{ kg/m}} = 10.0$ m/s

(b) $\lambda = v/f = (10.0 \text{ m/s})/(40.0 \text{ Hz}) = 0.250$ m

(c) $y(x,t) = A \cos(kx - \omega t)$. $k = 2\pi/\lambda = 8.00\pi$ rad/m; $\omega = 2\pi f = 80.0\pi$ rad/s.

$y(x,t) = (3.00 \text{ cm})\cos[\pi(8.00 \text{ rad/m})x - (80.0\pi \text{ rad/s})t]$

(d) $v_y = +A\omega \sin(kx - \omega t)$ and $a_y = -A\omega^2 \cos(kx - \omega t)$. $a_{y,\text{ max}} = A\omega^2 = A(2\pi f)^2 = 1890 \text{ m/s}^2$.

(e) $a_{y,\text{max}}$ is much larger than g, so it is a reasonable approximation to ignore gravity.

EVALUATE: $y(x,t)$ in part (c) gives $y(0,0) = A$, which does correspond to the oscillator having maximum upward displacement at $t = 0$.

15.23. **IDENTIFY:** The average power carried by the wave depends on the mass density of the wire and the tension in it, as well as on the square of both the frequency and amplitude of the wave (the target variable).

SET UP: $P_{\text{av}} = \frac{1}{2}\sqrt{\mu F}\,\omega^2 A^2$, $v = \sqrt{\dfrac{F}{\mu}}$.

EXECUTE: Solving $P_{\text{av}} = \frac{1}{2}\sqrt{\mu F}\,\omega^2 A^2$ for A gives $A = \left(\dfrac{2P_{\text{av}}}{\omega^2\sqrt{\mu F}}\right)^{1/2}$. $P_{\text{av}} = 0.365$ W.

$\omega = 2\pi f = 2\pi(69.0 \text{ Hz}) = 433.5$ rad/s. The tension is $F = 94.0$ N and $v = \sqrt{\dfrac{F}{\mu}}$ so

$\mu = \dfrac{F}{v^2} = \dfrac{94.0 \text{ N}}{(492 \text{ m/s})^2} = 3.883\times10^{-4}$ kg/m.

$A = \left(\dfrac{2(0.365 \text{ W})}{(433.5 \text{ rad/s})^2\sqrt{(3.883\times10^{-4} \text{ kg/m})(94.0 \text{ N})}}\right)^{1/2} = 4.51\times10^{-3} \text{ m} = 4.51 \text{ mm}$

EVALUATE: Vibrations of strings and wires normally have small amplitudes, which this wave does.

15.25. **IDENTIFY:** For a point source, $I = \dfrac{P}{4\pi r^2}$ and $\dfrac{I_1}{I_2} = \dfrac{r_2^2}{r_1^2}$.

SET UP: $1\,\mu\text{W} = 10^{-6}$ W

EXECUTE: **(a)** $r_2 = r_1\sqrt{\dfrac{I_1}{I_2}} = (30.0 \text{ m})\sqrt{\dfrac{10.0 \text{ W/m}^2}{1\times10^{-6} \text{ W/m}^2}} = 95$ km

(b) $\dfrac{I_2}{I_3} = \dfrac{r_3^2}{r_2^2}$, with $I_2 = 1.0\,\mu\text{W/m}^2$ and $r_3 = 2r_2$. $I_3 = I_2\left(\dfrac{r_2}{r_3}\right)^2 = I_2/4 = 0.25\,\mu\text{W/m}^2$.

(c) $P = I(4\pi r^2) = (10.0 \text{ W/m}^2)(4\pi)(30.0 \text{ m})^2 = 1.1\times10^5$ W

EVALUATE: These are approximate calculations, that assume the sound is emitted uniformly in all directions and that ignore the effects of reflection, for example reflections from the ground.

15.27. **IDENTIFY:** and **SET UP:** Apply Eq. (15.26) to relate I and r.

Power is related to intensity at a distance r by $P = I(4\pi r^2)$. Energy is power times time.

EXECUTE: **(a)** $I_1 r_1^2 = I_2 r_2^2$

$I_2 = I_1(r_1/r_2)^2 = (0.026 \text{ W/m}^2)(4.3 \text{ m}/3.1 \text{ m})^2 = 0.050 \text{ W/m}^2$

(b) $P = 4\pi r^2 I = 4\pi(4.3 \text{ m})^2(0.026 \text{ W/m}^2) = 6.04$ W

Energy $= Pt = (6.04 \text{ W})(3600 \text{ s}) = 2.2\times10^4$ J

EVALUATE: We could have used $r = 3.1$ m and $I = 0.050 \text{ W/m}^2$ in $P = 4\pi r^2 I$ and would have obtained the same P. Intensity becomes less as r increases because the radiated power spreads over a sphere of larger area.

15.41. **IDENTIFY:** Use Eq. (15.1) for v and Eq. (15.13) for the tension F. $v_y = \partial y/\partial t$ and $a_y = \partial v_y/\partial t$.

(a) SET UP: The fundamental standing wave is sketched in Figure 15.41.

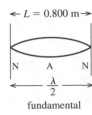

$f = 60.0$ Hz

From the sketch,

$\lambda/2 = L$ so

$\lambda = 2L = 1.60$ m

Figure 15.41

EXECUTE: $v = f\lambda = (60.0 \text{ Hz})(1.60 \text{ m}) = 96.0$ m/s

(b) The tension is related to the wave speed by Eq. (15.13):

$v = \sqrt{F/\mu}$ so $F = \mu v^2$.

$\mu = m/L = 0.0400 \text{ kg}/0.800 \text{ m} = 0.0500$ kg/m

$F = \mu v^2 = (0.0500 \text{ kg/m})(96.0 \text{ m/s})^2 = 461$ N.

(c) $\omega = 2\pi f = 377$ rad/s and $y(x, t) = A_{SW} \sin kx \sin \omega t$

$v_y = \omega A_{SW} \sin kx \cos \omega t;$ $a_y = -\omega^2 A_{SW} \sin kx \sin \omega t$

$(v_y)_{max} = \omega A_{SW} = (377 \text{ rad/s})(0.300 \text{ cm}) = 1.13$ m/s.

$(a_y)_{max} = \omega^2 A_{SW} = (377 \text{ rad/s})^2(0.300 \text{ cm}) = 426 \text{ m/s}^2.$

EVALUATE: The transverse velocity is different from the wave velocity. The wave velocity and tension are similar in magnitude to the values in the examples in the text. Note that the transverse acceleration is quite large.

15.43.   **IDENTIFY:**   Compare $y(x, t)$ given in the problem to Eq. (15.28). From the frequency and wavelength for the third harmonic find these values for the eighth harmonic.

(a) SET UP: The third harmonic standing wave pattern is sketched in Figure 15.43.

Figure 15.43

EXECUTE: **(b)** Eq. (15.28) gives the general equation for a standing wave on a string:

$y(x, t) = (A_{SW} \sin kx)\sin \omega t$

$A_{SW} = 2A,$ so $A = A_{SW}/2 = (5.60 \text{ cm})/2 = 2.80$ cm

(c) The sketch in part (a) shows that $L = 3(\lambda/2)$. $k = 2\pi/\lambda$, $\lambda = 2\pi/k$

Comparison of $y(x, t)$ given in the problem to Eq. (15.28) gives $k = 0.0340$ rad/cm. So,

$\lambda = 2\pi/(0.0340 \text{ rad/cm}) = 184.8$ cm

$L = 3(\lambda/2) = 277$ cm

(d) $\lambda = 185$ cm, from part (c)

$\omega = 50.0$ rad/s so $f = \omega/2\pi = 7.96$ Hz

period $T = 1/f = 0.126$ s

$v = f\lambda = 1470$ cm/s

(e) $v_y = \partial y/\partial t = \omega A_{SW} \sin kx \cos \omega t$

$v_{y, max} = \omega A_{SW} = (50.0 \text{ rad/s})(5.60 \text{ cm}) = 280$ cm/s

(f) $f_3 = 7.96$ Hz $= 3f_1$, so $f_1 = 2.65$ Hz is the fundamental

$f_8 = 8f_1 = 21.2$ Hz; $\omega_8 = 2\pi f_8 = 133$ rad/s

$\lambda = v/f = (1470 \text{ cm/s})/(21.2 \text{ Hz}) = 69.3$ cm and $k = 2\pi/\lambda = 0.0906$ rad/cm

$y(x,t) = (5.60 \text{ cm})\sin([0.0906 \text{ rad/cm}]x)\sin([133 \text{ rad/s}]t)$

EVALUATE: The wavelength and frequency of the standing wave equals the wavelength and frequency of the two traveling waves that combine to form the standing wave. In the 8th harmonic the frequency and wave number are larger than in the 3rd harmonic.

15.45. **(a) IDENTIFY and SET UP:** Use the angular frequency and wave number for the traveling waves in Eq. (15.28) for the standing wave.
EXECUTE: The traveling wave is $y(x,t) = (2.30 \text{ mm})\cos([6.98 \text{ rad/m}]x) + [742 \text{ rad/s}]t)$

$A = 2.30$ mm so $A_{SW} = 4.60$ mm; $k = 6.98$ rad/m and $\omega = 742$ rad/s

The general equation for a standing wave is $y(x,t) = (A_{SW}\sin kx)\sin \omega t$, so

$y(x,t) = (4.60 \text{ mm})\sin([6.98 \text{ rad/m}]x)\sin([742 \text{ rad/s}]t)$

(b) IDENTIFY and SET UP: Compare the wavelength to the length of the rope in order to identify the harmonic.
EXECUTE: $L = 1.35$ m (from Exercise 15.28)
$\lambda = 2\pi/k = 0.900$ m
$L = 3(\lambda/2)$, so this is the 3rd harmonic

(c) For this 3rd harmonic, $f = \omega/2\pi = 118$ Hz

$f_3 = 3f_1$ so $f_1 = (118 \text{ Hz})/3 = 39.3$ Hz

EVALUATE: The wavelength and frequency of the standing wave equals the wavelength and frequency of the two traveling waves that combine to form the standing wave. The nth harmonic has n node-to-node segments and the node-to-node distance is $\lambda/2$, so the relation between L and λ for the nth harmonic is $L = n(\lambda/2)$.

15.47. **IDENTIFY and SET UP:** Use the information given about the A_4 note to find the wave speed that depends on the linear mass density of the string and the tension. The wave speed isn't affected by the placement of the fingers on the bridge. Then find the wavelength for the D_5 note and relate this to the length of the vibrating portion of the string.
EXECUTE: **(a)** $f = 440$ Hz when a length $L = 0.600$ m vibrates; use this information to calculate the speed v of waves on the string. For the fundamental $\lambda/2 = L$ so $\lambda = 2L = 2(0.600 \text{ m}) = 1.20$ m. Then $v = f\lambda = (440 \text{ Hz})(1.20 \text{ m}) = 528$ m/s. Now find the length $L = x$ of the string that makes $f = 587$ Hz.

$\lambda = \dfrac{v}{f} = \dfrac{528 \text{ m/s}}{587 \text{ Hz}} = 0.900$ m

$L = \lambda/2 = 0.450$ m, so $x = 0.450$ m $= 45.0$ cm.

(b) No retuning means same wave speed as in part (a). Find the length of vibrating string needed to produce $f = 392$ Hz.

$\lambda = \dfrac{v}{f} = \dfrac{528 \text{ m/s}}{392 \text{ Hz}} = 1.35$ m

$L = \lambda/2 = 0.675$ m; string is shorter than this. No, not possible.

EVALUATE: Shortening the length of this vibrating string increases the frequency of the fundamental.

15.49. **IDENTIFY:** For the fundamental, $f_1 = \dfrac{v}{2L}$. $v = \sqrt{F/\mu}$. A standing wave on a string with frequency f produces a sound wave that also has frequency f.
SET UP: $f_1 = 245$ Hz. $L = 0.635$ m.
EXECUTE: **(a)** $v = 2f_1L = 2(245 \text{ Hz})(0.635 \text{ m}) = 311$ m/s.
(b) The frequency of the fundamental mode is proportional to the speed and hence to the square root of the tension; $(245 \text{ Hz})\sqrt{1.01} = 246$ Hz.

(c) The frequency will be the same, 245 Hz. The wavelength will be

$\lambda_{air} = v_{air}/f = (344 \text{ m/s})/(245 \text{ Hz}) = 1.40 \text{ m},$ which is larger than the wavelength of standing wave on the string by a factor of the ratio of the speeds.

EVALUATE: Increasing the tension increases the wave speed and this in turn increases the frequencies of the standing waves. The wavelength of each normal mode depends only on the length of the string and doesn't change when the tension changes.

15.51. **IDENTIFY and SET UP:** Calculate v, ω, and k from Eqs. (15.1), (15.5) and (15.6). Then apply Eq. (15.7) to obtain $y(x,t)$.

$A = 2.50\times10^{-3}$ m, $\lambda = 1.80$ m, $v = 36.0$ m/s

EXECUTE: **(a)** $v = f\lambda$ so $f = v/\lambda = (36.0 \text{ m/s})/1.80 \text{ m} = 20.0 \text{ Hz}$

$\omega = 2\pi f = 2\pi(20.0 \text{ Hz}) = 126 \text{ rad/s}$

$k = 2\pi/\lambda = 2\pi \text{ rad}/1.80 \text{ m} = 3.49 \text{ rad/m}$

(b) For a wave traveling to the right, $y(x,t) = A\cos(kx - \omega t)$. This equation gives that the $x = 0$ end of the string has maximum upward displacement at $t = 0$.

Put in the numbers: $y(x,t) = (2.50\times10^{-3} \text{ m})\cos((3.49 \text{ rad/m})x - (126 \text{ rad/s})t)$.

(c) The left-hand end is located at $x = 0$. Put this value into the equation of part (b):

$y(0,t) = +(2.50\times10^{-3} \text{ m})\cos((126 \text{ rad/s})t)$.

(d) Put $x = 1.35$ m into the equation of part (b):

$y(1.35 \text{ m}, t) = (2.50\times10^{-3} \text{ m})\cos((3.49 \text{ rad/m})(1.35 \text{ m}) - (126 \text{ rad/s})t)$.

$y(1.35 \text{ m}, t) = (2.50\times10^{-3} \text{ m})\cos(4.71 \text{ rad} - (126 \text{ rad/s})t)$

$4.71 \text{ rad} = 3\pi/2$ and $\cos(\theta) = \cos(-\theta)$, so $y(1.35 \text{ m}, t) = (2.50\times10^{-3} \text{ m})\cos((126 \text{ rad/s})t - 3\pi/2 \text{ rad})$

(e) $y = A\cos(kx - \omega t)$ (part (b))

The transverse velocity is given by $v_y = \dfrac{\partial y}{\partial t} = A\dfrac{\partial}{\partial t}\cos(kx - \omega t) = +A\omega\sin(kx - \omega t)$.

The maximum v_y is $A\omega = (2.50\times10^{-3} \text{ m})(126 \text{ rad/s}) = 0.315 \text{ m/s}$.

(f) $y(x,t) = (2.50\times10^{-3} \text{ m})\cos((3.49 \text{ rad/m})x - (126 \text{ rad/s})t)$

$t = 0.0625$ s and $x = 1.35$ m gives

$y = (2.50\times10^{-3} \text{ m})\cos((3.49 \text{ rad/m})(1.35 \text{ m}) - (126 \text{ rad/s})(0.0625 \text{ s})) = -2.50\times10^{-3}$ m.

$v_y = +A\omega\sin(kx - \omega t) = +(0.315 \text{ m/s})\sin((3.49 \text{ rad/m})x - (126 \text{ rad/s})t)$

$t = 0.0625$ s and $x = 1.35$ m gives

$v_y = (0.315 \text{ m/s})\sin((3.49 \text{ rad/m})(1.35 \text{ m}) - (126 \text{ rad/s})(0.0625 \text{ s})) = 0.0$

EVALUATE: The results of part (f) illustrate that $v_y = 0$ when $y = \pm A$, as we saw from SHM in Chapter 14.

15.53. **IDENTIFY:** The speed in each segment is $v = \sqrt{F/\mu}$. The time to travel through a segment is $t = L/v$.

SET UP: The travel times for each segment are $t_1 = L\sqrt{\dfrac{\mu_1}{F}}$, $t_2 = L\sqrt{\dfrac{4\mu_1}{F}}$, and $t_3 = L\sqrt{\dfrac{\mu_1}{4F}}$.

EXECUTE: **(a)** Adding the travel times gives $t_{total} = L\sqrt{\dfrac{\mu_1}{F}} + 2L\sqrt{\dfrac{\mu_1}{F}} + \dfrac{1}{2}L\sqrt{\dfrac{\mu_1}{F}} = \dfrac{7}{2}L\sqrt{\dfrac{\mu_1}{F}}$.

(b) No. The speed in a segment depends only on F and μ for that segment.

EVALUATE: The wave speed is greater and its travel time smaller when the mass per unit length of the segment decreases.

15.59. **IDENTIFY:** The frequency of the fundamental (the target variable) depends on the tension in the wire. The bar is in rotational equilibrium so the torques on it must balance.

SET UP: $v = \sqrt{\dfrac{F}{\mu}}$ and $f = \dfrac{v}{\lambda}$. $\Sigma\tau_z = 0$.

EXECUTE: $\lambda = 2L = 0.660$ m. The tension F in the wire is found by applying the rotational equilibrium methods of Chapter 11. Let l be the length of the bar. Then $\Sigma\tau_z = 0$ with the axis at the hinge gives

$$Fl\cos 30° = \frac{1}{2}lmg\sin 30°. \quad F = \frac{mg\tan 30°}{2} = \frac{(45.0\text{ kg})(9.80\text{ m/s}^2)\tan 30°}{2} = 127.3\text{ N}.$$

$$v = \sqrt{\frac{F}{\mu}} = \sqrt{\frac{127.3\text{ N}}{(0.0920\text{ kg}/0.330\text{ m})}} = 21.37\text{ m/s}. \quad f = \frac{v}{\lambda} = \frac{21.37\text{ m/s}}{0.660\text{ m}} = 32.4\text{ Hz}$$

EVALUATE: This is an audible frequency for humans.

15.61. **IDENTIFY:** The wavelengths of standing waves depend on the length of the string (the target variable), which in turn determine the frequencies of the waves.

SET UP: $f_n = nf_1$ where $f_1 = \dfrac{v}{2L}$.

EXECUTE: $f_n = nf_1$ and $f_{n+1} = (n+1)f_1$. We know the wavelengths of two adjacent modes, so

$$f_1 = f_{n+1} - f_n = 630\text{ Hz} - 525\text{ Hz} = 105\text{ Hz}. \text{ Solving } f_1 = \frac{v}{2L} \text{ for } L \text{ gives } L = \frac{v_1}{2f} = \frac{384\text{ m/s}}{2(105\text{ Hz})} = 1.83\text{ m}.$$

EVALUATE: The observed frequencies are both audible which is reasonable for a string that is about a half meter long.

15.63. **IDENTIFY:** The tension in the wires along with their lengths determine the fundamental frequency in each one (the target variables). These frequencies are different because the wires have different linear mass densities. The bar is in equilibrium, so the forces and torques on it balance.

SET UP: $T_a + T_c = w$, $\Sigma\tau_z = 0$, $v = \sqrt{\dfrac{F}{\mu}}$, $f_1 = v/2L$ and $\mu = \dfrac{m}{L}$, where $m = \rho V = \rho\pi r^2 L$. The densities of copper and aluminum are given in a table in the text.

EXECUTE: Using the subscript "a" for aluminum and "c" for copper, we have $T_a + T_c = w = 536$ N. $\Sigma\tau_z = 0$, with the axis at left-hand end of bar, gives $T_c(1.40\text{ m}) = w(0.90\text{ m})$, so $T_c = 344.6$ N.

$$T_a = 536\text{ N} - 344.6\text{ N} = 191.4\text{ N}. \quad f_1 = \frac{v}{2L}. \quad \mu = \frac{m}{L} = \frac{\rho\pi r^2 L}{L} = \rho\pi r^2.$$

For the copper wire: $F = 344.6$ N and $\mu = (8.90\times10^3\text{ kg/m}^3)\pi(0.280\times10^{-3}\text{ m})^2 = 2.19\times10^{-3}$ kg/m, so

$$v = \sqrt{\frac{F}{\mu}} = \sqrt{\frac{344.6\text{ N}}{2.19\times10^{-3}\text{ kg/m}}} = 396.7\text{ m/s}. \quad f_1 = \frac{v}{2L} = \frac{396.7\text{ m/s}}{2(0.600\text{ m})} = 330\text{ Hz}.$$

For the aluminum wire: $F = 191.4$ N and $\mu = (2.70\times10^3\text{ kg/m}^3)\pi(0.280\times10^{-3}\text{ m})^2 = 6.65\times10^{-4}$ kg/m,

so $v = \sqrt{\dfrac{F}{\mu}} = \sqrt{\dfrac{919.4\text{ N}}{6.65\times10^{-4}\text{ kg/m}}} = 536.5$ m/s, which gives $f_1 = \dfrac{536.5\text{ m/s}}{2(0.600\text{ m})} = 447$ Hz.

EVALUATE: The wires have different fundamental frequencies because they have different tensions and different linear mass densities.

15.69. **IDENTIFY and SET UP:** The average power is given by Eq. (15.25). Rewrite this expression in terms of v and λ in place of F and ω.

EXECUTE: (a) $P_{av} = \frac{1}{2}\sqrt{\mu F}\omega^2 A^2$

$v = \sqrt{F/\mu}$ so $\sqrt{F} = v\sqrt{\mu}$

$\omega = 2\pi f = 2\pi(v/\lambda)$

Using these two expressions to replace $\sqrt{F}$ and ω gives $P_{av} = 2\mu\pi^2 v^3 A^2/\lambda^2$;

$\mu = (6.00\times10^{-3}\text{ kg})/(8.00\text{ m})$

$$A = \left(\frac{2\lambda^2 P_{av}}{4\pi^2 v^3\mu}\right)^{1/2} = 7.07\text{ cm}$$

(b) EVALUATE: $P_{av} \sim v^3$ so doubling v increases P_{av} by a factor of 8.

$P_{av} = 8(50.0 \text{ W}) = 400.0 \text{ W}$

15.73.   **IDENTIFY and SET UP:** There is a node at the post and there must be a node at the clothespin. There could be additional nodes in between. The distance between adjacent nodes is $\lambda/2$, so the distance between *any* two nodes is $n(\lambda/2)$ for $n = 1, 2, 3, \ldots$ This must equal 45.0 cm, since there are nodes at the post and clothespin. Use this in Eq. (15.1) to get an expression for the possible frequencies f.
EXECUTE: $45.0 \text{ cm} = n(\lambda/2)$, $\lambda = v/f$, so $f = n[v/(90.0 \text{ cm})] = (0.800 \text{ Hz})n$, $n = 1, 2, 3, \ldots$

EVALUATE: Higher frequencies have smaller wavelengths, so more node-to-node segments fit between the post and clothespin.

15.77.   **IDENTIFY:** The standing wave frequencies are given by $f_n = n\left(\dfrac{v}{2L}\right)$. $v = \sqrt{F/\mu}$. Use the density of steel to calculate μ for the wire.

SET UP: For steel, $\rho = 7.8 \times 10^3 \text{ kg/m}^3$. For the first overtone standing wave, $n = 2$.

EXECUTE: $v = \dfrac{2Lf_2}{2} = (0.550 \text{ m})(311 \text{ Hz}) = 171 \text{ m/s}$. The volume of the wire is $V = (\pi r^2)L$. $m = \rho V$ so

$\mu = \dfrac{m}{L} = \dfrac{\rho V}{L} = \rho \pi r^2 = (7.8 \times 10^3 \text{ kg/m}^3)\pi(0.57 \times 10^{-3} \text{ m})^2 = 7.96 \times 10^{-3} \text{ kg/m}$. The tension is

$F = \mu v^2 = (7.96 \times 10^{-3} \text{ kg/m})(171 \text{ m/s})^2 = 233 \text{ N}$.

EVALUATE: The tension is not large enough to cause much change in length of the wire.

15.79.   **IDENTIFY:** At a node, $y(x,t) = 0$ for all t. $y_1 + y_2$ is a standing wave if the locations of the nodes don't depend on t.

SET UP: The string is fixed at each end so for all harmonics the ends are nodes. The second harmonic is the first overtone and has one additional node.
EXECUTE: **(a)** The fundamental has nodes only at the ends, $x = 0$ and $x = L$.
(b) For the second harmonic, the wavelength is the length of the string, and the nodes are at $x = 0, x = L/2$ and $x = L$.

(c) The graphs are sketched in Figure 15.79.
(d) The graphs in part (c) show that the locations of the nodes and antinodes between the ends vary in time.
EVALUATE: The sum of two standing waves of different frequencies is not a standing wave.

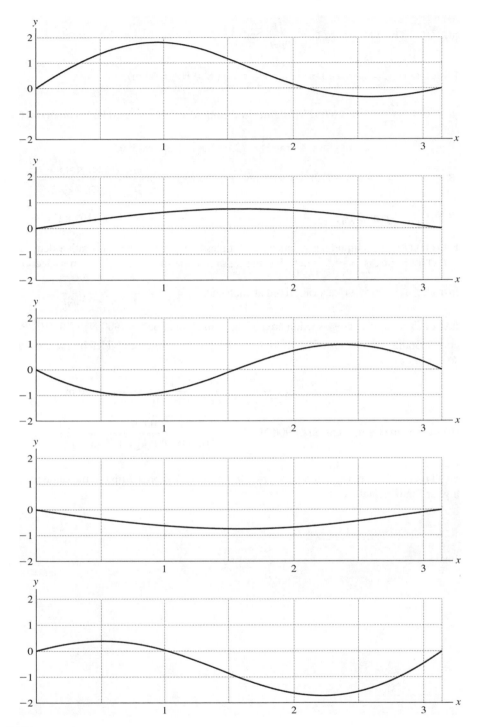

Figure 15.79

15.81.   **IDENTIFY:**   When the rock is submerged in the liquid, the buoyant force on it reduces the tension in the wire supporting it. This in turn changes the frequency of the fundamental frequency of the vibrations of the wire. The buoyant force depends on the density of the liquid (the target variable). The vertical forces on the rock balance in both cases, and the buoyant force is equal to the weight of the liquid displaced by the rock (Archimedes's principle).

SET UP: The wave speed is $v = \sqrt{\dfrac{F}{\mu}}$ and $v = f\lambda$. $B = \rho_{liq}V_{rock}g$. $\Sigma F_y = 0$.

EXECUTE: $\lambda = 2L = 6.00$ m. In air, $v = f\lambda = (42.0 \text{ Hz})(6.00 \text{ m}) = 252$ m/s. $v = \sqrt{\dfrac{F}{\mu}}$ so

$\mu = \dfrac{F}{v^2} = \dfrac{164.0 \text{ N}}{(252 \text{ m/s})^2} = 0.002583$ kg/m. In the liquid, $v = f\lambda = (28.0 \text{ Hz})(6.00 \text{ m}) = 168$ m/s.

$F = \mu v^2 = (0.002583 \text{ kg/m})(168 \text{ m/s})^2 = 72.90$ N. $F + B - mg = 0$.

$B = mg - F = 164.0 \text{ N} - 72.9 \text{ N} = 91.10$ N. For the rock, $V = \dfrac{m}{\rho} = \dfrac{(164.0 \text{ N}/9.8 \text{ m/s}^2)}{3200 \text{ kg/m}^3} = 5.230 \times 10^{-3}$ m^3.

$B = \rho_{liq}V_{rock}g$ and $\rho_{liq} = \dfrac{B}{V_{rock}g} = \dfrac{91.10 \text{ N}}{(5.230 \times 10^{-3} \text{ m}^3)(9.8 \text{ m/s}^2)} = 1.78 \times 10^3$ kg/m^3.

EVALUATE: This liquid has a density 1.78 times that of water, which is rather dense but not impossible.

15.83.   **IDENTIFY:** Stress is F/A, where F is the tension in the string and A is its cross-sectional area.

SET UP: $A = \pi r^2$. For a string fixed at each end, $f_1 = \dfrac{v}{2L} = \dfrac{1}{2L}\sqrt{\dfrac{F}{\mu}} = \dfrac{1}{2}\sqrt{\dfrac{F}{mL}}$.

EXECUTE: **(a)** The cross-section area of the string would be $A = (900 \text{ N})/(7.0 \times 10^8 \text{ Pa}) = 1.29 \times 10^{-6}$ m^2, corresponding to a radius of 0.640 mm. The length is the volume divided by the area, and the volume is $V = m/\rho$, so

$L = \dfrac{V}{A} = \dfrac{m/\rho}{A} = \dfrac{(4.00 \times 10^{-3} \text{ kg})}{(7.8 \times 10^3 \text{ kg/m}^3)(1.29 \times 10^{-6} \text{ m}^2)} = 0.40$ m.

(b) For the maximum tension of 900 N, $f_1 = \dfrac{1}{2}\sqrt{\dfrac{900 \text{ N}}{(4.00 \times 10^{-3} \text{ kg})(0.40 \text{ m})}} = 375$ Hz, or 380 Hz to two figures.

EVALUATE: The string could be shorter and thicker. A shorter string of the same mass would have a higher fundamental frequency.

16

SOUND AND HEARING

16.3. **IDENTIFY:** Use Eq. (16.5) to relate the pressure and displacement amplitudes.

SET UP: As stated in Example 16.1 the adiabatic bulk modulus for air is $B = 1.42 \times 10^5$ Pa. Use Eq. (15.1) to calculate λ from f, and then $k = 2\pi/\lambda$.

EXECUTE: (a) $f = 150$ Hz

Need to calculate k: $\lambda = v/f$ and $k = 2\pi/\lambda$ so $k = 2\pi f/v = (2\pi \text{ rad})(150 \text{ Hz})/344 \text{ m/s} = 2.74$ rad/m. Then

$p_{\max} = BkA = (1.42 \times 10^5 \text{ Pa})(2.74 \text{ rad/m})(0.0200 \times 10^{-3} \text{ m}) = 7.78$ Pa. This is below the pain threshold of 30 Pa.

(b) f is larger by a factor of 10 so $k = 2\pi f/v$ is larger by a factor of 10, and $p_{\max} = BkA$ is larger by a factor of 10. $p_{\max} = 77.8$ Pa, above the pain threshold.

(c) There is again an increase in f, k, and $p_{\max}$ of a factor of 10, so $p_{\max} = 778$ Pa, far above the pain threshold.

EVALUATE: When f increases, λ decreases so k increases and the pressure amplitude increases.

16.7. **IDENTIFY:** $d = vt$ for the sound waves in air and in water.

SET UP: Use $v_{\text{water}} = 1482$ m/s at 20°C, as given in Table 16.1. In air, $v = 344$ m/s.

EXECUTE: Since along the path to the diver the sound travels 1.2 m in air, the sound wave travels in water for the same time as the wave travels a distance $22.0 \text{ m} - 1.20 \text{ m} = 20.8 \text{ m}$ in air. The depth of the diver is

$(20.8 \text{ m})\dfrac{v_{\text{water}}}{v_{\text{air}}} = (20.8 \text{ m})\dfrac{1482 \text{ m/s}}{344 \text{ m/s}} = 89.6$ m. This is the depth of the diver; the distance from the horn is 90.8 m.

EVALUATE: The time it takes the sound to travel from the horn to the person on shore is

$t_1 = \dfrac{22.0 \text{ m}}{344 \text{ m/s}} = 0.0640$ s. The time it takes the sound to travel from the horn to the diver is

$t_2 = \dfrac{1.2 \text{ m}}{344 \text{ m/s}} + \dfrac{89.6 \text{ m}}{1482 \text{ m/s}} = 0.0035 \text{ s} + 0.0605 \text{ s} = 0.0640$ s. These times are indeed the same. For three figure accuracy the distance of the horn above the water can't be neglected.

16.9. **IDENTIFY:** $v = f\lambda$. The relation of v to gas temperature is given by $v = \sqrt{\dfrac{\gamma RT}{M}}$.

SET UP: Let $T = 22.0°\text{C} = 295.15$ K.

EXECUTE: At 22.0°C, $\lambda = \dfrac{v}{f} = \dfrac{325 \text{ m/s}}{1250 \text{ Hz}} = 0.260 \text{ m} = 26.0$ cm. $\lambda = \dfrac{v}{f} = \dfrac{1}{f}\sqrt{\dfrac{\gamma RT}{M}}$. $\dfrac{\lambda}{\sqrt{T}} = \dfrac{1}{f}\sqrt{\dfrac{\gamma R}{M}}$,

which is constant, so $\dfrac{\lambda_1}{\sqrt{T_1}} = \dfrac{\lambda_2}{\sqrt{T_2}}$. $T_2 = T_1\left(\dfrac{\lambda_2}{\lambda_1}\right)^2 = (295.15 \text{ K})\left(\dfrac{28.5 \text{ cm}}{26.0 \text{ cm}}\right)^2 = 354.6 \text{ K} = 81.4°\text{C}$.

EVALUATE: When T increases v increases and for fixed f, λ increases. Note that we did not need to know either γ or M for the gas.

16.11.   **IDENTIFY** and **SET UP:** Use $t =$ distance/speed. Calculate the time it takes each sound wave to travel the $L = 80.0$ m length of the pipe. Use Eq. (16.8) to calculate the speed of sound in the brass rod.

EXECUTE: wave in air: $t = 80.0$ m$/(344$ m/s$) = 0.2326$ s

wave in the metal: $v = \sqrt{\dfrac{Y}{\rho}} = \sqrt{\dfrac{9.0 \times 10^{10} \text{ Pa}}{8600 \text{ kg/m}^3}} = 3235$ m/s

$t = \dfrac{80.0 \text{ m}}{3235 \text{ m/s}} = 0.0247$ s

The time interval between the two sounds is $\Delta t = 0.2326$ s $- 0.0247$ s $= 0.208$ s

EVALUATE: The restoring forces that propagate the sound waves are much greater in solid brass than in air, so v is much larger in brass.

16.13.   **IDENTIFY** and **SET UP:**   Sound delivers energy (and hence power) to the ear. For a whisper, $I = 1 \times 10^{-10}$ W/m^2. The area of the tympanic membrane is $A = \pi r^2$, with $r = 4.2 \times 10^{-3}$ m. Intensity is energy per unit time per unit area.

EXECUTE: **(a)** $E = IAt = (1 \times 10^{-10}$ W/m$^2)\pi(4.2 \times 10^{-3}$ m$)^2(1$ s$) = 5.5 \times 10^{-15}$ J.

(b) $K = \frac{1}{2}mv^2$ so $v = \sqrt{\dfrac{2K}{m}} = \sqrt{\dfrac{2(5.5 \times 10^{-15} \text{ J})}{2.0 \times 10^{-6} \text{ kg}}} = 7.4 \times 10^{-5}$ m/s $= 0.074$ mm/s.

EVALUATE: Compared to the energy of ordinary objects, it takes only a very small amount of energy for hearing. As part (b) shows, a mosquito carries a lot more energy than is needed for hearing.

16.15.   **IDENTIFY:**   Apply Eq. (16.12) and solve for A. $\lambda = v/f$, with $v = \sqrt{B/\rho}$.

SET UP: $\omega = 2\pi f$. For air, $B = 1.42 \times 10^5$ Pa.

EXECUTE: **(a)** The amplitude is

$A = \sqrt{\dfrac{2I}{\sqrt{\rho B}\omega^2}} = \sqrt{\dfrac{2(3.00 \times 10^{-6} \text{ W/m}^2)}{\sqrt{(1000 \text{ kg/m}^3)(2.18 \times 10^9 \text{ Pa})}(2\pi(3400 \text{ Hz}))^2}} = 9.44 \times 10^{-11}$ m.

The wavelength is $\lambda = \dfrac{v}{f} = \dfrac{\sqrt{B/\rho}}{f} = \dfrac{\sqrt{(2.18 \times 10^9 \text{ Pa})/(1000 \text{ kg/m}^3)}}{3400 \text{ Hz}} = 0.434$ m.

(b) Repeating the above with $B = 1.42 \times 10^5$ Pa and the density of air gives $A = 5.66 \times 10^{-9}$ m and $\lambda = 0.100$ m.

EVALUATE: **(c)** The amplitude is larger in air, by a factor of about 60. For a given frequency, the much less dense air molecules must have a larger amplitude to transfer the same amount of energy.

16.17.   **IDENTIFY** and **SET UP:**   Apply Eqs. (16.5), (16.11) and (16.15).

EXECUTE: **(a)** $\omega = 2\pi f = (2\pi \text{ rad})(150 \text{ Hz}) = 942.5$ rad/s

$k = \dfrac{2\pi}{\lambda} = \dfrac{2\pi f}{v} = \dfrac{\omega}{v} = \dfrac{942.5 \text{ rad/s}}{344 \text{ m/s}} = 2.74$ rad/m

$B = 1.42 \times 10^5$ Pa (Example 16.1)

Then $p_{\max} = BkA = (1.42 \times 10^5 \text{ Pa})(2.74 \text{ rad/m})(5.00 \times 10^{-6} \text{ m}) = 1.95$ Pa.

(b) Eq. (16.11): $I = \frac{1}{2}\omega BkA^2$

$I = \frac{1}{2}(942.5 \text{ rad/s})(1.42 \times 10^5 \text{ Pa})(2.74 \text{ rad/m})(5.00 \times 10^{-6} \text{ m})^2 = 4.58 \times 10^{-3}$ W/m^2.

(c) Eq. (16.15): $\beta = (10 \text{ dB})\log(I/I_0)$, with $I_0 = 1 \times 10^{-12}$ W/m^2.

$\beta = (10 \text{ dB})\log((4.58 \times 10^{-3} \text{ W/m}^2)/(1 \times 10^{-12} \text{ W/m}^2)) = 96.6$ dB.

EVALUATE: Even though the displacement amplitude is very small, this is a very intense sound. Compare the sound intensity level to the values in Table 16.2.

16.21.   **IDENTIFY** and **SET UP:**   Let 1 refer to the mother and 2 to the father. Use the result derived in Example 16.9 for the difference in sound intensity level for the two sounds. Relate intensity to distance from the source using Eq. (15.26).

EXECUTE: From Example 16.9, $\beta_2 - \beta_1 = (10\ dB)\log(I_2/I_1)$

Eq. (15.26): $I_1/I_2 = r_2^2/r_1^2$ or $I_2/I_1 = r_1^2/r_2^2$

$\Delta\beta = \beta_2 - \beta_1 = (10\ dB)\log(I_2/I_1) = (10\ dB)\log(r_1/r_2)^2 = (20\ dB)\log(r_1/r_2)$

$\Delta\beta = (20\ dB)\log(1.50\ m/0.30\ m) = 14.0\ dB.$

EVALUATE: The father is 5 times closer so the intensity at his location is 25 times greater.

16.23. **IDENTIFY:** The intensity of sound obeys an inverse square law.

SET UP: $\dfrac{I_2}{I_1} = \dfrac{r_1^2}{r_2^2}.$ $\beta = (10\ dB)\log\left(\dfrac{I}{I_0}\right)$, with $I_0 = 1\times10^{-12}\ W/m^2$.

EXECUTE: (a) $\beta = 53\ dB$ gives $5.3 = \log\left(\dfrac{I}{I_0}\right)$ and $I = (10^{5.3})I_0 = 2.0\times10^{-7}\ W/m^2$.

(b) $r_2 = r_1\sqrt{\dfrac{I_1}{I_2}} = (3.0\ m)\sqrt{\dfrac{4}{1}} = 6.0\ m.$

(c) $\beta = \dfrac{53\ dB}{4} = 13.25\ dB$ gives $1.325 = \log\left(\dfrac{I}{I_0}\right)$ and $I = 2.1\times10^{-11}\ W/m^2$.

$r_2 = r_1\sqrt{\dfrac{I_1}{I_2}} = (3.0\ m)\sqrt{\dfrac{2.0\times10^{-7}\ W/m^2}{2.1\times10^{-11}\ W/m^2}} = 290\ m.$

EVALUATE: (d) Intensity obeys the inverse square law but noise level does not.

16.27. **IDENTIFY:** For a stopped pipe, the standing wave frequencies are given by Eq. (16.22).
SET UP: The first three standing wave frequencies correspond to $n = 1, 3$ and 5.

EXECUTE: $f_1 = \dfrac{(344\ m/s)}{4(0.17\ m)} = 506\ Hz$, $f_3 = 3f_1 = 1517\ Hz$, $f_5 = 5f_1 = 2529\ Hz.$

EVALUATE: All three of these frequencies are in the audible range, which is about 20 Hz to 20,000 Hz.

16.29. **IDENTIFY:** For either type of pipe, stopped or open, the fundamental frequency is proportional to the wave speed v. The wave speed is given in turn by Eq. (16.10).
SET UP: For He, $\gamma = 5/3$ and for air, $\gamma = 7/5$.

EXECUTE: (a) The fundamental frequency is proportional to the square root of the ratio $\dfrac{\gamma}{M}$, so

$f_{He} = f_{air}\sqrt{\dfrac{\gamma_{He}}{\gamma_{air}}\cdot\dfrac{M_{air}}{M_{He}}} = (262\ Hz)\sqrt{\dfrac{(5/3)}{(7/5)}\cdot\dfrac{28.8}{4.00}} = 767\ Hz.$

(b) No. In either case the frequency is proportional to the speed of sound in the gas.
EVALUATE: The frequency is much higher for helium, since the rms speed is greater for helium.

16.31. **IDENTIFY** and **SET UP:** Use the standing wave pattern to relate the wavelength of the standing wave to the length of the air column and then use Eq. (15.1) to calculate f. There is a displacement antinode at the top (open) end of the air column and a node at the bottom (closed) end, as shown in Figure 16.31.
EXECUTE: (a)

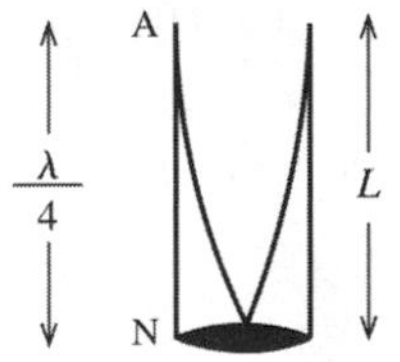

$\lambda/4 = L$
$\lambda = 4L = 4(0.140\ m) = 0.560\ m$
$f = \dfrac{v}{\lambda} = \dfrac{344\ m/s}{0.560\ m} = 614\ Hz$

Figure 16.31

(b) Now the length L of the air column becomes $\frac{1}{2}(0.140\ m) = 0.070\ m$ and $\lambda = 4L = 0.280\ m$.

$f = \dfrac{v}{\lambda} = \dfrac{344\ m/s}{0.280\ m} = 1230\ Hz$

EVALUATE: Smaller L means smaller λ which in turn corresponds to larger f.

16.33.

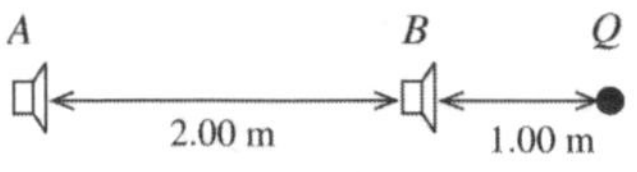

Figure 16.33

(a) IDENTIFY and SET UP: Path difference from points A and B to point Q is $3.00 \text{ m} - 1.00 \text{ m} = 2.00 \text{ m}$, as shown in Figure 16.33. Constructive interference implies path difference $= n\lambda$, $n = 1, 2, 3, \ldots$

EXECUTE: $2.00 \text{ m} = n\lambda$ so $\lambda = 2.00 \text{ m}/n$

$$f = \frac{v}{\lambda} = \frac{nv}{2.00 \text{ m}} = \frac{n(344 \text{ m/s})}{2.00 \text{ m}} = n(172 \text{ Hz}), \quad n = 1, 2, 3, \ldots$$

The lowest frequency for which constructive interference occurs is 172 Hz.

(b) IDENTIFY and SET UP: Destructive interference implies path difference $= (n/2)\lambda$, $n = 1, 3, 5, \ldots$

EXECUTE: $2.00 \text{ m} = (n/2)\lambda$ so $\lambda = 4.00 \text{ m}/n$

$$f = \frac{v}{\lambda} = \frac{nv}{4.00 \text{ m}} = \frac{n(344 \text{ m/s})}{(4.00 \text{ m})} = n(86 \text{ Hz}), \quad n = 1, 3, 5, \ldots$$

The lowest frequency for which destructive interference occurs is 86 Hz.

EVALUATE: As the frequency is slowly increased, the intensity at Q will fluctuate, as the interference changes between destructive and constructive.

16.35. **IDENTIFY:** For constructive interference the path difference is an integer number of wavelengths and for destructive interference the path difference is a half-integer number of wavelengths.

SET UP: $\lambda = v/f = (344 \text{ m/s})/(688 \text{ Hz}) = 0.500 \text{ m}$

EXECUTE: To move from constructive interference to destructive interference, the path difference must change by $\lambda/2$. If you move a distance x toward speaker B, the distance to B gets shorter by x and the distance to A gets longer by x so the path difference changes by $2x$. $2x = \lambda/2$ and $x = \lambda/4 = 0.125 \text{ m}$.

EVALUATE: If you walk an additional distance of 0.125 m farther, the interference again becomes constructive.

16.39. **IDENTIFY:** The beat is due to a difference in the frequencies of the two sounds.

SET UP: $f_{\text{beat}} = f_1 - f_2$. Tightening the string increases the wave speed for transverse waves on the string and this in turn increases the frequency.

EXECUTE: **(a)** If the beat frequency increases when she raises her frequency by tightening the string, it must be that her frequency is 433 Hz, 3 Hz above concert A.

(b) She needs to lower her frequency by loosening her string.

EVALUATE: The beat would only be audible if the two sounds are quite close in frequency. A musician with a good sense of pitch can come very close to the correct frequency just from hearing the tone.

16.41. **IDENTIFY:** $f_{\text{beat}} = |f_a - f_b|$. For a stopped pipe, $f_1 = \dfrac{v}{4L}$.

SET UP: $v = 344 \text{ m/s}$. Let $L_a = 1.14 \text{ m}$ and $L_b = 1.16 \text{ m}$. $L_b > L_a$ so $f_{1a} > f_{1b}$.

EXECUTE: $f_{1a} - f_{1b} = \dfrac{v}{4}\left(\dfrac{1}{L_a} - \dfrac{1}{L_b}\right) = \dfrac{v(L_b - L_a)}{4 L_a L_b} = \dfrac{(344 \text{ m/s})(2.00 \times 10^{-2} \text{ m})}{4(1.14 \text{ m})(1.16 \text{ m})} = 1.3 \text{ Hz}$. There are 1.3 beats per second.

EVALUATE: Increasing the length of the pipe increases the wavelength of the fundamental and decreases the frequency.

16.43. **IDENTIFY:** Apply the Doppler shift equation $f_L = \left(\dfrac{v + v_L}{v + v_S}\right) f_S$.

SET UP: The positive direction is from listener to source. $f_S = 1200 \text{ Hz}$. $f_L = 1240 \text{ Hz}$.

EXECUTE: $v_L = 0$. $v_S = -25.0$ m/s. $f_L = \left(\dfrac{v}{v + v_S}\right) f_S$ gives

$$v = \dfrac{v_S f_L}{f_S - f_L} = \dfrac{(-25 \text{ m/s})(1240 \text{ Hz})}{1200 \text{ Hz} - 1240 \text{ Hz}} = 780 \text{ m/s}.$$

EVALUATE: $f_L > f_S$ since the source is approaching the listener.

16.45.   **IDENTIFY:**   Apply the Doppler shift equation $f_L = \left(\dfrac{v + v_L}{v + v_S}\right) f_S$.

SET UP: The positive direction is from listener to source. $f_S = 392$ Hz.

(a) $v_S = 0$. $v_L = -15.0$ m/s. $f_L = \left(\dfrac{v + v_L}{v + v_S}\right) f_S = \left(\dfrac{344 \text{ m/s} - 15.0 \text{ m/s}}{344 \text{ m/s}}\right)(392 \text{ Hz}) = 375$ Hz

(b) $v_S = +35.0$ m/s. $v_L = +15.0$ m/s. $f_L = \left(\dfrac{v + v_L}{v + v_S}\right) f_S = \left(\dfrac{344 \text{ m/s} + 15.0 \text{ m/s}}{344 \text{ m/s} + 35.0 \text{ m/s}}\right)(392 \text{ Hz}) = 371$ Hz

(c) $f_{\text{beat}} = f_1 - f_2 = 4$ Hz

EVALUATE: The distance between whistle A and the listener is increasing, and for whistle A $f_L < f_S$. The distance between whistle B and the listener is also increasing, and for whistle B $f_L < f_S$.

16.47.   **IDENTIFY:**   The distance between crests is λ. In front of the source $\lambda = \dfrac{v - v_S}{f_S}$ and behind the source

$\lambda = \dfrac{v + v_S}{f_S}$. $f_S = 1/T$.

SET UP: $T = 1.6$ s. $v = 0.32$ m/s. The crest to crest distance is the wavelength, so $\lambda = 0.12$ m.

EXECUTE: **(a)** $f_S = 1/T = 0.625$ Hz. $\lambda = \dfrac{v - v_S}{f_S}$ gives

$v_S = v - \lambda f_S = 0.32 \text{ m/s} - (0.12 \text{ m})(0.625 \text{ Hz}) = 0.25$ m/s.

(b) $\lambda = \dfrac{v + v_S}{f_S} = \dfrac{0.32 \text{ m/s} + 0.25 \text{ m/s}}{0.625 \text{ Hz}} = 0.91$ m

EVALUATE: If the duck was held at rest but still paddled its feet, it would produce waves of wavelength $\lambda = \dfrac{0.32 \text{ m/s}}{0.625 \text{ Hz}} = 0.51$ m. In front of the duck the wavelength is decreased and behind the duck the wavelength is increased. The speed of the duck is 78% of the wave speed, so the Doppler effects are large.

16.51.   **IDENTIFY:**   Each bird is a moving source of sound and a moving observer, so each will experience a Doppler shift.

SET UP: Let one bird be the listener and the other be the source. Use coordinates as shown in Figure 16.51, with the positive direction from listener to source. $f_L = \left(\dfrac{v + v_L}{v + v_S}\right) f_S$.

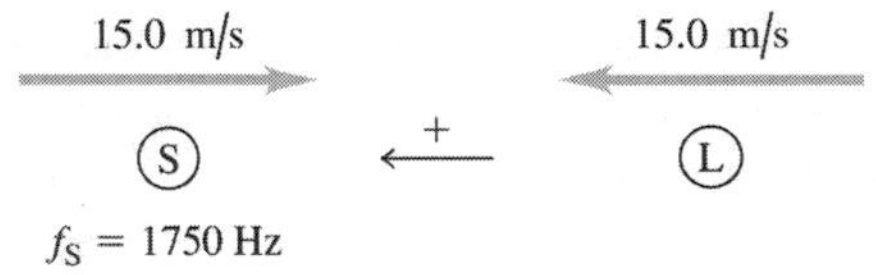

Figure 16.51

EXECUTE: **(a)** $f_S = 1750$ Hz, $v_S = -15.0$ m/s, and $v_L = +15.0$ m/s.

$f_L = \left(\dfrac{v + v_L}{v + v_S}\right) f_S = \left(\dfrac{344 \text{ m/s} + 15.0 \text{ m/s}}{344 \text{ m/s} - 15.0 \text{ m/s}}\right)(1750 \text{ Hz}) = 1910$ Hz.

(b) One canary hears a frequency of 1910 Hz and the waves move past it at $344 \text{ m/s} + 15 \text{ m/s}$, so the

wavelength it detects is $\lambda = \dfrac{344 \text{ m/s} + 15 \text{ m/s}}{1910 \text{ Hz}} = 0.188$ m. For a stationary bird, $\lambda = \dfrac{344 \text{ m/s}}{1750 \text{ Hz}} = 0.197$ m.

EVALUATE: The approach of the two birds raises the frequency, and the motion of the source toward the listener decreases the wavelength.

16.53. **IDENTIFY:** Apply Eq. (16.30).

SET UP: Require $f_R = 1.100 f_S$. Since $f_R > f_S$ the star would be moving toward us and $v < 0$, so

$v = -|v|$. $c = 3.00 \times 10^8$ m/s.

EXECUTE: $f_R = \sqrt{\dfrac{c + |v|}{c - |v|}} f_S$. $f_R = 1.100 f_S$ gives $\dfrac{c + |v|}{c - |v|} = (1.100)^2$. Solving for $|v|$ gives

$|v| = \dfrac{[(1.100)^2 - 1]c}{1 + (1.100)^2} = 0.0950c = 2.85 \times 10^7$ m/s.

EVALUATE: $\dfrac{v}{c} = 9.5\%$. $\dfrac{\Delta f}{f_S} = \dfrac{f_R - f_S}{f_S} = 10.0\%$. $\dfrac{v}{c}$ and $\dfrac{\Delta f}{f_S}$ are approximately equal.

16.55. **IDENTIFY:** Apply Eq. (16.31) to calculate α. Use the method of Example 16.19 to calculate t.

SET UP: Mach 1.70 means $v_S/v = 1.70$.

EXECUTE: **(a)** In Eq. (16.31), $v/v_S = 1/1.70 = 0.588$ and $\alpha = \arcsin(0.588) = 36.0°$.

(b) As in Example 16.19, $t = \dfrac{(950 \text{ m})}{(1.70)(344 \text{ m/s})(\tan(36.0°))} = 2.23$ s.

EVALUATE: The angle α decreases when the speed v_S of the plane increases.

16.59. **IDENTIFY:** The sound intensity level is $\beta = (10 \text{ dB})\log(I/I_0)$, so the same sound intensity level β means the same intensity I. The intensity is related to pressure amplitude by Eq. (16.13) and to the displacement amplitude by Eq. (16.12).

SET UP: $v = 344$ m/s. $\omega = 2\pi f$. Each octave higher corresponds to a doubling of frequency, so the note sung by the bass has frequency $(932 \text{ Hz})/8 = 116.5$ Hz. Let 1 refer to the note sung by the soprano and 2 refer to the note sung by the bass. $I_0 = 1 \times 10^{-12}$ W/m^2.

EXECUTE: **(a)** $I = \dfrac{v p_{max}^2}{2B}$ and $I_1 = I_2$ gives $p_{max,1} = p_{max,2}$; the ratio is 1.00.

(b) $I = \frac{1}{2}\sqrt{\rho B}\,\omega^2 A^2 = \frac{1}{2}\sqrt{\rho B}\,4\pi^2 f^2 A^2$. $I_1 = I_2$ gives $f_1 A_1 = f_2 A_2$. $\dfrac{A_2}{A_1} = \dfrac{f_1}{f_2} = 8.00$.

(c) $\beta = 72.0$ dB gives $\log(I/I_0) = 7.2$. $\dfrac{I}{I_0} = 10^{7.2}$ and $I = 1.585 \times 10^{-5}$ W/m^2. $I = \frac{1}{2}\sqrt{\rho B}\,4\pi^2 f^2 A^2$.

$A = \dfrac{1}{2\pi f}\sqrt{\dfrac{2I}{\sqrt{\rho B}}} = \dfrac{1}{2\pi(932 \text{ Hz})}\sqrt{\dfrac{2(1.585 \times 10^{-5} \text{ W/m}^2)}{\sqrt{(1.20 \text{ kg/m}^3)(1.42 \times 10^5 \text{ Pa})}}} = 4.73 \times 10^{-8}$ m $= 47.3$ nm.

EVALUATE: Even for this loud note the displacement amplitude is very small. For a given intensity, the displacement amplitude depends on the frequency of the sound wave but the pressure amplitude does not.

16.63. **IDENTIFY:** The flute acts as a stopped pipe and its harmonic frequencies are given by Eq. (16.23). The resonant frequencies of the string are $f_n = n f_1, n = 1, 2, 3, \ldots$ The string resonates when the string frequency equals the flute frequency.

SET UP: For the string $f_{1s} = 600.0$ Hz. For the flute, the fundamental frequency is

$f_{1f} = \dfrac{v}{4L} = \dfrac{344.0 \text{ m/s}}{4(0.1075 \text{ m})} = 800.0$ Hz. Let n_f label the harmonics of the flute and let n_s label the harmonics of the string.

EXECUTE: For the flute and string to be in resonance, $n_f f_{1f} = n_s f_{1s}$, where $f_{1s} = 600.0$ Hz is the fundamental frequency for the string. $n_s = n_f(f_{1f}/f_{1s}) = \frac{4}{3}n_f$. n_s is an integer when $n_f = 3N, N = 1, 3, 5, \ldots$ (the flute has only odd harmonics). $n_f = 3N$ gives $n_s = 4N$.

Flute harmonic $3N$ resonates with string harmonic $4N, N = 1, 3, 5, \ldots$

EVALUATE: We can check our results for some specific values of N. For $N = 1$, $n_f = 3$ and $f_{3f} = 2400$ Hz. For this N, $n_s = 4$ and $f_{4s} = 2400$ Hz. For $N = 3$, $n_f = 9$ and $f_{9f} = 7200$ Hz, and $n_s = 12$, $f_{12s} = 7200$ Hz. Our general results do give equal frequencies for the two objects.

16.67. **IDENTIFY** and **SET UP:** There is a node at the piston, so the distance the piston moves is the node to node distance, $\lambda/2$. Use Eq. (15.1) to calculate v and Eq. (16.10) to calculate γ from v.

EXECUTE: **(a)** $\lambda/2 = 37.5$ cm, so $\lambda = 2(37.5$ cm$) = 75.0$ cm $= 0.750$ m.

$v = f\lambda = (500$ Hz$)(0.750$ m$) = 375$ m/s

(b) $v = \sqrt{\gamma RT/M}$ (Eq. 16.10)

$\gamma = \dfrac{Mv^2}{RT} = \dfrac{(28.8\times10^{-3}\text{ kg/mol})(375\text{ m/s})^2}{(8.3145\text{ J/mol}\cdot\text{K})(350\text{ K})} = 1.39.$

(c) EVALUATE: There is a node at the piston so when the piston is 18.0 cm from the open end the node is inside the pipe, 18.0 cm from the open end. The node to antinode distance is $\lambda/4 = 18.8$ cm, so the antinode is 0.8 cm beyond the open end of the pipe.

The value of γ we calculated agrees with the value given for air in Example 16.4.

16.69. **IDENTIFY:** $v = f\lambda$. $v = \sqrt{\dfrac{\gamma RT}{M}}$. Solve for γ.

SET UP: The wavelength is twice the separation of the nodes, so $\lambda = 2L$, where $L = 0.200$ m.

EXECUTE: $v = \lambda f = 2Lf = \sqrt{\dfrac{\gamma RT}{M}}$. Solving for γ,

$$\gamma = \dfrac{M}{RT}(2Lf)^2 = \dfrac{(16.0\times10^{-3}\text{ kg/mol})}{(8.3145\text{ J/mol}\cdot\text{K})(293.15\text{ K})}(2(0.200\text{ m})(1100\text{ Hz}))^2 = 1.27.$$

EVALUATE: This value of γ is smaller than that of air. We will see in Chapter 19 that this value of γ is a typical value for polyatomic gases.

16.71. **IDENTIFY:** Apply $f_L = \left(\dfrac{v + v_L}{v + v_S}\right)f_S$.

SET UP: The positive direction is from the listener to the source. (a) The wall is the listener. $v_S = -30$ m/s. $v_L = 0$. $f_L = 600$ Hz. (b) The wall is the source and the car is the listener. $v_S = 0$. $v_L = +30$ m/s. $f_S = 600$ Hz.

EXECUTE: **(a)** $f_L = \left(\dfrac{v+v_L}{v+v_S}\right)f_S$. $f_S = \left(\dfrac{v+v_S}{v+v_L}\right)f_L = \left(\dfrac{344\text{ m/s} - 30\text{ m/s}}{344\text{ m/s}}\right)(600\text{ Hz}) = 548$ Hz

(b) $f_L = \left(\dfrac{v+v_L}{v+v_S}\right)f_S = \left(\dfrac{344\text{ m/s} + 30\text{ m/s}}{344\text{ m/s}}\right)(600\text{ Hz}) = 652$ Hz

EVALUATE: Since the singer and wall are moving toward each other the frequency received by the wall is greater than the frequency sung by the soprano, and the frequency she hears from the reflected sound is larger still.

16.73. **IDENTIFY:** For the sound coming directly to the observer at the top of the well, the source is moving away from the listener. For the reflected sound, the water at the bottom of the well is the "listener" so the source is moving toward the listener. The water reflects the same frequency sound it receives.

SET UP: $f_L = \left(\dfrac{v + v_L}{v + v_S} \right) f_S.$ Take the positive direction to be from the listener to the source. For reflection off the bottom of the well the water surface first serves as a listener and then as a source. The falling siren has constant downward acceleration of g and obeys the equation $v_y^2 = v_{0y}^2 + 2a_y(y - y_0).$

EXECUTE: For the falling siren, $v_y^2 = v_{0y}^2 + 2a_y(y - y_0),$ so the speed of the siren just before it hits the water is $\sqrt{2(9.80 \text{ m/s}^2)(125 \text{ m})} = 49.5 \text{ m/s}.$

(a) The situation is shown in Figure 16.73a.

Figure 16.73

$f_L = \left(\dfrac{v}{v + 49.5 \text{ m/s}} \right) f_S = \left(\dfrac{344 \text{ m/s}}{344 \text{ m/s} + 49.5 \text{ m/s}} \right)(2500 \text{ Hz}) = 2186 \text{ Hz}.$ $\lambda_L = \dfrac{v}{f_L} = \dfrac{344 \text{ m/s}}{2186 \text{ Hz}} = 0.157 \text{ m}.$

(b) The water serves as a listener (Figure 16.73b). $f_L = \left(\dfrac{v}{v - 49.5 \text{ m/s}} \right) f_S = 2920 \text{ Hz}.$ The source and listener are approaching and the frequency is raised. $\lambda_L = \dfrac{v}{f_L} = 0.118 \text{ m}.$ Both the person and the water are at rest so there is no Doppler effect when the water serves as a source and the person is the listener. The person detects sound with frequency 2920 Hz and wavelength 0.118 m.

(c) $f_{\text{beat}} = f_1 - f_2 = 2920 \text{ Hz} - 2186 \text{ Hz} = 734 \text{ Hz}.$

EVALUATE: In (a), the source is moving away from the listener and the frequency is lowered. In (b) the source is moving toward the "listener" so the frequency is increased.

16.75.   **(a) IDENTIFY** and **SET UP:** Use Eq. (15.1) to calculate $\lambda.$

EXECUTE: $\lambda = \dfrac{v}{f} = \dfrac{1482 \text{ m/s}}{22.0 \times 10^3 \text{ Hz}} = 0.0674 \text{ m}$

(b) IDENTIFY: Apply the Doppler effect equation, Eq. (16.29). The Problem-Solving Strategy in the text (Section 16.8) describes how to do this problem. The frequency of the directly radiated waves is $f_S = 22,000 \text{ Hz}.$ The moving whale first plays the role of a moving listener, receiving waves with frequency $f_L'.$ The whale then acts as a moving source, emitting waves with the same frequency, $f_S' = f_L'$ with which they are received. Let the speed of the whale be $v_W.$

SET UP: whale receives waves (Figure 16.75a)

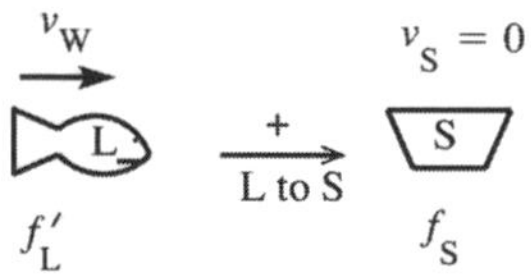

$v_W \longrightarrow$ L $\xrightarrow[\text{L to S}]{+}$ S $v_S = 0$ **EXECUTE:** $v_L = +v_W$

f_L' f_S $f_L' = f_S \left(\dfrac{v + v_L}{v + v_S} \right) = f_S \left(\dfrac{v + v_W}{v} \right)$

Figure 16.75a

SET UP: whale re-emits the waves (Figure 16.75b)

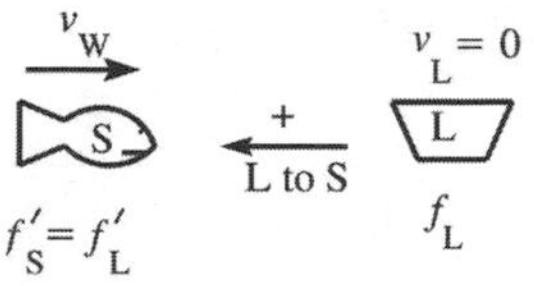

$$v_W \rightarrow \qquad v_L = 0 \qquad \text{EXECUTE:} \quad v_S = -v_W$$

$$f'_S = f'_L \qquad \xleftarrow[\text{L to S}]{+} \qquad f_L \qquad\qquad f_L = f_S\left(\frac{v+v_L}{v+v_S}\right) = f'_S\left(\frac{v}{v-v_W}\right)$$

Figure 16.75b

But $f'_S = f'_L$ so $f_L = f_S\left(\dfrac{v+v_W}{v}\right)\left(\dfrac{v}{v-v_W}\right) = f_S\left(\dfrac{v+v_W}{v-v_W}\right)$.

Then $\Delta f = f_S - f_L = f_S\left(1 - \dfrac{v+v_W}{v-v_W}\right) = f_S\left(\dfrac{v-v_W-v-v_W}{v-v_W}\right) = \dfrac{-2f_S v_W}{v-v_W}$.

$$\Delta f = \frac{-2(2.20\times10^4\ \text{Hz})(4.95\ \text{m/s})}{1482\ \text{m/s} - 4.95\ \text{m/s}} = 147\ \text{Hz}.$$

EVALUATE: Listener and source are moving toward each other so frequency is raised.

16.79. **IDENTIFY:** Apply the result derived in part (b) of Problem 16.78. The radius of the nebula is $R = vt$, where t is the time since the supernova explosion.

SET UP: When the source and receiver are moving toward each other, v is negative and $f_R > f_S$. The light from the explosion reached earth 952 years ago, so that is the amount of time the nebula has expanded. $1\ \text{ly} = 9.46\times10^{15}$ m.

EXECUTE: **(a)** $v = c\dfrac{f_S - f_R}{f_S} = (3.00\times10^8\ \text{m/s})\dfrac{-0.018\times10^{14}\ \text{Hz}}{4.568\times10^{14}\ \text{Hz}} = -1.2\times10^6\ \text{m/s}$, with the minus sign indicating that the gas is approaching the earth, as is expected since $f_R > f_S$.

(b) The radius is $(952\ \text{yr})(3.156\times10^7\ \text{s/yr})(1.2\times10^6\ \text{m/s}) = 3.6\times10^{16}\ \text{m} = 3.8\ \text{ly}$.

(c) The ratio of the width of the nebula to 2π times the distance from the earth is the ratio of the angular width (taken as 5 arc minutes) to an entire circle, which is 60×360 arc minutes. The distance to the nebula is then $\left(\dfrac{2}{2\pi}\right)(3.75\ \text{ly})\dfrac{(60)(360)}{5} = 5.2\times10^3\ \text{ly}$. The time it takes light to travel this distance is 5200 yr, so the explosion actually took place 5200 yr before 1054 C.E., or about 4100 B.C.E.

EVALUATE: $\left|\dfrac{v}{c}\right| = 4.0\times10^{-3}$, so even though $|v|$ is very large the approximation required for $v = c\dfrac{\Delta f}{f}$ is accurate.

16.81. **IDENTIFY:** Follow the method of Example 16.18 and apply the Doppler shift formula twice, once for the wall as a listener and then again with the wall as a source.

SET UP: In each application of the Doppler formula, the positive direction is from the listener to the source

EXECUTE: **(a)** The wall will receive and reflect pulses at a frequency $\dfrac{v}{v-v_w}f_0$, and the woman will hear this reflected wave at a frequency $\dfrac{v+v_w}{v}\dfrac{v}{v-v_w}f_0 = \dfrac{v+v_w}{v-v_w}f_0$. The beat frequency is

$$f_{\text{beat}} = f_0\left(\frac{v+v_w}{v-v_w} - 1\right) = f_0\left(\frac{2v_w}{v-v_w}\right).$$

(b) In this case, the sound reflected from the wall will have a lower frequency, and using $f_0(v-v_w)/(v+v_w)$ as the detected frequency, v_w is replaced by $-v_w$ in the calculation of part (a) and

$$f_{\text{beat}} = f_0\left(1 - \frac{v-v_w}{v+v_w}\right) = f_0\left(\frac{2v_w}{v+v_w}\right).$$

EVALUATE: The beat frequency is larger when she runs toward the wall, even though her speed is the same in both cases.

17

TEMPERATURE AND HEAT

17.9. **IDENTIFY** and **SET UP:** Fit the data to a straight line for $p(T)$ and use this equation to find T when $p = 0$.

EXECUTE: **(a)** If the pressure varies linearly with temperature, then $p_2 = p_1 + \gamma(T_2 - T_1)$.

$$\gamma = \frac{p_2 - p_1}{T_2 - T_1} = \frac{6.50 \times 10^4 \text{ Pa} - 4.80 \times 10^4 \text{ Pa}}{100°\text{C} - 0.01°\text{C}} = 170.0 \text{ Pa/C}°$$

Apply $p = p_1 + \gamma(T - T_1)$ with $T_1 = 0.01°\text{C}$ and $p = 0$ to solve for T.

$$0 = p_1 + \gamma(T - T_1)$$

$$T = T_1 - \frac{p_1}{\gamma} = 0.01°\text{C} - \frac{4.80 \times 10^4 \text{ Pa}}{170 \text{ Pa/C}°} = -282°\text{C}.$$

(b) Let $T_1 = 100°\text{C}$ and $T_2 = 0.01°\text{C}$; use Eq. (17.4) to calculate p_2. Eq. (17.4) says $T_2/T_1 = p_2/p_1$, where T is in kelvins.

$$p_2 = p_1\left(\frac{T_2}{T_1}\right) = 6.50 \times 10^4 \text{ Pa}\left(\frac{0.01 + 273.15}{100 + 273.15}\right) = 4.76 \times 10^4 \text{ Pa}; \text{ this differs from the } 4.80 \times 10^4 \text{ Pa that was}$$

measured so Eq. (17.4) is not precisely obeyed.

EVALUATE: The answer to part (a) is in reasonable agreement with the accepted value of $-273°\text{C}$.

17.15. **IDENTIFY:** Find the change ΔL in the diameter of the lid. The diameter of the lid expands according to Eq. (17.6).

SET UP: Assume iron has the same α as steel, so $\alpha = 1.2 \times 10^{-5}$ $(\text{C}°)^{-1}$.

EXECUTE: $\Delta L = \alpha L_0 \Delta T = (1.2 \times 10^{-5} \text{ } (\text{C}°)^{-1})(725 \text{ mm})(30.0 \text{ C}°) = 0.26 \text{ mm}$

EVALUATE: In Eq. (17.6), ΔL has the same units as L.

17.17. **IDENTIFY:** Apply $\Delta V = V_0 \beta \Delta T$.

SET UP: For copper, $\beta = 5.1 \times 10^{-5}$ $(\text{C}°)^{-1}$. $\Delta V/V_0 = 0.150 \times 10^{-2}$.

EXECUTE: $\Delta T = \frac{\Delta V/V_0}{\beta} = \frac{0.150 \times 10^{-2}}{5.1 \times 10^{-5} \text{ } (\text{C}°)^{-1}} = 29.4 \text{ C}°$. $T_f = T_i + \Delta T = 49.4°\text{C}$.

EVALUATE: The volume increases when the temperature increases.

17.19. **IDENTIFY:** Apply $\Delta V = V_0 \beta \Delta T$ to the volume of the flask and to the mercury. When heated, both the volume of the flask and the volume of the mercury increase.

SET UP: For mercury, $\beta_{\text{Hg}} = 18 \times 10^{-5}$ $(\text{C}°)^{-1}$.

EXECUTE: 8.95 cm^3 of mercury overflows, so $\Delta V_{\text{Hg}} - \Delta V_{\text{glass}} = 8.95 \text{ cm}^3$.

EXECUTE: $\Delta V_{\text{Hg}} = V_0 \beta_{\text{Hg}} \Delta T = (1000.00 \text{ cm}^3)(18 \times 10^{-5} \text{ } (\text{C}°)^{-1})(55.0 \text{ C}°) = 9.9 \text{ cm}^3$.

$\Delta V_{\text{glass}} = \Delta V_{\text{Hg}} - 8.95 \text{ cm}^3 = 0.95 \text{ cm}^3$. $\beta_{\text{glass}} = \frac{\Delta V_{\text{glass}}}{V_0 \Delta T} = \frac{0.95 \text{ cm}^3}{(1000.00 \text{ cm}^3)(55.0 \text{ C}°)} = 1.7 \times 10^{-5} \text{ } (\text{C}°)^{-1}$.

EVALUATE: The coefficient of volume expansion for the mercury is larger than for glass. When they are heated, both the volume of the mercury and the inside volume of the flask increase. But the increase for the mercury is greater and it no longer all fits inside the flask.

17.21.   IDENTIFY and SET UP: Apply the result of Exercise 17.20a to calculate ΔA for the plate, and then $A = A_0 + \Delta A$.

EXECUTE: (a) $A_0 = \pi r_0^2 = \pi(1.350 \text{ cm}/2)^2 = 1.431 \text{ cm}^2$

(b) Exercise 17.20 says $\Delta A = 2\alpha A_0 \Delta T$, so

$\Delta A = 2(1.2 \times 10^{-5} \text{ C}^{\circ-1})(1.431 \text{ cm}^2)(175^\circ\text{C} - 25^\circ\text{C}) = 5.15 \times 10^{-3} \text{ cm}^2$

$A = A_0 + \Delta A = 1.436 \text{ cm}^2$

EVALUATE: A hole in a flat metal plate expands when the metal is heated just as a piece of metal the same size as the hole would expand.

17.25.   IDENTIFY: Apply $\Delta L = L_0 \alpha \Delta T$ and stress $= F/A = -Y\alpha\Delta T$.

SET UP: For steel, $\alpha = 1.2 \times 10^{-5} \text{ (C}^\circ)^{-1}$ and $Y = 2.0 \times 10^{11} \text{ Pa}$.

EXECUTE: (a) $\Delta L = L_0 \alpha \Delta T = (12.0 \text{ m})(1.2 \times 10^{-5} \text{ (C}^\circ)^{-1})(35.0 \text{ C}^\circ) = 5.0 \text{ mm}$

(b) stress $= -Y\alpha\Delta T = -(2.0 \times 10^{11} \text{ Pa})(1.2 \times 10^{-5} \text{ (C}^\circ)^{-1})(35.0 \text{ C}^\circ) = -8.4 \times 10^7 \text{ Pa}$. The minus sign means the stress is compressive.

EVALUATE: Commonly occurring temperature changes result in very small fractional changes in length but very large stresses if the length change is prevented from occurring.

17.27.   IDENTIFY and SET UP: Apply Eq. (17.13) to the kettle and water.

EXECUTE: kettle

$Q = mc\Delta T$, $c = 910 \text{ J/kg} \cdot \text{K}$ (from Table 17.3)

$Q = (1.50 \text{ kg})(910 \text{ J/kg} \cdot \text{K})(85.0^\circ\text{C} - 20.0^\circ\text{C}) = 8.873 \times 10^4 \text{ J}$

water

$Q = mc\Delta T$, $c = 4190 \text{ J/kg} \cdot \text{K}$ (from Table 17.3)

$Q = (1.80 \text{ kg})(4190 \text{ J/kg} \cdot \text{K})(85.0^\circ\text{C} - 20.0^\circ\text{C}) = 4.902 \times 10^5 \text{ J}$

Total $Q = 8.873 \times 10^4 \text{ J} + 4.902 \times 10^5 \text{ J} = 5.79 \times 10^5 \text{ J}$

EVALUATE: Water has a much larger specific heat capacity than aluminum, so most of the heat goes into raising the temperature of the water.

17.31.   IDENTIFY: Apply $Q = mc\Delta T$ to find the heat that would raise the temperature of the student's body 7 C$^\circ$.

SET UP: 1 W $= 1$ J/s

EXECUTE: Find Q to raise the body temperature from 37°C to 44°C.

$Q = mc\Delta T = (70 \text{ kg})(3480 \text{ J/kg} \cdot \text{K})(7 \text{ C}^\circ) = 1.7 \times 10^6 \text{ J}$.

$t = \dfrac{1.7 \times 10^6 \text{ J}}{1200 \text{ J/s}} = 1400 \text{ s} = 23 \text{ min}$.

EVALUATE: Heat removal mechanisms are essential to the well-being of a person.

17.33.   IDENTIFY: The work done by friction is the loss of mechanical energy. The heat input for a temperature change is $Q = mc\Delta T$.

SET UP: The crate loses potential energy mgh, with $h = (8.00 \text{ m})\sin 36.9^\circ$, and gains kinetic energy $\frac{1}{2}mv_2^2$.

EXECUTE: (a)

$W_f = -mgh + \frac{1}{2}mv_2^2 = -(35.0 \text{ kg})((9.80 \text{ m/s}^2)(8.00 \text{ m})\sin 36.9^\circ + \frac{1}{2}(2.50 \text{ m/s})^2) = -1.54 \times 10^3 \text{ J}$.

(b) Using the results of part (a) for Q gives $\Delta T = (1.54 \times 10^3 \text{ J})/((35.0 \text{ kg})(3650 \text{ J/kg} \cdot \text{K})) = 1.21 \times 10^{-2} \text{ C}^\circ$.

EVALUATE: The temperature rise is very small.

17.37.   IDENTIFY: Some of the kinetic energy of the bullet is transformed through friction into heat, which raises the temperature of the water in the tank.

SET UP: Set the loss of kinetic energy of the bullet equal to the heat energy Q transferred to the water. $Q = mc\Delta T$. From Table 17.3, the specific heat of water is 4.19×10^3 J/kg·C°.

SOLVE: The kinetic energy lost by the bullet is

$K_i - K_f = \frac{1}{2}m(v_i^2 - v_f^2) = \frac{1}{2}(15.0 \times 10^{-3}$ kg$)[(865$ m/s$)^2 - (534$ m/s$)^2] = 3.47 \times 10^3$ J, so for the water

$Q = 3.47 \times 10^3$ J. $Q = mc\Delta T$ gives $\Delta T = \dfrac{Q}{mc} = \dfrac{3.47 \times 10^3 \text{ J}}{(13.5 \text{ kg})(4.19 \times 10^3 \text{ J/kg} \cdot \text{C°})} = 0.0613$ C°.

EVALUATE: The heat energy required to change the temperature of ordinary-size objects is very large compared to the typical kinetic energies of moving objects.

17.39. **IDENTIFY and SET UP:** Heat comes out of the metal and into the water. The final temperature is in the range $0 < T < 100°$C, so there are no phase changes. $Q_{\text{system}} = 0$.

(a) EXECUTE: $Q_{\text{water}} + Q_{\text{metal}} = 0$

$m_{\text{water}} c_{\text{water}} \Delta T_{\text{water}} + m_{\text{metal}} c_{\text{metal}} \Delta T_{\text{metal}} = 0$

$(1.00$ kg$)(4190$ J/kg·K$)(2.0$ C°$) + (0.500$ kg$)(c_{\text{metal}})(-78.0$ C°$) = 0$

$c_{\text{metal}} = 215$ J/kg·K

(b) EVALUATE: Water has a larger specific heat capacity so stores more heat per degree of temperature change.

(c) If some heat went into the styrofoam then Q_{metal} should actually be larger than in part (a), so the true c_{metal} is larger than we calculated; the value we calculated would be smaller than the true value.

17.41. **IDENTIFY:** The heat lost by the cooling copper is absorbed by the water and the pot, which increases their temperatures.

SET UP: For copper, $c_c = 390$ J/kg·K. For iron, $c_i = 470$ J/kg·K. For water, $c_w = 4.19 \times 10^3$ J/kg·K.

EXECUTE: For the copper pot,

$Q_c = m_c c_c \; \Delta T_c = (0.500$ kg$)(390$ J/kg·K$)(T - 20.0°$C$) = (195$ J/K$)T - 3900$ J. For the block of iron,

$Q_i = m_i c_i \; \Delta T_i = (0.250$ kg$)(470$ J/kg·K$)(T - 85.0°$C$) = (117.5$ J/K$)T - 9988$ J. For the water,

$Q_w = m_w c_w \; \Delta T_w = (0.170$ kg$)(4190$ J/kg·K$)(T - 20.0°$C$) = (712.3$ J/K$)T - 1.425 \times 10^4$ J. $\Sigma Q = 0$ gives

$(195$ J/K$)T - 3900$ J $+ (117.5$ J/K$)T - 9988$ J $+ (712.3$ J/K$)T - 1.425 \times 10^4$ J. $T = \dfrac{2.814 \times 10^4 \text{ J}}{1025 \text{ J/K}} = 27.5°$C.

EVALUATE: The basic principle behind this problem is conservation of energy: no energy is lost; it is only transferred.

17.45. **IDENTIFY:** By energy conservation, the heat lost by the copper is gained by the ice. This heat must first increase the temperature of the ice from $-20.0°$C to the melting point of $0.00°$C, then melt some of the ice. At the final thermal equilibrium state, there is ice and water, so the temperature must be $0.00°$C. The target variable is the initial temperature of the copper.

SET UP: For temperature changes, $Q = mc\Delta T$ and for a phase change from solid to liquid $Q = mL_F$.

EXECUTE: For the ice,

$Q_{\text{ice}} = (2.00$ kg$)[2100$ J/(kg·C°$)](20.0$C°$) + (0.80$ kg$)(3.34 \times 10^5$ J/kg$) = 3.512 \times 10^5$ J. For the copper, using the specific heat from the table in the text gives

$Q_{\text{copper}} = (6.00$ kg$)[390$ J/(kg·C°$)](0°$C$- T) = -(2.34 \times 10^3$ J/C°$)T$. Setting the sum of the two heats equal to zero gives 3.512×10^5 J $= (2.34 \times 10^3$ J/C°$)T$, which gives $T = 150°$C.

EVALUATE: Since the copper has a smaller specific heat than that of ice, it must have been quite hot initially to provide the amount of heat needed.

17.49. **IDENTIFY and SET UP:** Use Eq. (17.13) for the temperature changes and Eq. (17.20) for the phase changes.

EXECUTE: Heat must be added to do the following:

ice at $-10.0°$C $\rightarrow$ ice at $0°$C

$Q_{\text{ice}} = mc_{\text{ice}}\Delta T = (12.0 \times 10^{-3}$ kg$)(2100$ J/kg·K$)(0°$C$-(-10.0°$C$)) = 252$ J

phase transition ice $(0°$C$) \rightarrow$ liquid water $(0°$C$)$(melting)

$Q_{\text{melt}} = +mL_{\text{f}} = (12.0 \times 10^{-3}\text{ kg})(334 \times 10^{3}\text{ J/kg}) = 4.008 \times 10^{3}$ J

water at $0°C$ (from melted ice) $\rightarrow$ water at $100°C$

$Q_{\text{water}} = mc_{\text{water}}\Delta T = (12.0 \times 10^{-3}\text{ kg})(4190\text{ J/kg} \cdot \text{K})(100°C - 0°C) = 5.028 \times 10^{3}$ J

phase transition water $(100°C) \rightarrow$ steam $(100°C)$(boiling)

$Q_{\text{boil}} = +mL_{\text{v}} = (12.0 \times 10^{-3}\text{ kg})(2256 \times 10^{3}\text{ J/kg}) = 2.707 \times 10^{4}$ J

The total Q is $Q = 252\text{ J} + 4.008 \times 10^{3}\text{ J} + 5.028 \times 10^{3}\text{ J} + 2.707 \times 10^{4}\text{ J} = 3.64 \times 10^{4}$ J

$(3.64 \times 10^{4}\text{ J})(1\text{ cal}/4.186\text{ J}) = 8.70 \times 10^{3}$ cal

$(3.64 \times 10^{4}\text{ J})(1\text{ Btu}/1055\text{ J}) = 34.5$ Btu

EVALUATE: Q is positive and heat must be added to the material. Note that more heat is needed for the liquid to gas phase change than for the temperature changes.

17.51.   **IDENTIFY** and **SET UP:** The heat that must be added to a lead bullet of mass m to melt it is $Q = mc\Delta T + mL_{\text{f}}$ ($mc\Delta T$ is the heat required to raise the temperature from $25°C$ to the melting point of $327.3°C$; mL_{f} is the heat required to make the solid $\rightarrow$ liquid phase change.) The kinetic energy of the bullet if its speed is v is $K = \frac{1}{2}mv^{2}$.

EXECUTE: $K = Q$ says $\frac{1}{2}mv^{2} = mc\Delta T + mL_{\text{f}}$

$v = \sqrt{2(c\Delta T + L_{\text{f}})}$

$v = \sqrt{2[(130\text{ J/kg} \cdot \text{K})(327.3°C - 25°C) + 24.5 \times 10^{3}\text{ J/kg}]} = 357$ m/s

EVALUATE: This is a typical speed for a rifle bullet. A bullet fired into a block of wood does partially melt, but in practice not all of the initial kinetic energy is converted to heat that remains in the bullet.

17.55.   **IDENTIFY:** The asteroid's kinetic energy is $K = \frac{1}{2}mv^{2}$. To boil the water, its temperature must be raised to $100.0°C$ and the heat needed for the phase change must be added to the water.

SET UP: For water, $c = 4190\text{ J/kg} \cdot \text{K}$ and $L_{\text{v}} = 2256 \times 10^{3}$ J/kg.

EXECUTE: $K = \frac{1}{2}(2.60 \times 10^{15}\text{ kg})(32.0 \times 10^{3}\text{ m/s})^{2} = 1.33 \times 10^{24}$ J. $Q = mc\Delta T + mL_{\text{v}}$.

$m = \dfrac{Q}{c\Delta T + L_{\text{v}}} = \dfrac{1.33 \times 10^{22}\text{ J}}{(4190\text{ J/kg} \cdot \text{K})(90.0\text{ K}) + 2256 \times 10^{3}\text{ J/kg}} = 5.05 \times 10^{15}$ kg.

EVALUATE: The mass of water boiled is 2.5 times the mass of water in Lake Superior.

17.57.   **IDENTIFY** and **SET UP:** Heat flows out of the water and into the ice. The net heat flow for the system is zero. The ice warms to $0°C$, melts, and then the water from the melted ice warms from $0°C$ to the final temperature.

EXECUTE: $Q_{\text{system}} = 0$; calculate Q for each component of the system: (Beaker has small mass says that $Q = mc\Delta T$ for beaker can be neglected.)

0.250 kg of water: cools from $75.0°C$ to $40.0°C$

$Q_{\text{water}} = mc\Delta T = (0.250\text{ kg})(4190\text{ J/kg} \cdot \text{K})(40.0°C - 75.0°C) = -3.666 \times 10^{4}$ J.

ice: warms to $0°C$; melts; water from melted ice warms to $40.0°C$

$Q_{\text{ice}} = mc_{\text{ice}}\Delta T + mL_{\text{f}} + mc_{\text{water}}\Delta T$.

$Q_{\text{ice}} = m[(2100\text{ J/kg} \cdot \text{K})(0°C - (-20.0°C)) + 334 \times 10^{3}\text{ J/kg} + (4190\text{ J/kg} \cdot \text{K})(40.0°C - 0°C)]$.

$Q_{\text{ice}} = (5.436 \times 10^{5}\text{ J/kg})m$. $Q_{\text{system}} = 0$ says $Q_{\text{water}} + Q_{\text{ice}} = 0$. $-3.666 \times 10^{4}\text{ J} + (5.436 \times 10^{5}\text{ J/kg})m = 0$.

$m = \dfrac{3.666 \times 10^{4}\text{ J}}{5.436 \times 10^{5}\text{ J/kg}} = 0.0674$ kg.

EVALUATE: Since the final temperature is $40.0°C$ we know that all the ice melts and the final system is all liquid water. The mass of ice added is much less than the mass of the $75°C$ water; the ice requires a large heat input for the phase change.

17.61. **IDENTIFY:** Set $Q_{\text{system}} = 0$, for the system of water, ice and steam. $Q = mc\Delta T$ for a temperature change and $Q = \pm mL$ for a phase transition.

SET UP: For water, $c = 4190$ J/kg·K, $L_f = 334 \times 10^3$ J/kg and $L_v = 2256 \times 10^3$ J/kg.

EXECUTE: The steam both condenses and cools, and the ice melts and heats up along with the original water. $m_i L_f + m_i c(28.0\ \text{C}°) + m_w c(28.0\ \text{C}°) - m_{\text{steam}} L_v + m_{\text{steam}} c(-72.0\ \text{C}°) = 0$. The mass of steam needed is

$$m_{\text{steam}} = \frac{(0.450\ \text{kg})(334 \times 10^3\ \text{J/kg}) + (2.85\ \text{kg})(4190\ \text{J/kg·K})(28.0\ \text{C}°)}{2256 \times 10^3\ \text{J/kg} + (4190\ \text{J/kg·K})(72.0\ \text{C}°)} = 0.190\ \text{kg}.$$

EVALUATE: Since the final temperature is greater than 0.0°C, we know that all the ice melts.

17.65. **IDENTIFY and SET UP:** Call the temperature at the interface between the wood and the styrofoam T. The heat current in each material is given by $H = kA(T_H - T_C)/L$.

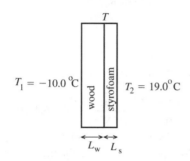

See Figure 17.65.

Heat current through the wood: $H_w = k_w A(T - T_1)L_w$

Heat current through the styrofoam: $H_s = k_s A(T_2 - T)/L_s$

Figure 17.65

In steady-state heat does not accumulate in either material. The same heat has to pass through both materials in succession, so $H_w = H_s$.

EXECUTE: **(a)** This implies $k_w A(T - T_1)/L_w = k_s A(T_2 - T)/L_s$

$k_w L_s (T - T_1) = k_s L_w (T_2 - T)$

$$T = \frac{k_w L_s T_1 + k_s L_w T_2}{k_w L_s + k_s L_w} = \frac{-0.0176\ \text{W·°C/K} + 00057\ \text{W·°C/K}}{0.00206\ \text{W/K}} = -5.8°\text{C}$$

EVALUATE: The temperature at the junction is much closer in value to T_1 than to T_2. The styrofoam has a very small k, so a larger temperature gradient is required for than for wood to establish the same heat current.

(b) IDENTIFY and SET UP: Heat flow per square meter is $\dfrac{H}{A} = k\left(\dfrac{T_H - T_C}{L}\right)$. We can calculate this either for the wood or for the styrofoam; the results must be the same.

EXECUTE: wood

$$\frac{H_w}{A} = k_w \frac{T - T_1}{L_w} = (0.080\ \text{W/m·K})\frac{(-5.8°\text{C} - (-10.0°\text{C}))}{0.030\ \text{m}} = 11\ \text{W/m}^2.$$

styrofoam

$$\frac{H_s}{A} = k_s \frac{T_2 - T}{L_s} = (0.010\ \text{W/m·K})\frac{(19.0°\text{C} - (-5.8°\text{C}))}{0.022\ \text{m}} = 11\ \text{W/m}^2.$$

EVALUATE: H must be the same for both materials and our numerical results show this. Both materials are good insulators and the heat flow is very small.

17.69. **IDENTIFY and SET UP:** The heat conducted through the bottom of the pot goes into the water at 100°C to convert it to steam at 100°C. We can calculate the amount of heat flow from the mass of material that changes phase. Then use Eq. (17.21) to calculate T_H, the temperature of the lower surface of the pan.

EXECUTE: $Q = mL_v = (0.390 \text{ kg})(2256 \times 10^3 \text{ J/kg}) = 8.798 \times 10^5 \text{ J}$

$H = Q/t = 8.798 \times 10^5 \text{ J}/180 \text{ s} = 4.888 \times 10^3 \text{ J/s}$

Then $H = k A(T_H - T_C)/L$ says that $T_H - T_C = \dfrac{HL}{kA} = \dfrac{(4.888 \times 10^3 \text{ J/s})(8.50 \times 10^{-3} \text{ m})}{(50.2 \text{ W/m} \cdot \text{K})(0.150 \text{ m}^2)} = 5.52 \text{ C}°$

$T_H = T_C + 5.52 \text{ C}° = 100°\text{C} + 5.52 \text{ C}° = 105.5°\text{C}$

EVALUATE: The larger $T_H - T_C$ is the larger H is and the faster the water boils.

17.71.   **IDENTIFY:**   Assume the temperatures of the surfaces of the window are the outside and inside temperatures. Use the concept of thermal resistance. For part (b) use the fact that when insulating materials are in layers, the R values are additive.
SET UP: From Table 17.5, $k = 0.8 \text{ W/m} \cdot \text{K}$ for glass. $R = L/k$.

EXECUTE: **(a)** For the glass, $R_{\text{glass}} = \dfrac{5.20 \times 10^{-3} \text{ m}}{0.8 \text{ W/m} \cdot \text{K}} = 6.50 \times 10^{-3} \text{ m}^2 \cdot \text{K/W}$.

$H = \dfrac{A(T_H - T_C)}{R} = \dfrac{(1.40 \text{ m})(2.50 \text{ m})(39.5 \text{ K})}{6.50 \times 10^{-3} \text{ m}^2 \cdot \text{K/W}} = 2.1 \times 10^4 \text{ W}$

(b) For the paper, $R_{\text{paper}} = \dfrac{0.750 \times 10^{-3} \text{ m}}{0.05 \text{ W/m} \cdot \text{K}} = 0.015 \text{ m}^2 \cdot \text{K/W}$. The total R is

$R = R_{\text{glass}} + R_{\text{paper}} = 0.0215 \text{ m}^2 \cdot \text{K/W}.$ $H = \dfrac{A(T_H - T_C)}{R} = \dfrac{(1.40 \text{ m})(2.50 \text{ m})(39.5 \text{ K})}{0.0215 \text{ m}^2 \cdot \text{K/W}} = 6.4 \times 10^3 \text{ W}.$

EVALUATE: The layer of paper decreases the rate of heat loss by a factor of about 3.

17.77.   **IDENTIFY and SET UP:**   Use the temperature difference in M° and in C° between the melting and boiling points of mercury to relate M° to C°. Also adjust for the different zero points on the two scales to get an equation for T_M in terms of T_C.
(a) EXECUTE: normal melting point of mercury: $-39°\text{C} = 0.0°\text{M}$
normal boiling point of mercury: $357°\text{C} = 100.0°\text{M}$
$100.0 \text{ M}° = 396 \text{ C}°$ so $1 \text{ M}° = 3.96 \text{ C}°$

Zero on the M scale is -39 on the C scale, so to obtain T_C multiply T_M by 3.96 and then subtract 39°:
$T_C = 3.96 T_M - 39°$

Solving for T_M gives $T_M = \frac{1}{3.96}(T_C + 39°)$

The normal boiling point of water is $100°\text{C}$; $T_M = \frac{1}{3.96}(100° + 39°) = 35.1°\text{M}$

(b) $10.0 \text{ M}° = 39.6 \text{ C}°$
EVALUATE: A M° is larger than a C° since it takes fewer of them to express the difference between the boiling and melting points for mercury.

17.79.   **IDENTIFY and SET UP:**   Use Eq. (17.8) for the volume expansion of the oil and of the cup. Both the volume of the cup and the volume of the olive oil increase when the temperature increases, but β is larger for the oil so it expands more. When the oil starts to overflow, $\Delta V_{\text{oil}} = \Delta V_{\text{glass}} + (2.00 \times 10^{-3} \text{ m})A$, where A is the cross-sectional area of the cup.
EXECUTE: $\Delta V_{\text{oil}} = V_{0,\text{oil}} \beta_{\text{oil}} \Delta T = (9.8 \text{ cm}) A \beta_{\text{oil}} \Delta T.$ $\Delta V_{\text{glass}} = V_{0,\text{glass}} \beta_{\text{glass}} \Delta T = (10.0 \text{ cm}) A \beta_{\text{glass}} \Delta T.$

$(9.8 \text{ cm}) A \beta_{\text{oil}} \Delta T = (10.0 \text{ cm}) A \beta_{\text{glass}} \Delta T + (0.200 \text{ cm})A.$ The A divides out. Solving for ΔT gives

$\Delta T = 31.3 \text{ C}°.$ $T_2 = T_1 + \Delta T = 53.3°\text{C}.$

EVALUATE: If the expansion of the cup is neglected, the olive oil will have expanded to fill the cup when $(0.200 \text{ cm})A = (9.8 \text{ cm}) A \beta_{\text{oil}} \Delta T,$ so $\Delta T = 30.0 \text{ C}°$ and $T_2 = 52.0°\text{C}.$ Our result is slightly higher than this. The cup also expands but not very much since $\beta_{\text{glass}} \ll \beta_{\text{oil}}.$

17.81.   **IDENTIFY:**   Use Eq. (17.6) to find the change in diameter of the sphere and the change in length of the cable. Set the sum of these two increases in length equal to 2.00 mm.
SET UP: $\alpha_{\text{brass}} = 2.0 \times 10^{-5} \text{ K}^{-1}$ and $\alpha_{\text{steel}} = 1.2 \times 10^{-5} \text{ K}^{-1}$.

EXECUTE: $\Delta L = (\alpha_{\text{brass}}L_{0,\text{brass}} + \alpha_{\text{steel}}L_{0,\text{steel}})\Delta T.$

$\Delta T = \dfrac{2.00\times10^{-3}\text{ m}}{(2.0\times10^{-5}\text{ K}^{-1})(0.350\text{ m}) + (1.2\times10^{-5}\text{ K}^{-1})(10.5\text{ m})} = 15.0\text{ C}°.\quad T_2 = T_1 + \Delta T = 35.0°\text{C}.$

EVALUATE: The change in diameter of the brass sphere is 0.10 mm. This is small, but should not be neglected.

17.83.   **IDENTIFY** and **SET UP:**   Call the metals A and B. Use the data given to calculate α for each metal.

EXECUTE: $\Delta L = L_0\alpha\Delta T$ so $\alpha = \Delta L/(L_0\Delta T)$

metal A: $\alpha_A = \dfrac{\Delta L}{L_0\Delta T} = \dfrac{0.0650\text{ cm}}{(30.0\text{ cm})(100\text{ C}°)} = 2.167\times10^{-5}\ (\text{C}°)^{-1}$

metal B: $\alpha_B = \dfrac{\Delta L}{L_0\Delta T} = \dfrac{0.0350\text{ cm}}{(30.0\text{ cm})(100\text{ C}°)} = 1.167\times10^{-5}\ (\text{C}°)^{-1}$

EVALUATE: L_0 and ΔT are the same, so the rod that expands the most has the larger α.

IDENTIFY and **SET UP:**   Now consider the composite rod (Figure 17.83). Apply Eq. (17.6). The target variables are L_A and L_B, the lengths of the metals A and B in the composite rod.

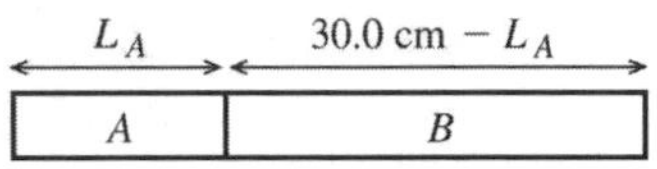

Figure 17.83

EXECUTE: $\Delta L = \Delta L_A + \Delta L_B = (\alpha_A L_A + \alpha_B L_B)\Delta T$

$\Delta L/\Delta T = \alpha_A L_A + \alpha_B(0.300\text{ m} - L_A)$

$L_A = \dfrac{\Delta L/\Delta T - (0.300\text{ m})\alpha_B}{\alpha_A - \alpha_B} = \dfrac{(0.058\times10^{-2}\text{ m}/100\text{ C}°) - (0.300\text{ m})(1.167\times10^{-5}(\text{C}°)^{-1})}{1.00\times10^{-5}\ (\text{C}°)^{-1}} = 23.0\text{ cm}$

$L_B = 30.0\text{ cm} - L_A = 30.0\text{ cm} - 23.0\text{ cm} = 7.0\text{ cm}$

EVALUATE: The expansion of the composite rod is similar to that of rod A, so the composite rod is mostly metal A.

17.87.   **IDENTIFY:**   For a string, $f_n = \dfrac{n}{2L}\sqrt{\dfrac{F}{\mu}}.$

SET UP: For the fundamental, $n = 1$. Solving for F gives $F = \mu 4L^2 f^2$. Note that $\mu = \pi r^2\rho$, so $\mu = \pi(0.203\times10^{-3}\text{ m})^2(7800\text{ kg/m}^3) = 1.01\times10^{-3}\text{ kg/m}.$

EXECUTE: **(a)** $F = (1.01\times10^{-3}\text{ kg/m})4(0.635\text{ m})^2(247.0\text{ Hz})^2 = 99.4\text{ N}$

(b) To find the fractional change in the frequency we must take the ratio of Δf to f: $f = \dfrac{1}{2L}\sqrt{\dfrac{F}{\mu}}$ and

$\Delta f = \Delta\left(\dfrac{1}{2L}\sqrt{\dfrac{F}{\mu}}\right) = \Delta\left(\dfrac{1}{2L\sqrt{\mu}}F^{\frac{1}{2}}\right) = \dfrac{1}{2L\sqrt{\mu}}\Delta\left(F^{\frac{1}{2}}\right) = \dfrac{1}{2L\sqrt{\mu}}\dfrac{1}{2}\dfrac{\Delta F}{\sqrt{F}}$

Now divide both sides by the original equation for f and cancel terms: $\dfrac{\Delta f}{f} = \dfrac{\dfrac{1}{2L\sqrt{\mu}}\dfrac{1}{2}\dfrac{\Delta F}{\sqrt{F}}}{\dfrac{1}{2L}\sqrt{\dfrac{F}{\mu}}} = \dfrac{1}{2}\dfrac{\Delta F}{F}.$

(c) The coefficient of thermal expansion α is defined by $\Delta l = l_0\alpha\Delta T$. Combining this with $Y = \dfrac{F/A}{\Delta l/l_0}$ gives

$\Delta F = -Y\alpha A\Delta T.\quad \Delta F = -(2.00\times10^{11}\text{ Pa})(1.20\times10^{-5}/\text{C}°)\pi(0.203\times10^{-3}\text{ m})^2(11\text{ C}°) = -3.4\text{ N}.$ Then $\Delta F/F - 0.034,\ \Delta f/f = -0.017$ and $\Delta f = -4.2\text{ Hz}$. The pitch falls. This also explains the constant tuning in the string sections of symphonic orchestras.

EVALUATE: An increase in temperature causes a decrease in tension of the string, and this lowers the frequency of each standing wave.

17.89. **(a) IDENTIFY and SET UP:** The diameter of the ring undergoes linear expansion (increases with T) just like a solid steel disk of the same diameter as the hole in the ring. Heat the ring to make its diameter equal to 2.5020 in.

EXECUTE: $\Delta L = \alpha L_0 \Delta T$ so $\Delta T = \dfrac{\Delta L}{L_0 \alpha} = \dfrac{0.0020 \text{ in.}}{(2.5000 \text{ in.})(1.2 \times 10^{-5} (\text{C}°)^{-1})} = 66.7 \text{ C}°$

$T = T_0 + \Delta T = 20.0°\text{C} + 66.7 \text{ C}° = 87°\text{C}$

(b) IDENTIFY and SET UP: Apply the linear expansion equation to the diameter of the brass shaft and to the diameter of the hole in the steel ring.

EXECUTE: $L = L_0(1 + \alpha \Delta T)$

Want L_s (steel) $= L_b$ (brass) for the same ΔT for both materials: $L_{0s}(1 + \alpha_s \Delta T) = L_{0b}(1 + \alpha_b \Delta T)$ so

$L_{0s} + L_{0s}\alpha_s \Delta T = L_{0b} + L_{0b}\alpha_b \Delta T$.

$\Delta T = \dfrac{L_{0b} - L_{0s}}{L_{0s}\alpha_s - L_{0b}\alpha_b} = \dfrac{2.5020 \text{ in.} - 2.5000 \text{ in.}}{(2.5000 \text{ in.})(1.2 \times 10^{-5} (\text{C}°)^{-1}) - (2.5050 \text{ in.})(2.0 \times 10^{-5}(\text{C}°)^{-1})}$

$\Delta T = \dfrac{0.0020}{3.00 \times 10^{-5} - 5.00 \times 10^{-5}} \text{ C}° = -100 \text{ C}°$

$T = T_0 + \Delta T = 20.0°\text{C} - 100 \text{ C}° = -80°\text{C}$

EVALUATE: Both diameters decrease when the temperature is lowered but the diameter of the brass shaft decreases more since $\alpha_b > \alpha_s$; $|\Delta L_b| - |\Delta L_s| = 0.0020$ in.

17.91. **IDENTIFY:** Apply Eq. (11.14) to the volume increase of the liquid due to the pressure decrease. Eq. (17.8) gives the volume decrease of the cylinder and liquid when they are cooled. Can think of the liquid expanding when the pressure is reduced and then contracting to the new volume of the cylinder when the temperature is reduced.

SET UP: Let β_l and β_m be the coefficients of volume expansion for the liquid and for the metal. Let ΔT be the (negative) change in temperature when the system is cooled to the new temperature.

EXECUTE: Change in volume of cylinder when cool: $\Delta V_m = \beta_m V_0 \Delta T$ (negative)

Change in volume of liquid when cool: $\Delta V_l = \beta_l V_0 \Delta T$ (negative)

The difference $\Delta V_l - \Delta V_m$ must be equal to the negative volume change due to the increase in pressure, which is $-\Delta p V_0 / B = -k \Delta p V_0$. Thus $\Delta V_l - \Delta V_m = -k \Delta p V_0$.

$\Delta T = -\dfrac{k \Delta p}{\beta_l - \beta_m}$

$\Delta T = -\dfrac{(8.50 \times 10^{-10} \text{ Pa}^{-1})(50.0 \text{ atm})(1.013 \times 10^5 \text{ Pa/1 atm})}{4.80 \times 10^{-4} \text{ K}^{-1} - 3.90 \times 10^{-5} \text{ K}^{-1}} = -9.8 \text{ C}°$

$T = T_0 + \Delta T = 30.0°\text{C} - 9.8 \text{ C}° = 20.2°\text{C}.$

EVALUATE: A modest temperature change produces the same volume change as a large change in pressure; $B \gg \beta$ for the liquid.

17.99. **IDENTIFY:** Use $Q = mL_f$ to find the heat that goes into the ice to melt it. This amount of heat must be conducted through the walls of the box; $Q = Ht$. Assume the surfaces of the styrofoam have temperatures of 5.00°C and 21.0°C.

SET UP: For water $L_f = 334 \times 10^3 \text{ J/kg}$. For styrofoam $k = 0.01 \text{ W/m} \cdot \text{K}$. One week is 6.048×10^5 s. The surface area of the box is $4(0.500 \text{ m})(0.800 \text{ m}) + 2(0.500 \text{ m})^2 = 2.10 \text{ m}^2$.

EXECUTE: $Q = mL_f = (24.0 \text{ kg})(334 \times 10^3 \text{ J/kg}) = 8.016 \times 10^6 \text{ J}$. $H = kA \dfrac{T_H - T_C}{L}$. $Q = Ht$ gives

$L = \dfrac{tkA(T_H - T_C)}{Q} = \dfrac{(6.048 \times 10^5 \text{ s})(0.01 \text{ W/m} \cdot \text{K})(2.10 \text{ m}^2)(21.0°\text{C} - 5.00°\text{C})}{8.016 \times 10^6 \text{ J}} = 2.5 \text{ cm}$

EVALUATE: We have assumed that the liquid water that is produced by melting the ice remains in thermal equilibrium with the ice so has a temperature of 0°C. The interior of the box and the ice are not in thermal equilibrium, since they have different temperatures.

17.101.   **(a) IDENTIFY and SET UP:**   Assume that all the ice melts and that all the steam condenses. If we calculate a final temperature T that is outside the range 0°C to 100°C then we know that this assumption is incorrect. Calculate Q for each piece of the system and then set the total $Q_{system} = 0$.

EXECUTE: copper can (changes temperature from $0.0°$ to T; no phase change)

$Q_{can} = mc\Delta T = (0.446 \text{ kg})(390 \text{ J/kg} \cdot \text{K})(T - 0.0°\text{C}) = (173.9 \text{ J/K})T$

ice (melting phase change and then the water produced warms to T)

$Q_{ice} = +mL_f + mc\Delta T = (0.0950 \text{ kg})(334 \times 10^3 \text{ J/kg}) + (0.0950 \text{ kg})(4190 \text{ J/kg} \cdot \text{K})(T - 0.0°\text{C})$

$Q_{ice} = 3.173 \times 10^4 \text{ J} + (398.0 \text{ J/K})T$.

steam (condenses to liquid and then water produced cools to T)

$Q_{steam} = -mL_v + mc\Delta T = -(0.0350 \text{ kg})(2256 \times 10^3 \text{ J/kg}) + (0.0350 \text{ kg})(4190 \text{ J/kg} \cdot \text{K})(T - 100.0°\text{C})$

$Q_{steam} = -7.896 \times 10^4 \text{ J} + (146.6 \text{ J/K})T - 1.466 \times 10^4 \text{ J} = -9.362 \times 10^4 \text{ J} + (146.6 \text{ J/K})T$

$Q_{system} = 0$ implies $Q_{can} + Q_{ice} + Q_{steam} = 0$.

$(173.9 \text{ J/K})T + 3.173 \times 10^4 \text{ J} + (398.0 \text{ J/K})T - 9.362 \times 10^4 \text{ J} + (146.6 \text{ J/K})T = 0$

$(718.5 \text{ J/K})T = 6.189 \times 10^4 \text{ J}$

$T = \dfrac{6.189 \times 10^4 \text{ J}}{718.5 \text{ J/K}} = 86.1°\text{C}$

EVALUATE: This is between 0°C and 100°C so our assumptions about the phase changes being complete were correct.

(b) No ice, no steam and $0.0950 \text{ kg} + 0.0350 \text{ kg} = 0.130 \text{ kg}$ of liquid water.

17.103.   **IDENTIFY and SET UP:**   Heat comes out of the steam when it changes phase and heat goes into the water and causes its temperature to rise. $Q_{system} = 0$. First determine what phases are present after the system has come to a uniform final temperature.

(a) EXECUTE: Heat that must be removed from steam if all of it condenses is

$Q = -mL_v = -(0.0400 \text{ kg})(2256 \times 10^3 \text{ J/kg}) = -9.02 \times 10^4 \text{ J}$

Heat absorbed by the water if it heats all the way to the boiling point of 100°C:

$Q = mc\Delta T = (0.200 \text{ kg})(4190 \text{ J/kg} \cdot \text{K})(50.0 \text{ C}°) = 4.19 \times 10^4 \text{ J}$

EVALUATE: The water can't absorb enough heat for all the steam to condense. Steam is left and the final temperature then must be 100°C.

(b) EXECUTE: Mass of steam that condenses is $m = Q/L_v = 4.19 \times 10^4 \text{ J}/2256 \times 10^3 \text{ J/kg} = 0.0186 \text{ kg}$. Thus there is $0.0400 \text{ kg} - 0.0186 \text{ kg} = 0.0214 \text{ kg}$ of steam left. The amount of liquid water is $0.0186 \text{ kg} + 0.200 \text{ kg} = 0.219 \text{ kg}$.

17.105.   **IDENTIFY:**   Heat Q_l comes out of the lead when it solidifies and the solid lead cools to T_f. If mass m_s of steam is produced, the final temperature is $T_f = 100°\text{C}$ and the heat that goes into the water is

$Q_w = m_w c_w (25.0 \text{ C}°) + m_s L_{v,w}$, where $m_w = 0.5000 \text{ kg}$. Conservation of energy says $Q_l + Q_w = 0$. Solve for m_s. The mass that remains is $1.250 \text{ kg} + 0.5000 \text{ kg} - m_s$.

SET UP: For lead, $L_{f,l} = 24.5 \times 10^3 \text{ J/kg}$, $c_l = 130 \text{ J/kg} \cdot \text{K}$ and the normal melting point of lead is 327.3°C.

For water, $c_w = 4190 \text{ J/kg} \cdot \text{K}$ and $L_{v,w} = 2256 \times 10^3 \text{ J/kg}$.

EXECUTE: $Q_l + Q_w = 0$. $-m_l L_{f,l} + m_l c_l (-227.3 \text{ C}°) + m_w c_w (25.0 \text{ C}°) + m_s L_{v,w} = 0$.

$m_s = \dfrac{m_l L_{f,l} + m_l c_l (+227.3 \text{ C}°) - m_w c_w (25.0 \text{ C}°)}{L_{v,w}}$.

$$m_s = \frac{+(1.250 \text{ kg})(24.5 \times 10^3 \text{ J/kg}) + (1.250 \text{ kg})(130 \text{ J/kg} \cdot \text{K})(227.3 \text{ K}) - (0.5000 \text{ kg})(4190 \text{ J/kg} \cdot \text{K})(25.0 \text{ K})}{2256 \times 10^3 \text{ J/kg}}$$

$$m_s = \frac{1.519 \times 10^4 \text{ J}}{2256 \times 10^3 \text{ J/kg}} = 0.0067 \text{ kg}. \text{ The mass of water and lead that remains is } 1.743 \text{ kg}.$$

EVALUATE: The magnitude of heat that comes out of the lead when it goes from liquid at 327.3°C to solid at 100.0°C is 6.76×10^4 J. The heat that goes into the water to warm it to 100°C is 5.24×10^4 J. The additional heat that goes into the water, $6.76 \times 10^4 \text{ J} - 5.24 \times 10^4 \text{ J} = 1.52 \times 10^4 \text{ J}$ converts 0.0067 kg of water at 100°C to steam.

17.107.   **IDENTIFY:**   Apply $H = kA \dfrac{\Delta T}{L}$.

SET UP: For the glass use $L = 12.45$ cm, to account for the thermal resistance of the air films on either side of the glass.

EXECUTE: **(a)** $H = (0.120 \text{ W/m} \cdot \text{K})(2.00 \times 0.95 \text{ m}^2)\left(\dfrac{28.0 \text{ C}°}{5.0 \times 10^{-2} \text{ m} + 1.8 \times 10^{-2} \text{ m}} \right) = 93.9 \text{ W}.$

(b) The heat flow through the wood part of the door is reduced by a factor of $1 - \dfrac{(0.50)^2}{(2.00 \times 0.95)} = 0.868$, so it becomes 81.5 W. The heat flow through the glass is

$H_{\text{glass}} = (0.80 \text{ W/m} \cdot \text{K})(0.50 \text{ m})^2 \left(\dfrac{28.0 \text{ C}°}{12.45 \times 10^{-2} \text{ m}} \right) = 45.0 \text{ W}$, and so the ratio is $\dfrac{81.5 + 45.0}{93.9} = 1.35$.

EVALUATE: The single-pane window produces a significant increase in heat loss through the door. (See Problem 17.109).

17.109.   **IDENTIFY** and **SET UP:**   Use H written in terms of the thermal resistance R: $H = A\Delta T/R$, where $R = L/k$ and $R = R_1 + R_2 + \dots$ (additive).

EXECUTE: single pane $R_s = R_{\text{glass}} + R_{\text{film}}$, where $R_{\text{film}} = 0.15 \text{ m}^2 \cdot \text{K/W}$ is the combined thermal resistance of the air films on the room and outdoor surfaces of the window.

$R_{\text{glass}} = L/k = (4.2 \times 10^{-3} \text{ m})/(0.80 \text{ W/m} \cdot \text{K}) = 0.00525 \text{ m}^2 \cdot \text{K/W}$

Thus $R_s = 0.00525 \text{ m}^2 \cdot \text{K/W} + 0.15 \text{ m}^2 \cdot \text{K/W} = 0.1553 \text{ m}^2 \cdot \text{K/W}.$

double pane $R_d = 2R_{\text{glass}} + R_{\text{air}} + R_{\text{film}}$, where R_{air} is the thermal resistance of the air space between the panes. $R_{\text{air}} = L/k = (7.0 \times 10^{-3} \text{ m})/(0.024 \text{ W/m} \cdot \text{K}) = 0.2917 \text{ m}^2 \cdot \text{K/W}$

Thus $R_d = 2(0.00525 \text{ m}^2 \cdot \text{K/W}) + 0.2917 \text{ m}^2 \cdot \text{K/W} + 0.15 \text{ m}^2 \cdot \text{K/W} = 0.4522 \text{ m}^2 \cdot \text{K/W}$

$H_s = A\Delta T/R_s, H_d = A\Delta T/R_d$, so $H_s/H_d = R_d/R_s$ (since A and ΔT are same for both)

$H_s/H_d = (0.4522 \text{ m}^2 \cdot \text{K/W})/(0.1553 \text{ m}^2 \cdot \text{K/W}) = 2.9$

EVALUATE: The heat loss is about a factor of 3 less for the double-pane window. The increase in R for a double-pane is due mostly to the thermal resistance of the air space between the panes.

17.111.   **(a) EXECUTE:**   Heat must be conducted from the water to cool it to 0°C and to cause the phase transition. The entire volume of water is not at the phase transition temperature, just the upper surface that is in contact with the ice sheet.

(b) IDENTIFY: The heat that must leave the water in order for it to freeze must be conducted through the layer of ice that has already been formed.

SET UP: Consider a section of ice that has area A. At time t let the thickness be h. Consider a short time interval t to $t + dt$. Let the thickness that freezes in this time be dh. The mass of the section that freezes in the time interval dt is $dm = \rho \, dV = \rho A \, dh$. The heat that must be conducted away from this mass of water to freeze it is $dQ = dm L_f = (\rho A L_f)dh$. $H = dQ/dt = kA(\Delta T/h)$, so the heat dQ conducted in time dt

throughout the thickness h that is already there is $dQ = kA\left(\dfrac{T_H - T_C}{h}\right)dt$. Solve for dh in terms of dt and

integrate to get an expression relating h and t.

EXECUTE: Equate these expressions for dQ.

$$\rho A L_f dh = kA\left(\frac{T_H - T_C}{h}\right)dt$$

$$h\,dh = \left(\frac{k(T_H - T_C)}{\rho L_f}\right)dt$$

Integrate from $t = 0$ to time t. At $t = 0$ the thickness h is zero.

$$\int_0^h h\,dh = [k(T_H - T_C)/\rho L_f]\int_0^t dt$$

$$\tfrac{1}{2}h^2 = \frac{k(T_H - T_C)}{\rho L_f}t \quad \text{and} \quad h = \sqrt{\frac{2k(T_H - T_C)}{\rho L_f}}\sqrt{t}$$

The thickness after time t is proportional to $\sqrt{t}$.

(c) The expression in part (b) gives $t = \dfrac{h^2 \rho L_f}{2k(T_H - T_C)} = \dfrac{(0.25\text{ m})^2(920\text{ kg/m}^3)(334\times10^3\text{ J/kg})}{2(1.6\text{ W/m}\cdot\text{K})(0°\text{C} - (-10°\text{C}))} = 6.0\times10^5\text{ s}$

$t = 170$ h.

(d) Find t for $h = 40$ m. t is proportional to h^2, so $t = (40\text{ m}/0.25\text{ m})^2(6.00\times10^5\text{ s}) = 1.5\times10^{10}$ s. This is about 500 years. With our current climate this will not happen.

EVALUATE: As the ice sheet gets thicker, the rate of heat conduction through it decreases. Part (d) shows that it takes a very long time for a moderately deep lake to totally freeze.

17.115.   **IDENTIFY and SET UP:** Use Eq. (17.26) to find the net heat current into the can due to radiation. Use $Q = Ht$ to find the heat that goes into the liquid helium, set this equal to mL and solve for the mass m of helium that changes phase.

EXECUTE: Calculate the net rate of radiation of heat from the can. $H_{\text{net}} = Ae\sigma(T^4 - T_s^4)$.

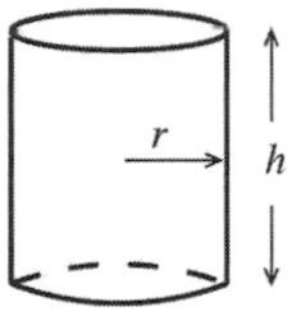

The surface area of the cylindrical can is
$A = 2\pi rh + 2\pi r^2$. (See Figure 17.115.)

Figure 17.115

$A = 2\pi r(h + r) = 2\pi(0.045\text{ m})(0.250\text{ m} + 0.045\text{ m}) = 0.08341\text{ m}^2$.

$H_{\text{net}} = (0.08341\text{ m}^2)(0.200)(5.67\times10^{-8}\text{ W/m}^2\cdot\text{K}^4)((4.22\text{ K})^4 - (77.3\text{ K})^4)$

$H_{\text{net}} = -0.0338$ W (the minus sign says that the net heat current is into the can). The heat that is put into the can by radiation in one hour is $Q = -(H_{\text{net}})t = (0.0338\text{ W})(3600\text{ s}) = 121.7$ J. This heat boils a mass m

of helium according to the equation $Q = mL_f$, so $m = \dfrac{Q}{L_f} = \dfrac{121.7\text{ J}}{2.09\times10^4\text{ J/kg}} = 5.82\times10^{-3}$ kg $= 5.82$ g.

EVALUATE: In the expression for the net heat current into the can the temperature of the surroundings is raised to the fourth power. The rate at which the helium boils away increases by about a factor of $(293/77)^4 = 210$ if the walls surrounding the can are at room temperature rather than at the temperature of the liquid nitrogen.

17.117.   **IDENTIFY:** The jogger radiates heat but the air radiates heat back into the jogger.

SET UP: The emissivity of a human body is taken to be 1.0. In the equation for the radiation heat current, $H_{\text{net}} = Ae\sigma(T^4 - T_s^4)$, the temperatures must be in kelvins.

EXECUTE: **(a)** $P_{jog} = (0.80)(1300 \text{ W}) = 1.04 \times 10^3 \text{ J/s}.$

(b) $H_{net} = Ae\sigma(T^4 - T_s^4),$ which gives

$H_{net} = (1.85 \text{ m}^2)(1.00)(5.67 \times 10^{-8} \text{ W/m}^2 \cdot \text{K}^4)([306 \text{ K}]^4 - [313 \text{ K}]^4) = -87.1 \text{ W}.$ The person gains 87.1 J of heat each second by radiation.

(c) The total excess heat per second is $1040 \text{ J/s} + 87 \text{ J/s} = 1130 \text{ J/s}.$

(d) In 1 min = 60 s, the runner must dispose of $(60 \text{ s})(1130 \text{ J/s}) = 6.78 \times 10^4 \text{ J}.$ If this much heat goes to evaporate water, the mass m of water that evaporates in one minute is given by $Q = mL_v,$ so

$$m = \frac{Q}{L_v} = \frac{6.78 \times 10^4 \text{ J}}{2.42 \times 10^6 \text{ J/kg}} = 0.028 \text{ kg} = 28 \text{ g}.$$

(e) In a half-hour, or 30 minutes, the runner loses $(30 \text{ min})(0.028 \text{ kg/min}) = 0.84 \text{ kg}.$ The runner must

drink 0.84 L, which is $\dfrac{0.84 \text{ L}}{0.750 \text{ L/bottle}} = 1.1$ bottles.

EVALUATE: The person *gains* heat by radiation since the air temperature is greater than his skin temperature.

17.119. **IDENTIFY:** For the water, $Q = mc\Delta T.$

SET UP: For water, $c = 4190 \text{ J/kg} \cdot \text{K}.$

EXECUTE: **(a)** At steady state, the input power all goes into heating the water, so $P = \dfrac{Q}{t} = \dfrac{mc\Delta T}{t}$ and

$\Delta T = \dfrac{Pt}{cm} = \dfrac{(1800 \text{ W})(60 \text{ s/min})}{(4190 \text{ J/kg} \cdot \text{K})(0.500 \text{ kg/min})} = 51.6 \text{ K},$ and the output temperature is

$18.0°\text{C} + 51.6 \text{ C}° = 69.6°\text{C}.$

EVALUATE: **(b)** At steady state, the temperature of the apparatus is constant and the apparatus will neither remove heat from nor add heat to the water.

18

THERMAL PROPERTIES OF MATTER

18.5. **IDENTIFY:** We know the pressure and temperature and want to find the density of the gas. The ideal gas law applies.

SET UP: $M_{CO_2} = (12 + 2[16])$ g/mol $= 44$ g/mol. $M_{N_2} = 28$ g/mol. $\rho = \dfrac{pM}{RT}$.

$R = 8.315$ J/mol·K. T must be in kelvins. Express M in kg/mol and p in Pa. 1 atm $= 1.013 \times 10^5$ Pa.

EXECUTE: **(a)** *Mars:* $\rho = \dfrac{(650\ \text{Pa})(44 \times 10^{-3}\ \text{kg/mol})}{(8.315\ \text{J/mol·K})(253\ \text{K})} = 0.0136$ kg/m^3.

Venus: $\rho = \dfrac{(92\ \text{atm})(1.013 \times 10^5\ \text{Pa/atm})(44 \times 10^{-3}\ \text{kg/mol})}{(8.315\ \text{J/mol·K})(730\ \text{K})} = 67.6$ kg/m^3.

Titan: $T = -178 + 273 = 95$ K.

$\rho = \dfrac{(1.5\ \text{atm})(1.013 \times 10^5\ \text{Pa/atm})(28 \times 10^{-3}\ \text{kg/mol})}{(8.315\ \text{J/mol·K})(95\ \text{K})} = 5.39$ kg/m^3.

EVALUATE: **(b)** Table 12.1 gives the density of air at 20°C and $p = 1$ atm to be 1.20 kg/m^3. The density of the atmosphere of Mars is much less, the density for Venus is much greater and the density for Titan is somewhat greater.

18.7. **IDENTIFY:** We are asked to compare two states. Use the ideal gas law to obtain T_2 in terms of T_1 and ratios of pressures and volumes of the gas in the two states.

SET UP: $pV = nRT$ and n, R constant implies $pV/T = nR = $ constant and $p_1 V_1 / T_1 = p_2 V_2 / T_2$

EXECUTE: $T_1 = (27 + 273)$ K $= 300$ K

$p_1 = 1.01 \times 10^5$ Pa

$p_2 = 2.72 \times 10^6$ Pa $+ 1.01 \times 10^5$ Pa $= 2.82 \times 10^6$ Pa (in the ideal gas equation the pressures must be absolute, not gauge, pressures)

$T_2 = T_1 \left(\dfrac{p_2}{p_1} \right) \left(\dfrac{V_2}{V_1} \right) = 300$ K $\left(\dfrac{2.82 \times 10^6\ \text{Pa}}{1.01 \times 10^5\ \text{Pa}} \right) \left(\dfrac{46.2\ \text{cm}^3}{499\ \text{cm}^3} \right) = 776$ K

$T_2 = (776 - 273)$°C $= 503$°C

EVALUATE: The units cancel in the V_2 / V_1 volume ratio, so it was not necessary to convert the volumes in cm^3 to m^3. It was essential, however, to use T in kelvins.

18.9. **IDENTIFY:** $pV = nRT$.

SET UP: $T_1 = 300$ K, $T_2 = 430$ K.

EXECUTE: **(a)** n, R are constant so $\dfrac{pV}{T} = nR = $ constant. $\dfrac{p_1 V_1}{T_1} = \dfrac{p_2 V_2}{T_2}$.

$$p_2 = p_1 \left(\frac{V_1}{V_2} \right) \left(\frac{T_2}{T_1} \right) = (7.50 \times 10^3 \text{ Pa}) \left(\frac{0.750 \text{ m}^3}{0.480 \text{ m}^3} \right) \left(\frac{430 \text{ K}}{300 \text{ K}} \right) = 1.68 \times 10^4 \text{ Pa.}$$

EVALUATE: Since the temperature increased while the volume decreased, the pressure must have increased. In $pV = nRT$, T must be in kelvins, even if we use a ratio of temperatures.

18.13. **IDENTIFY:** We know the volume of the gas at STP on the earth and want to find the volume it would occupy on Venus where the pressure and temperature are much greater.
SET UP: STP is $T = 273$ K and $p = 1$ atm. Set up a ratio using $pV = nRT$ with nR constant.

$T_V = 1003 + 273 = 1276$ K.

EXECUTE: $pV = nRT$ gives $\frac{pV}{T} = nR = $ constant, so $\frac{p_E V_E}{T_E} = \frac{p_V V_V}{T_V}$.

$$V_V = V_E \left(\frac{p_E}{p_V} \right) \left(\frac{T_V}{T_E} \right) = V \left(\frac{1 \text{ atm}}{92 \text{ atm}} \right) \left(\frac{1276 \text{ K}}{273 \text{ K}} \right) = 0.0508V.$$

EVALUATE: Even though the temperature on Venus is higher than it is on Earth, the pressure there is much greater than on Earth, so the volume of the gas on Venus is only about 5% what it is on Earth.

18.15. **IDENTIFY:** We are asked to compare two states. First use $pV = nRT$ to calculate p_1. Then use it to obtain T_2 in terms of T_1 and the ratio of pressures in the two states.

(a) SET UP: $pV = nRT$. Find the initial pressure p_1.

EXECUTE: $p_1 = \dfrac{nRT_1}{V} = \dfrac{(11.0 \text{ mol})(8.3145 \text{ J/mol} \cdot \text{K})(23.0 + 273.15)\text{K}}{3.10 \times 10^{-3} \text{ m}^3} = 8.737 \times 10^6$ Pa

SET UP: $p_2 = 100 \text{ atm}(1.013 \times 10^5 \text{ Pa/1 atm}) = 1.013 \times 10^7$ Pa

$p/T = nR/V = $ constant, so $p_1/T_1 = p_2/T_2$

EXECUTE: $T_2 = T_1 \left(\dfrac{p_2}{p_1} \right) = (296.15 \text{ K}) \left(\dfrac{1.013 \times 10^7 \text{ Pa}}{8.737 \times 10^6 \text{ Pa}} \right) = 343.4 \text{ K} = 70.2°C$

(b) EVALUATE: The coefficient of volume expansion for a gas is much larger than for a solid, so the expansion of the tank is negligible.

18.19. **IDENTIFY:** We know the volume, pressure and temperature of the gas and want to find its mass and density.
SET UP: $V = 3.00 \times 10^{-3} \text{ m}^3$. $T = 295$ K. $p = 2.03 \times 10^{-8}$ Pa. The ideal gas law, $pV = nRT$, applies.
EXECUTE: **(a)** $pV = nRT$ gives

$n = \dfrac{pV}{RT} = \dfrac{(2.03 \times 10^{-8} \text{ Pa})(3.00 \times 10^{-3} \text{ m}^3)}{(8.315 \text{ J/mol} \cdot \text{K})(295 \text{ K})} = 2.48 \times 10^{-14}$ mol. The mass of this amount of gas is

$m = nM = (2.48 \times 10^{-14} \text{ mol})(28.0 \times 10^{-3} \text{ kg/mol}) = 6.95 \times 10^{-16}$ kg.

(b) $\rho = \dfrac{m}{V} = \dfrac{6.95 \times 10^{-16} \text{ kg}}{3.00 \times 10^{-3} \text{ m}^3} = 2.32 \times 10^{-13} \text{ kg/m}^3.$

EVALUATE: The density at this level of vacuum is 13 orders of magnitude less than the density of air at STP, which is 1.20 kg/m^3.

18.21. **IDENTIFY:** Use Eq. (18.5) and solve for p.
SET UP: $\rho = pM/RT$ and $p = RT\rho/M$

$T = (-56.5 + 273.15) \text{ K} = 216.6 \text{ K}$

For air $M = 28.8 \times 10^{-3}$ kg/mol (Example 18.3)

EXECUTE: $p = \dfrac{(8.3145 \text{ J/mol} \cdot \text{K})(216.6 \text{ K})(0.364 \text{ kg/m}^3)}{28.8 \times 10^{-3} \text{ kg/mol}} = 2.28 \times 10^4$ Pa

EVALUATE: The pressure is about one-fifth the pressure at sea-level.

18.23.   **IDENTIFY:** The mass m_{tot} is related to the number of moles n by $m_{tot} = nM$. Mass is related to volume by $\rho = m/V$.

SET UP: For gold, $M = 196.97$ g/mol and $\rho = 19.3 \times 10^3$ kg/m^3. The volume of a sphere of radius r is $V = \frac{4}{3}\pi r^3$.

EXECUTE: **(a)** $m_{tot} = nM = (3.00 \text{ mol})(196.97 \text{ g/mol}) = 590.9$ g. The value of this mass of gold is $(590.9 \text{ g})(\$14.75/\text{g}) = \$8720$.

(b) $V = \dfrac{m}{\rho} = \dfrac{0.5909 \text{ kg}}{19.3 \times 10^3 \text{ kg/m}^3} = 3.06 \times 10^{-5}$ m^3. $V = \frac{4}{3}\pi r^3$ gives

$$r = \left(\frac{3V}{4\pi}\right)^{1/3} = \left(\frac{3[3.06 \times 10^{-5} \text{ m}^3]}{4\pi}\right)^{1/3} = 0.0194 \text{ m} = 1.94 \text{ cm}.$$ The diameter is $2r = 3.88$ cm.

EVALUATE: The mass and volume are directly proportional to the number of moles.

18.25.   **IDENTIFY:** We are asked about a single state of the system.

SET UP: Use the ideal-gas law. Write n in terms of the number of molecules N.

(a) EXECUTE: $pV = nRT$, $n = N/N_A$ so $pV = (N/N_A)RT$

$$p = \left(\frac{N}{V}\right)\left(\frac{R}{N_A}\right)T$$

$$p = \left(\frac{80 \text{ molecules}}{1 \times 10^{-6} \text{ m}^3}\right)\left(\frac{8.3145 \text{ J/mol} \cdot \text{K}}{6.022 \times 10^{23} \text{ molecules/mol}}\right)(7500 \text{ K}) = 8.28 \times 10^{-12} \text{ Pa}$$

$p = 8.2 \times 10^{-17}$ atm. This is much lower than the laboratory pressure of 9×10^{-14} atm in Exercise 18.24.

(b) EVALUATE: The Lagoon Nebula is a very rarefied low pressure gas. The gas would exert *very* little force on an object passing through it.

18.29.   **(a) IDENTIFY** and **SET UP:** Use the density and the mass of 5.00 mol to calculate the volume. $\rho = m/V$ implies $V = m/\rho$, where $m = m_{tot}$, the mass of 5.00 mol of water.

EXECUTE: $m_{tot} = nM = (5.00 \text{ mol})(18.0 \times 10^{-3} \text{ kg/mol}) = 0.0900$ kg

Then $V = \dfrac{m}{\rho} = \dfrac{0.0900 \text{ kg}}{1000 \text{ kg/m}^3} = 9.00 \times 10^{-5}$ m^3

(b) One mole contains $N_A = 6.022 \times 10^{23}$ molecules, so the volume occupied by one molecule is

$$\frac{9.00 \times 10^{-5} \text{ m}^3/\text{mol}}{(5.00 \text{ mol})(6.022 \times 10^{23} \text{ molecules/mol})} = 2.989 \times 10^{-29} \text{ m}^3/\text{molecule}$$

$V = a^3$, where a is the length of each side of the cube occupied by a molecule. $a^3 = 2.989 \times 10^{-29}$ m^3, so $a = 3.1 \times 10^{-10}$ m.

(c) EVALUATE: Atoms and molecules are on the order of 10^{-10} m in diameter, in agreement with the above estimates.

18.35.   **IDENTIFY:** $v_{rms} = \sqrt{\dfrac{3kT}{m}}$

SET UP: The mass of a deuteron is $m = m_p + m_n = 1.673 \times 10^{-27}$ kg $+ 1.675 \times 10^{-27}$ kg $= 3.35 \times 10^{-27}$ kg. $c = 3.00 \times 10^8$ m/s. $k = 1.381 \times 10^{-23}$ J/molecule $\cdot$ K.

EXECUTE: **(a)** $v_{rms} = \sqrt{\dfrac{3(1.381 \times 10^{-23} \text{ J/molecule} \cdot \text{K})(300 \times 10^6 \text{ K})}{3.35 \times 10^{-27} \text{ kg}}} = 1.93 \times 10^6$ m/s. $\dfrac{v_{rms}}{c} = 6.43 \times 10^{-3}$.

(b) $T = \left(\dfrac{m}{3k}\right)(v_{rms})^2 = \left(\dfrac{3.35 \times 10^{-27} \text{ kg}}{3(1.381 \times 10^{-23} \text{ J/molecule} \cdot \text{K})}\right)(3.0 \times 10^7 \text{ m/s})^2 = 7.3 \times 10^{10}$ K.

EVALUATE: Even at very high temperatures and for this light nucleus, v_{rms} is a small fraction of the speed of light.

18.37. **IDENTIFY** and **SET UP:** Apply the analysis of Section 18.3.

EXECUTE: (a) $\frac{1}{2}m(v^2)_{av} = \frac{3}{2}kT = \frac{3}{2}(1.38\times10^{-23} \text{ J/molecule} \cdot \text{K})(300 \text{ K}) = 6.21\times10^{-21} \text{ J}$

(b) We need the mass m of one molecule:

$$m = \frac{M}{N_A} = \frac{32.0\times10^{-3} \text{ kg/mol}}{6.022\times10^{23} \text{ molecules/mol}} = 5.314\times10^{-26} \text{ kg/molecule}$$

Then $\frac{1}{2}m(v^2)_{av} = 6.21\times10^{-21}$ J (from part (a)) gives

$$(v^2)_{av} = \frac{2(6.21\times10^{-21} \text{ J})}{m} = \frac{2(6.21\times10^{-21} \text{ J})}{5.314\times10^{-26} \text{ kg}} = 2.34\times10^5 \text{ m}^2/\text{s}^2$$

(c) $v_{rms} = \sqrt{(v^2)_{rms}} = \sqrt{2.34\times10^4 \text{ m}^2/\text{s}^2} = 484 \text{ m/s}$

(d) $p = mv_{rms} = (5.314\times10^{-26} \text{ kg})(484 \text{ m/s}) = 2.57\times10^{-23} \text{ kg} \cdot \text{m/s}$

(e) Time between collisions with one wall is $t = \dfrac{0.20 \text{ m}}{v_{rms}} = \dfrac{0.20 \text{ m}}{484 \text{ m/s}} = 4.13\times10^{-4} \text{ s}$

In a collision $\vec{v}$ changes direction, so $\Delta p = 2mv_{rms} = 2(2.57\times10^{-23} \text{ kg} \cdot \text{m/s}) = 5.14\times10^{-23} \text{ kg} \cdot \text{m/s}$

$$F = \frac{dp}{dt} \text{ so } F_{av} = \frac{\Delta p}{\Delta t} = \frac{5.14\times10^{-23} \text{ kg} \cdot \text{m/s}}{4.13\times10^{-4} \text{ s}} = 1.24\times10^{-19} \text{ N}$$

(f) pressure $= F/A = 1.24\times10^{-19} \text{ N}/(0.10 \text{ m})^2 = 1.24\times10^{-17}$ Pa (due to one molecule)

(g) pressure $= 1$ atm $= 1.013\times10^5$ Pa

Number of molecules needed is 1.013×10^5 Pa$/(1.24\times10^{-17}$ Pa/molecule$) = 8.17\times10^{21}$ molecules

(h) $pV = NkT$ (Eq. 18.18), so $N = \dfrac{pV}{kT} = \dfrac{(1.013\times10^5 \text{ Pa})(0.10 \text{ m})^3}{(1.381\times10^{-23} \text{ J/molecule} \cdot \text{K})(300 \text{ K})} = 2.45\times10^{22}$ molecules

(i) From the factor of $\frac{1}{3}$ in $(v_x^2)_{av} = \frac{1}{3}(v^2)_{av}$.

EVALUATE: This exercise shows that the pressure exerted by a gas arises from collisions of the molecules of the gas with the walls.

18.41. **IDENTIFY:** Use Eq. (18.24), applied to a finite temperature change.
SET UP: $C_V = 5R/2$ for a diatomic ideal gas and $C_V = 3R/2$ for a monatomic ideal gas.

EXECUTE: (a) $Q = nC_V\Delta T = n\left(\frac{5}{2}R\right)\Delta T$. $Q = (2.5 \text{ mol})(\frac{5}{2})(8.3145 \text{ J/mol} \cdot \text{K})(50.0 \text{ K}) = 2600 \text{ J}$.

(b) $Q = nC_V\Delta T = n(\frac{3}{2}R)\Delta T$. $Q = (2.5 \text{ mol})(\frac{3}{2})(8.3145 \text{ J/mol} \cdot \text{K})(50.0 \text{ K}) = 1560 \text{ J}$.

EVALUATE: More heat is required for the diatomic gas; not all the heat that goes into the gas appears as translational kinetic energy, some goes into energy of the internal motion of the molecules (rotations).

18.43. **IDENTIFY:** $C = Mc$, where C is the molar heat capacity and c is the specific heat capacity.

$$pV = nRT = \frac{m}{M}RT.$$

SET UP: $M_{N_2} = 2(14.007 \text{ g/mol}) = 28.014\times10^{-3}$ kg/mol. For water, $c_w = 4190$ J/kg $\cdot$ K. For N_2, $C_V = 20.76$ J/mol $\cdot$ K.

EXECUTE: (a) $c_{N_2} = \dfrac{C}{M} = \dfrac{20.76 \text{ J/mol} \cdot \text{K}}{28.014\times10^{-3} \text{ kg/mol}} = 741$ J/kg $\cdot$ K. $\dfrac{c_w}{c_{N_2}} = 5.65$; c_w is over five time larger.

(b) To warm the water, $Q = mc_w \Delta T = (1.00 \text{ kg})(4190 \text{ J/mol} \cdot \text{K})(10.0 \text{ K}) = 4.19 \times 10^4 \text{ J}$. For air,

$$m = \frac{Q}{c_{N_2} \Delta T} = \frac{4.19 \times 10^4 \text{ J}}{(741 \text{ J/kg} \cdot \text{K})(10.0 \text{ K})} = 5.65 \text{ kg}.$$

$$V = \frac{mRT}{Mp} = \frac{(5.65 \text{ kg})(8.314 \text{ J/mol} \cdot \text{K})(293 \text{ K})}{(28.014 \times 10^{-3} \text{ kg/mol})(1.013 \times 10^5 \text{ Pa})} = 4.85 \text{ m}^3.$$

EVALUATE: c is smaller for N_2, so less heat is needed for 1.0 kg of N_2 than for 1.0 kg of water.

18.47. **IDENTIFY:** Apply Eqs. (18.34), (18.35) and (18.36).

SET UP: Note that $\frac{k}{m} = \frac{R/N_A}{M/N_A} = \frac{R}{M}$. $M = 44.0 \times 10^{-3}$ kg/mol.

EXECUTE: **(a)** $v_{mp} = \sqrt{2(8.3145 \text{ J/mol} \cdot \text{K})(300 \text{ K})/(44.0 \times 10^{-3} \text{ kg/mol})} = 3.37 \times 10^2$ m/s.

(b) $v_{av} = \sqrt{8(8.3145 \text{ J/mol} \cdot \text{K})(300 \text{ K})/(\pi (44.0 \times 10^{-3} \text{ kg/mol}))} = 3.80 \times 10^2$ m/s.

(c) $v_{rms} = \sqrt{3(8.3145 \text{ J/mol} \cdot \text{K})(300 \text{ K})/(44.0 \times 10^{-3} \text{ kg/mol})} = 4.12 \times 10^2$ m/s.

EVALUATE: The average speed is greater than the most probable speed and the rms speed is greater than the average speed.

18.51. **IDENTIFY:** Figure 18.24 in the textbook shows that there is no liquid phase below the triple point pressure.

SET UP: Table 18.3 gives the triple point pressure to be 610 Pa for water and 5.17×10^5 Pa for CO_2.

EXECUTE: The atmospheric pressure is below the triple point pressure of water, and there can be no liquid water on Mars. The same holds true for CO_2.

EVALUATE: On earth $p_{atm} = 1 \times 10^5$ Pa, so on the surface of the earth there can be liquid water but not liquid CO_2.

18.53. **IDENTIFY:** We can model the atmosphere as a fluid of constant density, so the pressure depends on the depth in the fluid, as we saw in Section 12.2.

SET UP: The pressure difference between two points in a fluid is $\Delta p = \rho g h$, where h is the difference in height of two points.

EXECUTE: **(a)** $\Delta p = \rho g h = (1.2 \text{ kg/m}^3)(9.80 \text{ m/s}^2)(1000 \text{ m}) = 1.18 \times 10^4$ Pa.

(b) At the bottom of the mountain, $p = 1.013 \times 10^5$ Pa. At the top, $p = 8.95 \times 10^4$ Pa.

$$pV = nRT = \text{constant so } p_b V_b = p_t V_t \text{ and } V_t = V_b \left(\frac{p_b}{p_t} \right) = (0.50 \text{ L}) \left(\frac{1.013 \times 10^5 \text{ Pa}}{8.95 \times 10^4 \text{ Pa}} \right) = 0.566 \text{ L}.$$

EVALUATE: The pressure variation with altitude is affected by changes in air density and temperature and we have neglected those effects. The pressure decreases with altitude and the volume increases. You may have noticed this effect: bags of potato chips "puff up" when taken to the top of a mountain.

18.55. **IDENTIFY:** The buoyant force on the balloon must be equal to the weight of the load plus the weight of the gas.

SET UP: The buoyant force is $F_B = \rho_{air} V g$. A lift of 290 kg means $\frac{F_B}{g} - m_{hot} = 290$ kg, where m_{hot} is the mass of hot air in the balloon. $m = \rho V$.

EXECUTE: $m_{hot} = \rho_{hot} V$. $\frac{F_B}{g} - m_{hot} = 290$ kg gives $(\rho_{air} - \rho_{hot})V = 290$ kg.

Solving for ρ_{hot} gives $\rho_{hot} = \rho_{air} - \frac{290 \text{ kg}}{V} = 1.23 \text{ kg/m}^3 - \frac{290 \text{ kg}}{500.0 \text{ m}^3} = 0.65 \text{ kg/m}^3$. $\rho_{hot} = \frac{pM}{RT_{hot}}$.

$\rho_{air} = \frac{pM}{RT_{air}}$. $\rho_{hot} T_{hot} = \rho_{air} T_{air}$ so

$$T_{hot} = T_{air}\left(\frac{\rho_{air}}{\rho_{hot}}\right) = (288 \text{ K})\left(\frac{1.23 \text{ kg/m}^3}{0.65 \text{ kg/m}^3}\right) = 545 \text{ K} = 272°\text{C}.$$

EVALUATE: This temperature is well above normal air temperatures, so the air in the balloon would need considerable heating.

18.57. **IDENTIFY:** We are asked to compare two states. Use the ideal-gas law to obtain m_2 in terms of m_1 and the ratio of pressures in the two states. Apply Eq. (18.4) to the initial state to calculate m_1.

SET UP: $pV = nRT$ can be written $pV = (m/M)RT$

T, V, M, R are all constant, so $p/m = RT/MV = $ constant.

So $p_1/m_1 = p_2/m_2$, where m is the mass of the gas in the tank.

EXECUTE: $p_1 = 1.30\times10^6 \text{ Pa} + 1.01\times10^5 \text{ Pa} = 1.40\times10^6 \text{ Pa}$

$p_2 = 2.50\times10^5 \text{ Pa} + 1.01\times10^5 \text{ Pa} = 3.51\times10^5 \text{ Pa}$

$m_1 = p_1VM/RT; \quad V = hA = h\pi r^2 = (1.00 \text{ m})\pi(0.060 \text{ m})^2 = 0.01131 \text{ m}^3$

$$m_1 = \frac{(1.40\times10^6 \text{ Pa})(0.01131 \text{ m}^3)(44.1\times10^{-3} \text{ kg/mol})}{(8.3145 \text{ J/mol}\cdot\text{K})((22.0+273.15)\text{K})} = 0.2845 \text{ kg}$$

Then $m_2 = m_1\left(\dfrac{p_2}{p_1}\right) = (0.2845 \text{ kg})\left(\dfrac{3.51\times10^5 \text{ Pa}}{1.40\times10^6 \text{ Pa}}\right) = 0.0713 \text{ kg}.$

m_2 is the mass that remains in the tank. The mass that has been used is

$m_1 - m_2 = 0.2845 \text{ kg} - 0.0713 \text{ kg} = 0.213 \text{ kg}.$

EVALUATE: Note that we have to use absolute pressures. The absolute pressure decreases by a factor of four and the mass of gas in the tank decreases by a factor of four.

18.59. **IDENTIFY:** $pV = NkT$ gives $\dfrac{N}{V} = \dfrac{p}{kT}$.

SET UP: $1 \text{ atm} = 1.013\times10^5 \text{ Pa}. \quad T_K = T_C + 273.15. \quad k = 1.381\times10^{-23} \text{ J/molecule}\cdot\text{K}.$

EXECUTE: **(a)** $T_C = T_K - 273.15 = 94 \text{ K} - 273.15 = -179°\text{C}$

(b) $\dfrac{N}{V} = \dfrac{p}{kT} = \dfrac{(1.5 \text{ atm})(1.013\times10^5 \text{ Pa/atm})}{(1.381\times10^{-23} \text{ J/molecule}\cdot\text{K})(94 \text{ K})} = 1.2\times10^{26} \text{ molecules/m}^3$

(c) For the earth, $p = 1.0 \text{ atm} = 1.013\times10^5 \text{ Pa}$ and $T = 22°\text{C} = 295 \text{ K}.$

$\dfrac{N}{V} = \dfrac{(1.0 \text{ atm})(1.013\times10^5 \text{ Pa/atm})}{(1.381\times10^{-23} \text{ J/molecule}\cdot\text{K})(295 \text{ K})} = 2.5\times10^{25} \text{ molecules/m}^3.$ The atmosphere of Titan is about

five times denser than earth's atmosphere.

EVALUATE: Though it is smaller than Earth and has weaker gravity at its surface, Titan can maintain a dense atmosphere because of the very low temperature of that atmosphere.

18.61. **IDENTIFY:** $pV = nRT$

SET UP: In $pV = nRT$ we must use the absolute pressure. $T_1 = 278 \text{ K}. \quad p_1 = 2.72 \text{ atm}. \quad T_2 = 318 \text{ K}.$

EXECUTE: n, R constant, so $\dfrac{pV}{T} = nR = $ constant. $\dfrac{p_1V_1}{T_1} = \dfrac{p_2V_2}{T_2}$ and

$p_2 = p_1\left(\dfrac{V_1}{V_2}\right)\left(\dfrac{T_2}{T_1}\right) = (2.72 \text{ atm})\left(\dfrac{0.0150 \text{ m}^3}{0.0159 \text{ m}^3}\right)\left(\dfrac{318 \text{ K}}{278 \text{ K}}\right) = 2.94 \text{ atm}.$ The final gauge pressure is

$2.94 \text{ atm} - 1.02 \text{ atm} = 1.92 \text{ atm}.$

EVALUATE: Since a ratio is used, pressure can be expressed in atm. But absolute pressures must be used. The ratio of gauge pressures is not equal to the ratio of absolute pressures.

18.63. **(a) IDENTIFY:** Consider the gas in one cylinder. Calculate the volume to which this volume of gas expands when the pressure is decreased from $(1.20\times10^6 \text{ Pa} + 1.01\times10^5 \text{ Pa}) = 1.30\times10^6 \text{ Pa}$ to

1.01×10^5 Pa. Apply the ideal-gas law to the two states of the system to obtain an expression for V_2 in terms of V_1 and the ratio of the pressures in the two states.

SET UP: $pV = nRT$

n, R, T constant implies $pV = nRT = $ constant, so $p_1 V_1 = p_2 V_2$.

EXECUTE: $V_2 = V_1(p_1/p_2) = (1.90 \text{ m}^3)\left(\dfrac{1.30 \times 10^6 \text{ Pa}}{1.01 \times 10^5 \text{ Pa}}\right) = 24.46 \text{ m}^3$

The number of cylinders required to fill a 750 m^3 balloon is $750 \text{ m}^3 / 24.46 \text{ m}^3 = 30.7$ cylinders.

EVALUATE: The ratio of the volume of the balloon to the volume of a cylinder is about 400. Fewer cylinders than this are required because of the large factor by which the gas is compressed in the cylinders.
(b) IDENTIFY: The upward force on the balloon is given by Archimedes's principle (Chapter 12): $B = $ weight of air displaced by balloon $ = \rho_{air} V g$. Apply Newton's second law to the balloon and solve for the weight of the load that can be supported. Use the ideal-gas equation to find the mass of the gas in the balloon.
SET UP: The free-body diagram for the balloon is given in Figure 18.63.

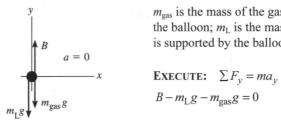

m_{gas} is the mass of the gas that is inside the balloon; m_L is the mass of the load that is supported by the balloon.

EXECUTE: $\sum F_y = ma_y$

$B - m_L g - m_{gas} g = 0$

Figure 18.63

$\rho_{air} V g - m_L g - m_{gas} g = 0$

$m_L = \rho_{air} V - m_{gas}$

Calculate m_{gas}, the mass of hydrogen that occupies 750 m^3 at $15°C$ and $p = 1.01 \times 10^5$ Pa.

$pV = nRT = (m_{gas}/M)RT$ gives

$$m_{gas} = pVM/RT = \dfrac{(1.01 \times 10^5 \text{ Pa})(750 \text{ m}^3)(2.02 \times 10^{-3} \text{ kg/mol})}{(8.3145 \text{ J/mol} \cdot \text{K})(288 \text{ K})} = 63.9 \text{ kg}$$

Then $m_L = (1.23 \text{ kg/m}^3)(750 \text{ m}^3) - 63.9 \text{ kg} = 859 \text{ kg}$, and the weight that can be supported is

$w_L = m_L g = (859 \text{ kg})(9.80 \text{ m/s}^2) = 8420 \text{ N}$.

(c) $m_L = \rho_{air} V - m_{gas}$

$m_{gas} = pVM/RT = (63.9 \text{ kg})((4.00 \text{ g/mol})/(2.02 \text{ g/mol})) = 126.5 \text{ kg}$ (using the results of part (b)).

Then $m_L = (1.23 \text{ kg/m}^3)(750 \text{ m}^3) - 126.5 \text{ kg} = 796 \text{ kg}$.

$w_L = m_L g = (796 \text{ kg})(9.80 \text{ m/s}^2) = 7800 \text{ N}$.

EVALUATE: A greater weight can be supported when hydrogen is used because its density is less.

18.65. **IDENTIFY:** Apply Bernoulli's equation to relate the efflux speed of water out the hose to the height of water in the tank and the pressure of the air above the water in the tank. Use the ideal-gas equation to relate the volume of the air in the tank to the pressure of the air.
(a) SET UP: Points 1 and 2 are shown in Figure 18.65.

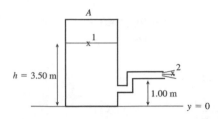

$p_1 = 4.20 \times 10^5$ Pa

$p_2 = p_{air} = 1.00 \times 10^5$ Pa

large tank implies $v_1 \approx 0$

Figure 18.65

EXECUTE: $p_1 + \rho g y_1 + \frac{1}{2}\rho v_1^2 = p_2 + \rho g y_2 + \frac{1}{2}\rho v_2^2$

$\frac{1}{2}\rho v_2^2 = p_1 - p_2 + \rho g(y_1 - y_2)$

$v_2 = \sqrt{(2/\rho)(p_1 - p_2) + 2g(y_1 - y_2)}$

$v_2 = 26.2$ m/s

(b) $h = 3.00$ m

The volume of the air in the tank increases so its pressure decreases. $pV = nRT = \text{constant}$, so $pV = p_0 V_0$

(p_0 is the pressure for $h_0 = 3.50$ m and p is the pressure for $h = 3.00$ m)

$p(4.00 \text{ m} - h)A = p_0(4.00 \text{ m} - h_0)A$

$p = p_0\left(\dfrac{4.00 \text{ m} - h_0}{4.00 \text{ m} - h}\right) = (4.20 \times 10^5 \text{ Pa})\left(\dfrac{4.00 \text{ m} - 3.50 \text{ m}}{4.00 \text{ m} - 3.00 \text{ m}}\right) = 2.10 \times 10^5$ Pa

Repeat the calculation of part (a), but now $p_1 = 2.10 \times 10^5$ Pa and $y_1 = 3.00$ m.

$v_2 = \sqrt{(2/\rho)(p_1 - p_2) + 2g(y_1 - y_2)}$

$v_2 = 16.1$ m/s

$h = 2.00$ m

$p = p_0\left(\dfrac{4.00 \text{ m} - h_0}{4.00 \text{ m} - h}\right) = (4.20 \times 10^5 \text{ Pa})\left(\dfrac{4.00 \text{ m} - 3.50 \text{ m}}{4.00 \text{ m} - 2.00 \text{ m}}\right) = 1.05 \times 10^5$ Pa

$v_2 = \sqrt{(2/\rho)(p_1 - p_2) + 2g(y_1 - y_2)}$

$v_2 = 5.44$ m/s

(c) $v_2 = 0$ means $(2/\rho)(p_1 - p_2) + 2g(y_1 - y_2) = 0$

$p_1 - p_2 = -\rho g(y_1 - y_2)$

$y_1 - y_2 = h - 1.00$ m

$p = p_0\left(\dfrac{0.50 \text{ m}}{4.00 \text{ m} - h}\right) = (4.20 \times 10^5 \text{ Pa})\left(\dfrac{0.50 \text{ m}}{4.00 \text{ m} - h}\right)$. This is p_1, so

$(4.20 \times 10^5 \text{ Pa})\left(\dfrac{0.50 \text{ m}}{4.00 \text{ m} - h}\right) - 1.00 \times 10^5 \text{ Pa} = (9.80 \text{ m/s}^2)(1000 \text{ kg/m}^3)(1.00 \text{ m} - h)$

$(210/(4.00 - h)) - 100 = 9.80 - 9.80h$, with h in meters.

$210 = (4.00 - h)(109.8 - 9.80h)$

$9.80h^2 - 149h + 229.2 = 0$ and $h^2 - 15.20h + 23.39 = 0$

quadratic formula: $h = \frac{1}{2}\left(15.20 \pm \sqrt{(15.20)^2 - 4(23.39)}\right) = (7.60 \pm 5.86)$ m

h must be less than 4.00 m, so the only acceptable value is $h = 7.60$ m $- 5.86$ m $= 1.74$ m

EVALUATE: The flow stops when $p + \rho g(y_1 - y_2)$ equals air pressure. For $h = 1.74$ m, $p = 9.3 \times 10^4$ Pa

and $\rho g(y_1 - y_2) = 0.7 \times 10^4$ Pa, so $p + \rho g(y_1 - y_2) = 1.0 \times 10^5$ Pa, which is air pressure.

18.71. **IDENTIFY:** The mass of one molecule is the molar mass, M, divided by the number of molecules in a mole, N_A. The average translational kinetic energy of a single molecule is $\frac{1}{2}m(v^2)_{av} = \frac{3}{2}kT$. Use $pV = NkT$ to calculate N, the number of molecules.

SET UP: $k = 1.381 \times 10^{-23}$ J/molecule$\cdot$K. $M = 28.0 \times 10^{-3}$ kg/mol. $T = 295.15$ K. The volume of the balloon is $V = \frac{4}{3}\pi(0.250 \text{ m})^3 = 0.0654 \text{ m}^3$. $p = 1.25$ atm $= 1.27 \times 10^5$ Pa.

EXECUTE: **(a)** $m = \dfrac{M}{N_A} = \dfrac{28.0 \times 10^{-3} \text{ kg/mol}}{6.022 \times 10^{23} \text{ molecules/mol}} = 4.65 \times 10^{-26}$ kg

(b) $\frac{1}{2}m(v^2)_{av} = \frac{3}{2}kT = \frac{3}{2}(1.381 \times 10^{-23} \text{ J/molecule}\cdot\text{K})(295.15 \text{ K}) = 6.11 \times 10^{-21}$ J

(c) $N = \dfrac{pV}{kT} = \dfrac{(1.27 \times 10^5 \text{ Pa})(0.0654 \text{ m}^3)}{(1.381 \times 10^{-23} \text{ J/molecule}\cdot\text{K})(295.15 \text{ K})} = 2.04 \times 10^{24}$ molecules

(d) The total average translational kinetic energy is

$N\left(\frac{1}{2}m(v^2)_{av}\right) = (2.04 \times 10^{24} \text{ molecules})(6.11 \times 10^{-21} \text{ J/molecule}) = 1.25 \times 10^4$ J.

EVALUATE: The number of moles is $n = \dfrac{N}{N_A} = \dfrac{2.04 \times 10^{24} \text{ molecules}}{6.022 \times 10^{23} \text{ molecules/mol}} = 3.39$ mol.

$K_{tr} = \frac{3}{2}nRT = \frac{3}{2}(3.39 \text{ mol})(8.314 \text{ J/mol}\cdot\text{K})(295.15 \text{ K}) = 1.25 \times 10^4$ J, which agrees with our results in part (d).

18.75. **IDENTIFY and SET UP:** Apply Eq. (18.19) for v_{rms}. The equation preceeding Eq. (18.12) relates v_{rms} and $(v_x)_{rms}$.

EXECUTE: **(a)** $v_{rms} = \sqrt{3RT/M}$

$v_{rms} = \sqrt{\dfrac{3(8.3145 \text{ J/mol}\cdot\text{K})(300 \text{ K})}{28.0 \times 10^{-3} \text{ kg/mol}}} = 517$ m/s

(b) $(v_x^2)_{av} = \frac{1}{3}(v^2)_{av}$ so $\sqrt{(v_x^2)_{av}} = \left(1/\sqrt{3}\right)\sqrt{(v^2)_{av}} = \left(1/\sqrt{3}\right)v_{rms} = \left(1/\sqrt{3}\right)(517 \text{ m/s}) = 298$ m/s

EVALUATE: The speed of sound is approximately equal to $(v_x)_{rms}$ since it is the motion along the direction of propagation of the wave that transmits the wave.

18.77. **(a) IDENTIFY and SET UP:** Apply conservation of energy $K_1 + U_1 + W_{other} = K_2 + U_2$, where $U = -Gmm_p/r$. Let point 1 be at the surface of the planet, where the projectile is launched, and let point 2 be far from the earth. Just barely escapes says $v_2 = 0$.

EXECUTE: Only gravity does work says $W_{other} = 0$.

$U_1 = -Gmm_p/R_p$; $r_2 \to \infty$ so $U_2 = 0$; $v_2 = 0$ so $K_2 = 0$.

The conservation of energy equation becomes $K_1 - Gmm_p/R_p = 0$ and $K_1 = Gmm_p/R_p$.

But $g = Gm_p/R_p^2$ so $Gm_p/R_p = R_pg$ and $K_1 = mgR_p$, as was to be shown.

EVALUATE: The greater gR_p is, the more initial kinetic energy is required for escape.

(b) IDENTIFY and SET UP: Set K_1 from part (a) equal to the average kinetic energy of a molecule as given by Eq. (18.16). $\frac{1}{2}m(v^2)_{av} = mgR_p$ (from part (a)). But also, $\frac{1}{2}m(v^2)_{av} = \frac{3}{2}kT$, so $mgR_p = \frac{3}{2}kT$

EXECUTE: $T = \dfrac{2mgR_p}{3k}$

nitrogen

$m_{N_2} = (28.0 \times 10^{-3} \text{ kg/mol})/(6.022 \times 10^{23} \text{ molecules/mol}) = 4.65 \times 10^{-26}$ kg/molecule

$T = \dfrac{2mgR_p}{3k} = \dfrac{2(4.65 \times 10^{-26} \text{ kg/molecule})(9.80 \text{ m/s}^2)(6.38 \times 10^6 \text{ m})}{3(1.381 \times 10^{-23} \text{ J/molecule}\cdot\text{K})} = 1.40 \times 10^5$ K hydrogen

$$m_{H_2} = (2.02 \times 10^{-3} \text{ kg/mol})/(6.022 \times 10^{23} \text{ molecules/mol}) = 3.354 \times 10^{-27} \text{ kg/molecule}$$

$$T = \frac{2mgR_p}{3k} = \frac{2(3.354 \times 10^{-27} \text{ kg/molecule})(9.80 \text{ m/s}^2)(6.38 \times 10^6 \text{ m})}{3(1.381 \times 10^{-23} \text{ J/molecule} \cdot \text{K})} = 1.01 \times 10^4 \text{ K}$$

(c) $T = \dfrac{2mgR_p}{3k}$

nitrogen

$$T = \frac{2(4.65 \times 10^{-26} \text{ kg/molecule})(1.63 \text{ m/s}^2)(1.74 \times 10^6 \text{ m})}{3(1.381 \times 10^{-23} \text{ J/molecule} \cdot \text{K})} = 6370 \text{ K}$$

hydrogen

$$T = \frac{2(3.354 \times 10^{-27} \text{ kg/molecule})(1.63 \text{ m/s}^2)(1.74 \times 10^6 \text{ m})}{3(1.381 \times 10^{-23} \text{ J/molecule} \cdot \text{K})} = 459 \text{ K}$$

(d) EVALUATE: The "escape temperatures" are much less for the moon than for the earth. For the moon a larger fraction of the molecules at a given temperature will have speeds in the Maxwell-Boltzmann distribution larger than the escape speed. After a long time most of the molecules will have escaped from the moon.

18.79. **IDENTIFY:** $v_{rms} = \sqrt{\dfrac{3kT}{m}}$. The number of molecules in an object of mass m is $N = nN_A = \dfrac{m}{M}N_A$.

SET UP: The volume of a sphere of radius r is $V = \dfrac{4}{3}\pi r^3$.

EXECUTE: **(a)** $m = \dfrac{3kT}{v_{rms}^2} = \dfrac{3(1.381 \times 10^{-23} \text{ J/K})(300 \text{ K})}{(0.0010 \text{ m/s})^2} = 1.24 \times 10^{-14} \text{kg}.$

(b) $N = mN_A/M = (1.24 \times 10^{-14} \text{ kg})(6.022 \times 10^{23} \text{ molecules/mol})/(18.0 \times 10^{-3} \text{ kg/mol})$

$N = 4.16 \times 10^{11}$ molecules.

(c) The diameter is $D = 2r = 2\left(\dfrac{3V}{4\pi}\right)^{1/3} = 2\left(\dfrac{3m/\rho}{4\pi}\right)^{1/3} = 2\left(\dfrac{3(1.24 \times 10^{-14} \text{ kg})}{4\pi(920 \text{ kg/m}^3)}\right)^{1/3} = 2.95 \times 10^{-6} \text{ m}$ which is

too small to see.

EVALUATE: v_{rms} decreases as m increases.

18.81. **IDENTIFY:** The equipartition principle says that each atom has an average kinetic energy of $\frac{1}{2}kT$ for each degree of freedom. There is an equal average potential energy.

SET UP: The atoms in a three-dimensional solid have three degrees of freedom and the atoms in a two-dimensional solid have two degrees of freedom.

EXECUTE: **(a)** In the same manner that Eq. (18.28) was obtained, the heat capacity of the two-dimensional solid would be $2R = 16.6 \text{ J/mol} \cdot \text{K}$.

(b) The heat capacity would behave qualitatively like those in Figure 18.21 in the textbook, and the heat capacity would decrease with decreasing temperature.

EVALUATE: At very low temperatures the equipartition theorem doesn't apply. Most of the atoms remain in their lowest energy states because the next higher energy level is not accessible.

18.83. **IDENTIFY:** $C_V = N\left(\frac{1}{2}R\right)$, where N is the number of degrees of freedom.

SET UP: There are three translational degrees of freedom.

EXECUTE: For CO_2, $N = 5$ and the contribution to C_V other than from vibration is

$\frac{5}{2}R = 20.79 \text{ J/mol} \cdot \text{K}$ and $C_V - \frac{5}{2}R = 0.270\, C_V$. So 27% of C_V is due to vibration. For both SO_2 and H_2S,

$N = 6$ and the contribution to C_V other than from vibration is $\frac{6}{2}R = 24.94$ J/mol·K. The respective fractions of C_V from vibration are 21% and 3.9%.

EVALUATE: The vibrational contribution is much less for H_2S. In H_2S the vibrational energy steps are larger because the two hydrogen atoms have small mass and $\omega = \sqrt{k/m}$.

18.89.   **IDENTIFY:**   The measurement gives the dew point. Relative humidity is defined in Problem 18.88.

SET UP: $\text{relative humidity} = \dfrac{\text{partial pressure of water vapor at temperature } T}{\text{vapor pressure of water at temperature } T}$

EXECUTE: The experiment shows that the dew point is $16.0°C$, so the partial pressure of water vapor at $30.0°C$ is equal to the vapor pressure at $16.0°C$, which is 1.81×10^3 Pa.

Thus the relative humidity $= \dfrac{1.81 \times 10^3 \text{ Pa}}{4.25 \times 10^3 \text{ Pa}} = 0.426 = 42.6\%.$

EVALUATE: The lower the dew point is compared to the air temperature, the smaller the relative humidity.

THE FIRST LAW OF THERMODYNAMICS

19.5. **IDENTIFY:** Example 19.1 shows that for an isothermal process $W = nRT \ln(p_1/p_2)$. Solve for p_1.

SET UP: For a compression (V decreases) W is negative, so $W = -468$ J. $T = 295.15$ K.

EXECUTE: **(a)** $\dfrac{W}{nRT} = \ln\left(\dfrac{p_1}{p_2}\right)$. $\dfrac{p_1}{p_2} = e^{W/nRT}$.

$$\frac{W}{nRT} = \frac{-468 \text{ J}}{(0.305 \text{ mol})(8.314 \text{ J/mol} \cdot \text{K})(295.15 \text{ K})} = -0.6253.$$

$p_1 = p_2 e^{W/nRT} = (1.76 \text{ atm}) e^{-0.6253} = 0.942$ atm.

(b) In the process the pressure increases and the volume decreases. The pV-diagram is sketched in Figure 19.5.

EVALUATE: W is the work done by the gas, so when the surroundings do work on the gas, W is negative. The gas was compressed at constant temperature, so its pressure must have increased, which means that $p_1 < p_2$, which is what we found.

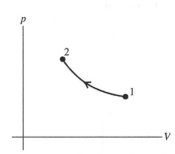

Figure 19.5

19.9. **IDENTIFY:** $\Delta U = Q - W$. For a constant pressure process, $W = p\Delta V$.

SET UP: $Q = +1.15 \times 10^5$ J, since heat enters the gas.

EXECUTE: **(a)** $W = p\Delta V = (1.65 \times 10^5 \text{ Pa})(0.320 \text{ m}^3 - 0.110 \text{ m}^3) = 3.47 \times 10^4$ J.

(b) $\Delta U = Q - W = 1.15 \times 10^5 \text{ J} - 3.47 \times 10^4 \text{ J} = 8.04 \times 10^4$ J.

EVALUATE: **(c)** $W = p\Delta V$ for a constant pressure process and $\Delta U = Q - W$ both apply to any material. The ideal gas law wasn't used and it doesn't matter if the gas is ideal or not.

19.11. **IDENTIFY:** Part ab is isochoric, but bc is not any of the familiar processes.

SET UP: $pV = nRT$ determines the Kelvin temperature of the gas. The work done in the process is the area under the curve in the pV diagram. Q is positive since heat goes into the gas. 1 atm $= 1.013 \times 10^5$ Pa. $1 \text{ L} = 1 \times 10^{-3} \text{ m}^3$. $\Delta U = Q - W$.

EXECUTE: **(a)** The lowest T occurs when pV has its smallest value. This is at point a, and

$$T_a = \frac{p_a V_a}{nR} = \frac{(0.20 \text{ atm})(1.013 \times 10^5 \text{ Pa/atm})(2.0 \text{ L})(1.0 \times 10^{-3} \text{ m}^3/\text{L})}{(0.0175 \text{ mol})(8.315 \text{ J/mol} \cdot \text{K})} = 278 \text{ K}.$$

(b) a to b: $\Delta V = 0$ so $W = 0$.

b to c: The work done by the gas is positive since the volume increases. The magnitude of the work is the area under the curve so $W = \frac{1}{2}(0.50 \text{ atm} + 0.30 \text{ atm})(6.0 \text{ L} - 2.0 \text{ L})$ and

$W = (1.6 \text{ L} \cdot \text{atm})(1 \times 10^{-3} \text{ m}^3/\text{L})(1.013 \times 10^5 \text{ Pa/atm}) = 162 \text{ J}.$

(c) For abc, $W = 162 \text{ J}$. $\Delta U = Q - W = 215 \text{ J} - 162 \text{ J} = 53 \text{ J}.$

EVALUATE: 215 J of heat energy went into the gas. 53 J of energy stayed in the gas as increased internal energy and 162 J left the gas as work done by the gas on its surroundings.

19.19. **IDENTIFY:** For a constant pressure process, $W = p\Delta V$, $Q = nC_p\Delta T$ and $\Delta U = nC_V\Delta T$. $\Delta U = Q - W$ and $C_p = C_V + R$. For an ideal gas, $p\Delta V = nR\Delta T$.

SET UP: From Table 19.1, $C_V = 28.46 \text{ J/mol} \cdot \text{K}$.

EXECUTE: **(a)** The pV diagram is given in Figure 19.19.

(b) $W = pV_2 - pV_1 = nR(T_2 - T_1) = (0.250 \text{ mol})(8.3145 \text{ J/mol} \cdot \text{K})(100.0 \text{ K}) = 208 \text{ J}.$

(c) The work is done on the piston.

(d) Since Eq. (19.13) holds for any process,

$\Delta U = nC_V\Delta T = (0.250 \text{ mol})(28.46 \text{ J/mol} \cdot \text{K})(100.0 \text{ K}) = 712 \text{ J}.$

(e) Either $Q = nC_p\Delta T$ or $Q = \Delta U + W$ gives $Q = 920 \text{ J}$ to three significant figures.

(f) The lower pressure would mean a correspondingly larger volume, and the net result would be that the work done would be the same as that found in part (b).

EVALUATE: $W = nR\Delta T$, so W, Q and ΔU all depend only on ΔT. When T increases at constant pressure, V increases and $W > 0$. ΔU and Q are also positive when T increases.

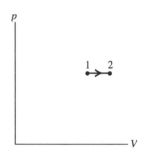

Figure 19.19

19.21. **IDENTIFY:** For constant volume, $Q = nC_V\Delta T$. For constant pressure, $Q = nC_p\Delta T$.

SET UP: From Table 19.1, $C_V = 20.76 \text{ J/mol} \cdot \text{K}$ and $C_p = 29.07 \text{ J/mol} \cdot \text{K}$.

EXECUTE: **(a)** Using Eq. (19.12), $\Delta T = \dfrac{Q}{nC_V} = \dfrac{645 \text{ J}}{(0.185 \text{ mol})(20.76 \text{ J/mol} \cdot \text{K})} = 167.9 \text{ K}$ and $T = 948 \text{ K}$.

The pV-diagram is sketched in Figure 19.21a.

(b) Using Eq. (19.14), $\Delta T = \dfrac{Q}{nC_p} = \dfrac{645 \text{ J}}{(0.185 \text{ mol})(29.07 \text{ J/mol} \cdot \text{K})} = 119.9 \text{ K}$ and $T = 900 \text{ K}$.

The pV-diagram is sketched in Figure 19.21b.

EVALUATE: At constant pressure some of the heat energy added to the gas leaves the gas as expansion work and the internal energy change is less than if the same amount of heat energy is added at constant volume. ΔT is proportional to ΔU.

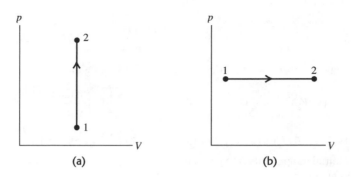

Figure 19.21

19.25.   **IDENTIFY:**   For a constant volume process, $Q = nC_V\Delta T$. For a constant pressure process, $Q = nC_p\Delta T$. For any process of an ideal gas, $\Delta U = nC_V\Delta T$.

SET UP: From Table 19.1, for N_2, $C_V = 20.76$ J/mol·K and $C_p = 29.07$ J/mol·K. Heat is added, so Q is positive and $Q = +1557$ J.

EXECUTE: **(a)** $\Delta T = \dfrac{Q}{nC_V} = \dfrac{1557\ \text{J}}{(3.00\ \text{mol})(20.76\ \text{J/mol·K})} = +25.0$ K

(b) $\Delta T = \dfrac{Q}{nC_p} = \dfrac{1557\ \text{J}}{(3.00\ \text{mol})(29.07\ \text{J/mol·K})} = +17.9$ K

(c) $\Delta U = nC_V\Delta T$ for either process, so ΔU is larger when ΔT is larger. The final internal energy is larger for the constant volume process in (a).

EVALUATE: For constant volume $W = 0$ and all the energy added as heat stays in the gas as internal energy. For the constant pressure process the gas expands and $W > 0$. Part of the energy added as heat leaves the gas as expansion work done by the gas.

19.27.   **IDENTIFY:**   Calculate W and ΔU and then use the first law to calculate Q.

(a) SET UP: $W = \int_{V_1}^{V_2} p\, dV$

$pV = nRT$ so $p = nRT/V$

$W = \int_{V_1}^{V_2} (nRT/V)\, dV = nRT \int_{V_1}^{V_2} dV/V = nRT \ln(V_2/V_1)$ (work done during an isothermal process).

EXECUTE: $W = (0.150\ \text{mol})(8.3145\ \text{J/mol·K})(350\ \text{K})\ln(0.25V_1/V_1) = (436.5\ \text{J})\ln(0.25) = -605$ J.

EVALUATE: W for the gas is negative, since the volume decreases.

(b) EXECUTE: $\Delta U = nC_V\Delta T$ for any ideal gas process.

$\Delta T = 0$ (isothermal) so $\Delta U = 0$.

EVALUATE: $\Delta U = 0$ for any ideal gas process in which T doesn't change.

(c) EXECUTE: $\Delta U = Q - W$

$\Delta U = 0$ so $Q = W = -605$ J. (Q is negative; the gas liberates 605 J of heat to the surroundings.)

EVALUATE: $Q = nC_V\Delta T$ is only for a constant volume process so doesn't apply here.

$Q = nC_p\Delta T$ is only for a constant pressure process so doesn't apply here.

19.29.   **IDENTIFY:**   For an adiabatic process of an ideal gas, $p_1V_1^{\gamma} = p_2V_2^{\gamma}$, $W = \dfrac{1}{\gamma - 1}(p_1V_1 - p_2V_2)$ and

$T_1V_1^{\gamma-1} = T_2V_2^{\gamma-1}$.

SET UP: For a monatomic ideal gas $\gamma = 5/3$.

EXECUTE: (a) $p_2 = p_1 \left(\dfrac{V_1}{V_2} \right)^{\gamma} = (1.50 \times 10^5 \text{ Pa}) \left(\dfrac{0.0800 \text{ m}^3}{0.0400 \text{ m}^3} \right)^{5/3} = 4.76 \times 10^5 \text{ Pa}.$

(b) This result may be substituted into Eq. (19.26), or, substituting the above form for p_2,

$$W = \frac{1}{\gamma - 1} p_1 V_1 \left(1 - (V_1/V_2)^{\gamma - 1} \right) = \frac{3}{2} (1.50 \times 10^5 \text{ Pa})(0.0800 \text{ m}^3) \left(1 - \left(\frac{0.0800}{0.0400} \right)^{2/3} \right) = -1.06 \times 10^4 \text{ J}.$$

(c) From Eq. (19.22), $(T_2/T_1) = (V_2/V_1)^{\gamma - 1} = (0.0800/0.0400)^{2/3} = 1.59,$ and since the final temperature is higher than the initial temperature, the gas is heated.

EVALUATE: In an adiabatic compression $W < 0$ since $\Delta V < 0.$ $Q = 0$ so $\Delta U = -W.$ $\Delta U > 0$ and the temperature increases.

19.31. IDENTIFY: For an adiabatic process of an ideal gas, $W = \dfrac{1}{\gamma - 1} (p_1 V_1 - p_2 V_2)$ and $p_1 V_1^{\gamma} = p_2 V_2^{\gamma}.$

SET UP: $\gamma = 1.40$ for an ideal diatomic gas. 1 atm $= 1.013 \times 10^5$ Pa and 1 L $= 10^{-3}$ m^3.

EXECUTE: $Q = \Delta U + W = 0$ for an adiabatic process, so $\Delta U = -W = \dfrac{1}{\gamma - 1} (p_2 V_2 - p_1 V_1).$

$p_1 = 1.22 \times 10^5$ Pa. $p_2 = p_1 (V_1/V_2)^{\gamma} = (1.22 \times 10^5 \text{ Pa})(3)^{1.4} = 5.68 \times 10^5$ Pa.

$W = \dfrac{1}{0.40} ([5.68 \times 10^5 \text{ Pa}][10 \times 10^{-3} \text{ m}^{-3}] - [1.22 \times 10^5 \text{ Pa}][30 \times 10^{-3} \text{ m}^{-3}]) = 5.05 \times 10^3$ J. The internal energy increases because work is done on the gas $(\Delta U > 0)$ and $Q = 0.$ The temperature increases because the internal energy has increased.

EVALUATE: In an adiabatic compression $W < 0$ since $\Delta V < 0.$ $Q = 0$ so $\Delta U = -W.$ $\Delta U > 0$ and the temperature increases.

19.35. IDENTIFY: Combine $T_1 V_1^{\gamma - 1} = T_2 V_2^{\gamma - 1}$ with $pV = nRT$ to obtain an expression relating T and p for an adiabatic process of an ideal gas.

SET UP: $T_1 = 299.15$ K

EXECUTE: $V = \dfrac{nRT}{p}$ so $T_1 \left(\dfrac{nRT_1}{p_1} \right)^{\gamma - 1} = T_2 \left(\dfrac{nRT_2}{p_2} \right)^{\gamma - 1}$ and $\dfrac{T_1^{\gamma}}{p_1^{\gamma - 1}} = \dfrac{T_2^{\gamma}}{p_2^{\gamma - 1}}.$

$T_2 = T_1 \left(\dfrac{p_2}{p_1} \right)^{(\gamma - 1)/\gamma} = (299.15 \text{ K}) \left(\dfrac{0.850 \times 10^5 \text{ Pa}}{1.01 \times 10^5 \text{ Pa}} \right)^{0.4/1.4} = 284.8 \text{ K} = 11.6°\text{C}$

EVALUATE: For an adiabatic process of an ideal gas, when the pressure decreases the temperature decreases.

19.39. IDENTIFY and SET UP: For an ideal gas, $pV = nRT.$ The work done is the area under the path in the pV-diagram.

EXECUTE: (a) The product pV increases and this indicates a temperature increase.

(b) The work is the area in the pV plane bounded by the blue line representing the process and the verticals at V_a and $V_b.$ The area of this trapezoid is $\frac{1}{2}(p_b + p_a)(V_b - V_a) = \frac{1}{2}(2.40 \times 10^5 \text{ Pa})(0.0400 \text{ m}^3) = 4800$ J.

EVALUATE: The work done is the average pressure, $\frac{1}{2}(p_1 + p_2),$ times the volume increase.

19.41. IDENTIFY: Use $\Delta U = Q - W$ and the fact that ΔU is path independent.

$W > 0$ when the volume increases, $W < 0$ when the volume decreases, and $W = 0$ when the volume is constant. $Q > 0$ if heat flows into the system.

SET UP: The paths are sketched in Figure 19.41.

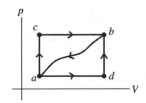

$Q_{acb} = +90.0 \text{ J}$ (positive since heat flows in)

$W_{acb} = +60.0 \text{ J}$ (positive since $\Delta V > 0$)

Figure 19.41

EXECUTE: **(a)** $\Delta U = Q - W$

ΔU is path independent; Q and W depend on the path.

$\Delta U = U_b - U_a$

This can be calculated for any path from a to b, in particular for path acb:

$\Delta U_{a \to b} = Q_{acb} - W_{acb} = 90.0 \text{ J} - 60.0 \text{ J} = 30.0 \text{ J}$.

Now apply $\Delta U = Q - W$ to path adb; $\Delta U = 30.0 \text{ J}$ for this path also.

$W_{adb} = +15.0 \text{ J}$ (positive since $\Delta V > 0$)

$\Delta U_{a \to b} = Q_{adb} - W_{adb}$ so $Q_{adb} = \Delta U_{a \to b} + W_{adb} = 30.0 \text{ J} + 15.0 \text{ J} = +45.0 \text{ J}$

(b) Apply $\Delta U = Q - W$ to path ba: $\Delta U_{b \to a} = Q_{ba} - W_{ba}$

$W_{ba} = -35.0 \text{ J}$ (negative since $\Delta V < 0$)

$\Delta U_{b \to a} = U_a - U_b = -(U_b - U_a) = -\Delta U_{a \to b} = -30.0 \text{ J}$

Then $Q_{ba} = \Delta U_{b \to a} + W_{ba} = -30.0 \text{ J} - 35.0 \text{ J} = -65.0 \text{ J}$.

($Q_{ba} < 0$; the system liberates heat.)

(c) $U_a = 0$, $U_d = 8.0 \text{ J}$

$\Delta U_{a \to b} = U_b - U_a = +30.0 \text{ J}$, so $U_b = +30.0 \text{ J}$.

process $a \to d$

$\Delta U_{a \to d} = Q_{ad} - W_{ad}$

$\Delta U_{a \to d} = U_d - U_a = +8.0 \text{ J}$

$W_{adb} = +15.0 \text{ J}$ and $W_{adb} = W_{ad} + W_{db}$. But the work W_{db} for the process $d \to b$ is zero since $\Delta V = 0$ for that process. Therefore $W_{ad} = W_{adb} = +15.0 \text{ J}$.

Then $Q_{ad} = \Delta U_{a \to d} + W_{ad} = +8.0 \text{ J} + 15.0 \text{ J} = +23.0 \text{ J}$ (positive implies heat absorbed).

process $d \to b$

$\Delta U_{d \to b} = Q_{db} - W_{db}$

$W_{db} = 0$, as already noted.

$\Delta U_{d \to b} = U_b - U_d = 30.0 \text{ J} - 8.0 \text{ J} = +22.0 \text{ J}$.

Then $Q_{db} = \Delta U_{d \to b} + W_{db} = +22.0 \text{ J}$ (positive; heat absorbed).

EVALUATE: The signs of our calculated Q_{ad} and Q_{db} agree with the problem statement that heat is absorbed in these processes.

19.43. **IDENTIFY:** Use $pV = nRT$ to calculate T_c/T_a. Calculate ΔU and W and use $\Delta U = Q - W$ to obtain Q.

SET UP: For path ac, the work done is the area under the line representing the process in the pV-diagram.

EXECUTE: **(a)** $\dfrac{T_c}{T_a} = \dfrac{p_c V_c}{p_a V_a} = \dfrac{(1.0 \times 10^5 \text{ J})(0.060 \text{ m}^3)}{(3.0 \times 10^5 \text{ J})(0.020 \text{ m}^3)} = 1.00$. $T_c = T_a$.

(b) Since $T_c = T_a$, $\Delta U = 0$ for process abc. For ab, $\Delta V = 0$ and $W_{ab} = 0$. For bc, p is constant and

$W_{bc} = p \Delta V = (1.0 \times 10^5 \text{ Pa})(0.040 \text{ m}^3) = 4.0 \times 10^3 \text{ J}$. Therefore, $W_{abc} = +4.0 \times 10^3 \text{ J}$. Since $\Delta U = 0$,

$Q = W = +4.0 \times 10^3 \text{ J}$. $4.0 \times 10^3 \text{ J}$ of heat flows into the gas during process abc.

(c) $W = \frac{1}{2}(3.0 \times 10^5 \text{ Pa} + 1.0 \times 10^5 \text{ Pa})(0.040 \text{ m}^3) = +8.0 \times 10^3 \text{ J}.$ $Q_{ac} = W_{ac} = +8.0 \times 10^3 \text{ J}.$

EVALUATE: The work done is path dependent and is greater for process ac than for process abc, even though the initial and final states are the same.

19.45.   **IDENTIFY:** Use the 1st law to relate Q_{tot} to W_{tot} for the cycle.

Calculate W_{ab} and W_{bc} and use what we know about W_{tot} to deduce W_{ca}.

(a) SET UP: We aren't told whether the pressure increases or decreases in process bc. The two possibilities for the cycle are sketched in Figure 19.45.

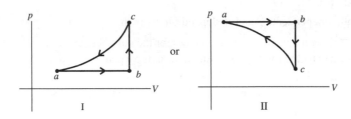

Figure 19.45

In cycle I, the total work is negative and in cycle II the total work is positive. For a cycle, $\Delta U = 0$, so $Q_{tot} = W_{tot}$.

The net heat flow for the cycle is out of the gas, so heat $Q_{tot} < 0$ and $W_{tot} < 0$. Sketch I is correct.

(b) EXECUTE: $W_{tot} = Q_{tot} = -800 \text{ J}$

$W_{tot} = W_{ab} + W_{bc} + W_{ca}$

$W_{bc} = 0$ since $\Delta V = 0$.

$W_{ab} = p\Delta V$ since p is constant. But since it is an ideal gas, $p\Delta V = nR\Delta T$.

$W_{ab} = nR(T_b - T_a) = 1660 \text{ J}$

$W_{ca} = W_{tot} - W_{ab} = -800 \text{ J} - 1660 \text{ J} = -2460 \text{ J}$

EVALUATE: In process ca the volume decreases and the work W is negative.

19.47.   **IDENTIFY:** Segment ab is isochoric, bc is isothermal, and ca is isobaric.

SET UP: For bc, $\Delta T = 0$, $\Delta U = 0$, and $Q = W = nRT \ln(V_c/V_b)$. For ideal H_2 (diatomic), $C_V = \frac{5}{2}R$ and

$C_p = \frac{7}{2}R$. $\Delta U = nC_V \Delta T$ for any process of an ideal gas.

EXECUTE: (a) $T_b = T_c$. For states b and c, $pV = nRT = $ constant so $p_b V_b = p_c V_c$ and

$$V_c = V_b\left(\frac{p_b}{p_c}\right) = (0.20 \text{ L})\left(\frac{2.0 \text{ atm}}{0.50 \text{ atm}}\right) = 0.80 \text{ L}.$$

(b) $T_a = \dfrac{p_a V_a}{nR} = \dfrac{(0.50 \text{ atm})(1.013 \times 10^5 \text{ Pa/atm})(0.20 \times 10^{-3} \text{ m}^3)}{(0.0040 \text{ mol})(8.315 \text{ J/mol} \cdot \text{K})} = 305 \text{ K}.$ $V_a = V_b$ so for states a and b,

$\dfrac{T}{p} = \dfrac{V}{nR} = $ constant so $\dfrac{T_a}{p_a} = \dfrac{T_b}{p_b}$. $T_b = T_c = T_a\left(\dfrac{p_b}{p_a}\right) = (305 \text{ K})\left(\dfrac{2.0 \text{ atm}}{0.50 \text{ atm}}\right) = 1220 \text{ K}$; $T_c = 1220 \text{ K}$.

(c) ab: $Q = nC_V \Delta T = n\left(\frac{5}{2}R\right)\Delta T$, which gives

$Q = (0.0040 \text{ mol})\left(\frac{5}{2}\right)(8.315 \text{ J/mol} \cdot \text{K})(1220 \text{ K} - 305 \text{ K}) = +76 \text{ J}.$ Q is positive and heat goes into the gas.

ca: $Q = nC_p \Delta T = n\left(\frac{7}{2}R\right)\Delta T$, which gives

$Q = (0.0040 \text{ mol})\left(\frac{7}{2}\right)(8.315 \text{ J/mol} \cdot \text{K})(305 \text{ K} - 1220 \text{ K}) = -107 \text{ J}.$ Q is negative and heat comes out of the gas.

bc: $Q = W = nRT \ln(V_c/V_b)$, which gives

$Q = (0.0040 \text{ mol})(8.315 \text{ J/mol} \cdot \text{K})(1220 \text{ K}) \ln(0.80 \text{ L}/0.20 \text{ L}) = 56 \text{ J}$. Q is positive and heat goes into the gas.

(d) *ab*: $\Delta U = nC_V \Delta T = n\left(\frac{5}{2}R\right)\Delta T$, which gives

$\Delta U = (0.0040 \text{ mol})\left(\frac{5}{2}\right)(8.315 \text{ J/mol} \cdot \text{K})(1220 \text{ K} - 305 \text{ K}) = +76 \text{ J}$. The internal energy increased.

bc: $\Delta T = 0$ so $\Delta U = 0$. The internal energy does not change.

ca: $\Delta U = nC_V \Delta T = n\left(\frac{5}{2}R\right)\Delta T$, which gives

$\Delta U = (0.0040 \text{ mol})\left(\frac{5}{2}\right)(8.315 \text{ J/mol} \cdot \text{K})(305 \text{ K} - 1220 \text{ K}) = -76 \text{ J}$. The internal energy decreased.

EVALUATE: The net internal energy change for the complete cycle $a \rightarrow b \rightarrow c \rightarrow a$ is

$\Delta U_{\text{tot}} = +76 \text{ J} + 0 + (-76 \text{ J}) = 0$. For any complete cycle the final state is the same as the initial state and the net internal energy change is zero. For the cycle the net heat flow is

$Q_{\text{tot}} = +76 \text{ J} + (-107 \text{ J}) + 56 \text{ J} = +25 \text{ J}$. $\Delta U_{\text{tot}} = 0$ so $Q_{\text{tot}} = W_{\text{tot}}$. The net work done in the cycle is positive and this agrees with our result that the net heat flow is positive.

19.49. **IDENTIFY:** The segments *ab* and *bc* are not any of the familiar ones, such as isothermal, isobaric or isochoric, but *ac* is isobaric.

SET UP: For helium, $C_V = 12.47 \text{ J/mol} \cdot \text{K}$ and $C_p = 20.78 \text{ J/mol} \cdot \text{K}$. $\Delta U = Q - W$. W is the area under the p versus V curve. $\Delta U = nC_V \Delta T$ for any process of an ideal gas.

EXECUTE: **(a)** $W = \frac{1}{2}(1.0 \times 10^5 \text{ Pa} + 3.5 \times 10^5 \text{ Pa})(0.0060 \text{ m}^3 - 0.0020 \text{ m}^3)$

$+ \frac{1}{2}(1.0 \times 10^5 \text{ Pa} + 3.5 \times 10^5 \text{ Pa})(0.0100 \text{ m}^3 - 0.0060 \text{ m}^3) = 1800 \text{ J}$.

Find $\Delta T = T_c - T_a$. p is constant so $\Delta T = \dfrac{p \Delta V}{nR} = \dfrac{(1.0 \times 10^5 \text{ Pa})(0.0100 \text{ m}^3 - 0.0020 \text{ m}^3)}{\left(\frac{1}{3} \text{ mol}\right)(8.315 \text{ J/mol} \cdot \text{K})} = 289 \text{ K}$. Then

$\Delta U = nC_V \Delta T = \left(\frac{1}{3} \text{ mol}\right)(12.47 \text{ J/mol} \cdot \text{K})(289 \text{ K}) = 1.20 \times 10^3 \text{ J}$.

$Q = \Delta U + W = 1.20 \times 10^3 \text{ J} + 1800 \text{ J} = 3.00 \times 10^3 \text{ J}$. $Q > 0$, so this heat is transferred into the gas.

(b) This process is isobaric, so $Q = nC_p \Delta T = \left(\frac{1}{3} \text{ mol}\right)(20.78 \text{ J/mol} \cdot \text{K})(289 \text{ K}) = 2.00 \times 10^3 \text{ J}$. $Q > 0$, so this heat is transferred into the gas.

(c) Q is larger in part (a).

EVALUATE: ΔU is the same in parts (a) and (b) because the initial and final states are the same, but in (a) more work is done.

19.51. **IDENTIFY:** Use $Q = nC_V \Delta T$ to calculate the temperature change in the constant volume process and use $pV = nRT$ to calculate the temperature change in the constant pressure process. The work done in the constant volume process is zero and the work done in the constant pressure process is $W = p \Delta V$. Use $Q = nC_p \Delta T$ to calculate the heat flow in the constant pressure process. $\Delta U = nC_V \Delta T$, or $\Delta U = Q - W$.

SET UP: For N_2, $C_V = 20.76 \text{ J/mol} \cdot \text{K}$ and $C_p = 29.07 \text{ J/mol} \cdot \text{K}$.

EXECUTE: **(a)** For process *ab*, $\Delta T = \dfrac{Q}{nC_V} = \dfrac{1.52 \times 10^4 \text{ J}}{(2.50 \text{ mol})(20.76 \text{ J/mol} \cdot \text{K})} = 293 \text{ K}$. $T_a = 293 \text{ K}$, so

$T_b = 586 \text{ K}$. $pV = nRT$ says T doubles when V doubles and p is constant, so $T_c = 2(586 \text{ K}) = 1172 \text{ K} = 899°\text{C}$.

(b) For process *ab*, $W_{ab} = 0$. For process *bc*,

$W_{bc} = p \Delta V = nR \Delta T = (2.50 \text{ mol})(8.314 \text{ J/mol} \cdot \text{K})(1172 \text{ K} - 586 \text{ K}) = 1.22 \times 10^4 \text{ J}$.

$W = W_{ab} + W_{bc} = 1.22 \times 10^4 \text{ J}$.

(c) For process bc, $Q = nC_p\Delta T = (2.50 \text{ mol})(29.07 \text{ J/mol} \cdot \text{K})(1172 \text{ K} - 586 \text{ K}) = 4.26 \times 10^4 \text{ J}.$

(d) $\Delta U = nC_V\Delta T = (2.50 \text{ mol})(20.76 \text{ J/mol} \cdot \text{K})(1172 \text{ K} - 293 \text{ K}) = 4.56 \times 10^4 \text{ J}.$

EVALUATE: The total Q is $1.52 \times 10^4 \text{ J} + 4.26 \times 10^4 \text{ J} = 5.78 \times 10^4 \text{ J}.$

$\Delta U = Q - W = 5.78 \times 10^4 \text{ J} - 1.22 \times 10^4 \text{ J} = 4.56 \times 10^4 \text{ J},$ which agrees with our results in part (d).

19.55. **IDENTIFY:** $\Delta V = V_0\beta\Delta T.$ $W = p\Delta V$ since the force applied to the piston is constant. $Q = mc_p\Delta T.$
$\Delta U = Q - W.$

SET UP: $m = \rho V$

EXECUTE: **(a)** The change in volume is

$\Delta V = V_0\beta\Delta T = (1.20 \times 10^{-2} \text{ m}^3)(1.20 \times 10^{-3} \text{ K}^{-1})(30.0 \text{ K}) = 4.32 \times 10^{-4} \text{ m}^3.$

(b) $W = p\Delta V = (F/A)\Delta V = ((3.00 \times 10^4 \text{ N})/(0.0200 \text{ m}^2))(4.32 \times 10^{-4} \text{ m}^3) = 648 \text{ J}.$

(c) $Q = mc_p\Delta T = V_0\rho c_p\Delta T = (1.20 \times 10^{-2} \text{ m}^3)(791 \text{ kg/m}^3)(2.51 \times 10^3 \text{ J/kg} \cdot \text{K})(30.0 \text{ K}).$

$Q = 7.15 \times 10^5 \text{ J}.$

(d) $\Delta U = Q - W = 7.15 \times 10^5 \text{ J}$ to three figures.

(e) Under these conditions W is much less than Q and there is no substantial difference between c_V and c_p.

EVALUATE: $\Delta U = Q - W$ is valid for any material. For liquids the expansion work is much less than Q.

19.57. **IDENTIFY and SET UP:** The heat produced from the reaction is $Q_{\text{reaction}} = mL_{\text{reaction}},$ where L_{reaction} is the
heat of reaction of the chemicals.

$Q_{\text{reaction}} = W + \Delta U_{\text{spray}}$

EXECUTE: For a mass m of spray, $W = \frac{1}{2}mv^2 = \frac{1}{2}m(19 \text{ m/s})^2 = (180.5 \text{ J/kg})m$ and

$\Delta U_{\text{spray}} = Q_{\text{spray}} = mc\Delta T = m(4190 \text{ J/kg} \cdot \text{K})(100°C - 20°C) = (335,200 \text{ J/kg})m.$

Then $Q_{\text{reaction}} = (180 \text{ J/kg} + 335,200 \text{ J/kg})m = (335,380 \text{ J/kg})m$ and $Q_{\text{reaction}} = mL_{\text{reaction}}$ implies

$mL_{\text{reaction}} = (335,380 \text{ J/kg})m.$

The mass m divides out and $L_{\text{reaction}} = 3.4 \times 10^5 \text{ J/kg}.$

EVALUATE: The amount of energy converted to work is negligible for the two significant figures to which
the answer should be expressed. Almost all of the energy produced in the reaction goes into heating the
compound.

19.59. **IDENTIFY:** For an adiabatic process of an ideal gas, $T_1V_1^{\gamma-1} = T_2V_2^{\gamma-1}.$ $pV = nRT.$

SET UP: For air, $\gamma = 1.40 = \frac{7}{5}.$

EXECUTE: **(a)** As the air moves to lower altitude its density increases; under an adiabatic compression,
the temperature rises. If the wind is fast-moving, Q is not as likely to be significant, and modeling the
process as adiabatic (no heat loss to the surroundings) is more accurate.

(b) $V = \dfrac{nRT}{p},$ so $T_1V_1^{\gamma-1} = T_2V_2^{\gamma-1}$ gives $T_1^\gamma p_1^{1-\gamma} = T_2^\gamma p_2^{1-\gamma}.$ The temperature at the higher pressure is

$T_2 = T_1(p_1/p_2)^{(\gamma-1)/\gamma} = (258.15 \text{ K})([8.12 \times 10^4 \text{ Pa}]/[5.60 \times 10^4 \text{ Pa}])^{2/7} = 287.1 \text{ K} = 13.9°C$ so the
temperature would rise by 11.9 C°.

EVALUATE: In an adiabatic compression, $Q = 0$ but the temperature rises because of the work done on
the gas.

19.61. **IDENTIFY:** Assume that the gas is ideal and that the process is adiabatic. Apply Eqs. (19.22) and (19.24)
to relate pressure and volume and temperature and volume. The distance the piston moves is related to the
volume of the gas. Use Eq. (19.25) to calculate $W.$

(a) SET UP: $\gamma = C_p/C_V = (C_V + R)/C_V = 1 + R/C_V = 1.40.$ The two positions of the piston are shown in
Figure 19.61.

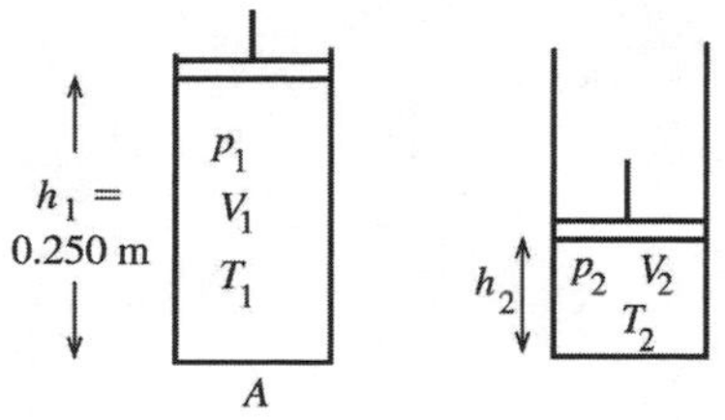

$p_1 = 1.01 \times 10^5$ Pa

$p_2 = 4.20 \times 10^5$ Pa $+ p_{air} = 5.21 \times 10^5$ Pa

$V_1 = h_1 A$

$V_2 = h_2 A$

Figure 19.61

EXECUTE: adiabatic process: $p_1 V_1^\gamma = p_2 V_2^\gamma$

$p_1 h_1^\gamma A^\gamma = p_2 h_2^\gamma A^\gamma$

$h_2 = h_1 \left(\dfrac{p_1}{p_2} \right)^{1/\gamma} = (0.250 \text{ m}) \left(\dfrac{1.01 \times 10^5 \text{ Pa}}{5.21 \times 10^5 \text{ Pa}} \right)^{1/1.40} = 0.0774 \text{ m}$

The piston has moved a distance $h_1 - h_2 = 0.250 \text{ m} - 0.0774 \text{ m} = 0.173 \text{ m}$.

(b) $T_1 V_1^{\gamma-1} = T_2 V_2^{\gamma-1}$

$T_1 h_1^{\gamma-1} A^{\gamma-1} = T_2 h_2^{\gamma-1} A^{\gamma-1}$

$T_2 = T_1 \left(\dfrac{h_1}{h_2} \right)^{\gamma-1} = 300.1 \text{ K} \left(\dfrac{0.250 \text{ m}}{0.0774 \text{ m}} \right)^{0.40} = 479.7 \text{ K} = 207°\text{C}$

(c) $W = n C_V (T_1 - T_2)$ (Eq. 19.25)

$W = (20.0 \text{ mol})(20.8 \text{ J/mol} \cdot \text{K})(300.1 \text{ K} - 479.7 \text{ K}) = -7.47 \times 10^4 \text{ J}$

EVALUATE: In an adiabatic compression of an ideal gas the temperature increases. In any compression the work W is negative.

THE SECOND LAW OF THERMODYNAMICS

20.3. **IDENTIFY** and **SET UP:** The problem deals with a heat engine. $W = +3700$ W and $Q_H = +16,100$ J. Use Eq. (20.4) to calculate the efficiency e and Eq. (20.2) to calculate $|Q_C|$. Power $= W/t$.

EXECUTE: (a) $e = \dfrac{\text{work output}}{\text{heat energy input}} = \dfrac{W}{Q_H} = \dfrac{3700 \text{ J}}{16,100 \text{ J}} = 0.23 = 23\%$.

(b) $W = Q = |Q_H| - |Q_C|$

Heat discarded is $|Q_C| = |Q_H| - W = 16,100 \text{ J} - 3700 \text{ J} = 12,400 \text{ J}$.

(c) Q_H is supplied by burning fuel; $Q_H = mL_c$ where L_c is the heat of combustion.

$m = \dfrac{Q_H}{L_c} = \dfrac{16,100 \text{ J}}{4.60 \times 10^4 \text{ J/g}} = 0.350$ g.

(d) $W = 3700$ J per cycle

In $t = 1.00$ s the engine goes through 60.0 cycles.

$P = W/t = 60.0(3700 \text{ J})/1.00 \text{ s} = 222$ kW

$P = (2.22 \times 10^5 \text{ W})(1 \text{ hp}/746 \text{ W}) = 298$ hp

EVALUATE: $Q_C = -12,400$ J. In one cycle $Q_{\text{tot}} = Q_C + Q_H = 3700$ J. This equals W_{tot} for one cycle.

20.5. **IDENTIFY:** This cycle involves adiabatic (ab), isobaric (bc), and isochoric (ca) processes.

SET UP: ca is at constant volume, ab has $Q = 0$, and bc is at constant pressure. For a constant pressure

process $W = p\Delta V$ and $Q = nC_p \Delta T$. $pV = nRT$ gives $n\Delta T = \dfrac{p\Delta V}{R}$, so $Q = \left(\dfrac{C_p}{R}\right) p\Delta V$. If $\gamma = 1.40$ the

gas is diatomic and $C_p = \frac{7}{2}R$. For a constant volume process $W = 0$ and $Q = nC_V \Delta T$. $pV = nRT$ gives

$n\Delta T = \dfrac{V\Delta p}{R}$, so $Q = \left(\dfrac{C_V}{R}\right) V\Delta p$. For a diatomic ideal gas $C_V = \frac{5}{2}R$. 1 atm $= 1.013 \times 10^5$ Pa.

EXECUTE: (a) $V_b = 9.0 \times 10^{-3}$ m^3, $p_b = 1.5$ atm and $V_a = 2.0 \times 10^{-3}$ m^3. For an adiabatic process

$p_a V_a^\gamma = p_b V_b^\gamma$. $p_a = p_b \left(\dfrac{V_b}{V_a}\right)^\gamma = (1.5 \text{ atm}) \left(\dfrac{9.0 \times 10^{-3} \text{ m}^3}{2.0 \times 10^{-3} \text{ m}^3}\right)^{1.4} = 12.3$ atm.

(b) Heat enters the gas in process ca, since T increases.

$Q = \left(\dfrac{C_V}{R}\right) V\Delta p = \left(\dfrac{5}{2}\right)(2.0 \times 10^{-3} \text{ m}^3)(12.3 \text{ atm} - 1.5 \text{ atm})(1.013 \times 10^5 \text{ Pa/atm}) = 5470$ J. $Q_H = 5470$ J.

(c) Heat leaves the gas in process bc, since T increases.

$Q = \left(\dfrac{C_p}{R}\right) p\Delta V = \left(\dfrac{7}{2}\right)(1.5 \text{ atm})(1.013 \times 10^5 \text{ Pa/atm})(-7.0 \times 10^{-3} \text{ m}^3) = -3723$ J. $Q_C = -3723$ J.

(d) $W = Q_H + Q_C = +5470 \text{ J} + (-3723 \text{ J}) = 1747$ J.

(e) $e = \dfrac{W}{Q_{\text{H}}} = \dfrac{1747 \text{ J}}{5470 \text{ J}} = 0.319 = 31.9\%$.

EVALUATE: We did not use the number of moles of the gas.

20.7. **IDENTIFY:** $e = 1 - r^{1-\gamma}$

SET UP: r is the compression ratio.

EXECUTE: **(a)** $e = 1 - (8.8)^{-0.40} = 0.581$, which rounds to 58%.

(b) $e = 1 - (9.6)^{-0.40} = 0.595$ an increase of 1.4%.

EVALUATE: An increase in r gives an increase in e.

20.11. **IDENTIFY:** The heat $Q = mc\Delta T$ that comes out of the water to cool it to 5.0°C is Q_{C} for the refrigerator.

SET UP: For water 1.0 L has a mass of 1.0 kg and $c = 4.19 \times 10^3$ J/kg·C°. $P = \dfrac{|W|}{t}$. The coefficient of

performance is $K = \dfrac{|Q_{\text{C}}|}{|W|}$.

EXECUTE: $Q = mc\Delta T = (12.0 \text{ kg})(4.19 \times 10^3 \text{ J/kg} \cdot \text{C}°)(5.0°\text{C} - 31°\text{C}) = -1.31 \times 10^6$ J. $|Q_{\text{C}}| = 1.31 \times 10^6$ J.

$K = \dfrac{|Q_{\text{C}}|}{|W|} = \dfrac{|Q_{\text{C}}|}{Pt}$ so $t = \dfrac{|Q_{\text{C}}|}{PK} = \dfrac{1.31 \times 10^6 \text{ J}}{(95 \text{ W})(2.25)} = 6129 \text{ s} = 102 \text{ min} = 1.7 \text{ h}$.

EVALUATE: 1.7 h seems like a reasonable time to cool down the dozen bottles.

20.17. **IDENTIFY:** $|Q_{\text{H}}| = |W| + |Q_{\text{C}}|$. $Q_{\text{H}} < 0$, $Q_{\text{C}} > 0$. $K = \dfrac{|Q_{\text{C}}|}{|W|}$. For a Carnot cycle, $\dfrac{Q_{\text{C}}}{Q_{\text{H}}} = -\dfrac{T_{\text{C}}}{T_{\text{H}}}$.

SET UP: $T_{\text{C}} = 270$ K, $T_{\text{H}} = 320$ K. $|Q_{\text{C}}| = 415$ J.

EXECUTE: **(a)** $Q_{\text{H}} = -\left(\dfrac{T_{\text{H}}}{T_{\text{C}}}\right)Q_{\text{C}} = -\left(\dfrac{320 \text{ K}}{270 \text{ K}}\right)(415 \text{ J}) = -492$ J.

(b) For one cycle, $|W| = |Q_{\text{H}}| - |Q_{\text{C}}| = 492 \text{ J} - 415 \text{ J} = 77$ J. $P = \dfrac{(165)(77 \text{ J})}{60 \text{ s}} = 212$ W.

(c) $K = \dfrac{|Q_{\text{C}}|}{|W|} = \dfrac{415 \text{ J}}{77 \text{ J}} = 5.4$.

EVALUATE: The amount of heat energy $|Q_{\text{H}}|$ delivered to the high-temperature reservoir is greater than the amount of heat energy $|Q_{\text{C}}|$ removed from the low-temperature reservoir.

20.19. **IDENTIFY:** $e = \dfrac{W}{Q_{\text{H}}} = 1 - \dfrac{Q_{\text{C}}}{Q_{\text{H}}}$. For a Carnot cycle, $\dfrac{Q_{\text{C}}}{Q_{\text{H}}} = -\dfrac{T_{\text{C}}}{T_{\text{H}}}$ and $e = 1 - \dfrac{T_{\text{C}}}{T_{\text{H}}}$.

SET UP: $T_{\text{H}} = 800$ K. $Q_{\text{C}} = -3000$ J.

EXECUTE: For a heat engine, $Q_{\text{H}} = -Q_{\text{C}}/(1-e) = -(-3000 \text{ J})/(1 - 0.600) = 7500$ J, and then $W = eQ_{\text{H}} = (0.600)(7500 \text{ J}) = 4500$ J.

EVALUATE: This does not make use of the given value of T_{H}. If T_{H} is used, then $T_{\text{C}} = T_{\text{H}}(1-e) = (800 \text{ K})(1 - 0.600) = 320$ K and $Q_{\text{H}} = -Q_{\text{C}}T_{\text{H}}/T_{\text{C}}$, which gives the same result.

20.21. **IDENTIFY:** The power output is $P = \dfrac{W}{t}$. The theoretical maximum efficiency is $e_{\text{Carnot}} = 1 - \dfrac{T_{\text{C}}}{T_{\text{H}}}$. $e = \dfrac{W}{Q_{\text{H}}}$.

SET UP: $Q_{\text{H}} = 1.50 \times 10^4$ J. $T_{\text{C}} = 350$ K. $T_{\text{H}} = 650$ K. 1 hp = 746 W.

EXECUTE: $e_{\text{Carnot}} = 1 - \dfrac{T_{\text{C}}}{T_{\text{H}}} = 1 - \dfrac{350 \text{ K}}{650 \text{ K}} = 0.4615$. $W = eQ_{\text{H}} = (0.4615)(1.50 \times 10^4 \text{ J}) = 6.923 \times 10^3$ J; this is

the work output in one cycle. $P = \dfrac{W}{t} = \dfrac{(240)(6.923 \times 10^3 \text{ J})}{60.0 \text{ s}} = 2.77 \times 10^4 \text{ W} = 37.1$ hp.

EVALUATE: We could also use $\frac{Q_C}{Q_H} = -\frac{T_C}{T_H}$ to calculate

$Q_C = -\left(\frac{T_C}{T_H}\right) Q_H = -\left(\frac{350 \text{ K}}{650 \text{ K}}\right)(1.50 \times 10^4 \text{ J}) = -8.08 \times 10^3 \text{ J}.$ Then $W = Q_C + Q_H = 6.92 \times 10^3 \text{ J},$ the same as

previously calculated.

20.25. **IDENTIFY:** Both the ice and the room are at a constant temperature, so $\Delta S = \frac{Q}{T}$. For the melting phase

transition, $Q = mL_f$. Conservation of energy requires that the quantity of heat that goes into the ice is the amount of heat that comes out of the room.

SET UP: For ice, $L_f = 334 \times 10^3$ J/kg. When heat flows into an object, $Q > 0$, and when heat flows out of an object, $Q < 0$.

EXECUTE: **(a)** Irreversible because heat will not spontaneously flow out of 15 kg of water into a warm room to freeze the water.

(b) $\Delta S = \Delta S_{ice} + \Delta S_{room} = \frac{mL_f}{T_{ice}} + \frac{-mL_f}{T_{room}} = \frac{(15.0 \text{ kg})(334 \times 10^3 \text{ J/kg})}{273 \text{ K}} + \frac{-(15.0 \text{ kg})(334 \times 10^3 \text{ J/kg})}{293 \text{ K}}.$

$\Delta S = +1250$ J/K.

EVALUATE: This result is consistent with the answer in (a) because $\Delta S > 0$ for irreversible processes.

20.31. **IDENTIFY:** No heat is transferred, but the entropy of the He increases because it occupies a larger volume and hence is more disordered. To calculate the entropy change, we need to find a reversible process that connects the same initial and final states.

SET UP: The reversible process that connects the same initial and final states is an isothermal expansion at $T = 293$ K, from $V_1 = 10.0$ L to $V_2 = 35.0$ L. For an isothermal expansion of an ideal gas $\Delta U = 0$ and $Q = W = nRT \ln(V_2/V_1)$.

EXECUTE:

(a) $Q = (3.20 \text{ mol})(8.315 \text{ J/mol} \cdot \text{K})(293 \text{ K}) \ln(35.0 \text{ L}/10.0 \text{ L}) = 9767 \text{ J}.$ $\Delta S = \frac{Q}{T} = \frac{9767 \text{ J}}{293 \text{ K}} = +33.3$ J/K.

(b) The isolated system has $\Delta S > 0$ so the process is irreversible.

EVALUATE: The reverse process, where all the gas in 35.0 L goes through the hole and into the tank does not ever occur.

20.35. **IDENTIFY:** The total work that must be done is $W_{tot} = mg\Delta y$. $|W| = |Q_H| - |Q_C|$. $Q_H > 0$, $W > 0$ and

$Q_C < 0$. For a Carnot cycle, $\frac{Q_C}{Q_H} = -\frac{T_C}{T_H}$,

SET UP: $T_C = 373$ K, $T_H = 773$ K. $|Q_H| = 250$ J.

EXECUTE: **(a)** $Q_C = -Q_H \left(\frac{T_C}{T_H}\right) = -(250 \text{ J})\left(\frac{373 \text{ K}}{773 \text{ K}}\right) = -121$ J.

(b) $|W| = 250 \text{ J} - 121 \text{ J} = 129 \text{ J}.$ This is the work done in one cycle.

$W_{tot} = (500 \text{ kg})(9.80 \text{ m/s}^2)(100 \text{ m}) = 4.90 \times 10^5$ J. The number of cycles required is

$\frac{W_{tot}}{|W|} = \frac{4.90 \times 10^5 \text{ J}}{129 \text{ J/cycle}} = 3.80 \times 10^3$ cycles.

EVALUATE: In $\frac{Q_C}{Q_H} = -\frac{T_C}{T_H}$, the temperatures must be in kelvins.

20.37. **IDENTIFY:** We know the efficiency of this Carnot engine, the heat it absorbs at the hot reservoir and the temperature of the hot reservoir.

SET UP: For a heat engine $e = \dfrac{W}{|Q_H|}$ and $Q_H + Q_C = W$. For a Carnot cycle, $\dfrac{Q_C}{Q_H} = -\dfrac{T_C}{T_H}$. $Q_C < 0$, $W > 0$, and $Q_H > 0$. $T_H = 135°C = 408$ K. In each cycle, $|Q_H|$ leaves the hot reservoir and $|Q_C|$ enters the cold reservoir. The work done on the water equals its increase in gravitational potential energy, mgh.

EXECUTE: **(a)** $e = \dfrac{W}{Q_H}$ so $W = eQ_H = (0.22)(150 \text{ J}) = 33$ J.

(b) $Q_C = W - Q_H = 33 \text{ J} - 150 \text{ J} = -117$ J.

(c) $\dfrac{Q_C}{Q_H} = -\dfrac{T_C}{T_H}$ so $T_C = -T_H\left(\dfrac{Q_C}{Q_H}\right) = -(408 \text{ K})\left(\dfrac{-117 \text{ J}}{150 \text{ J}}\right) = 318 \text{ K} = 45°C$.

(d) $\Delta S = \dfrac{-|Q_H|}{T_H} + \dfrac{|Q_C|}{T_C} = \dfrac{-150 \text{ J}}{408 \text{ K}} + \dfrac{117 \text{ J}}{318 \text{ K}} = 0$. The Carnot cycle is reversible and $\Delta S = 0$.

(e) $W = mgh$ so $m = \dfrac{W}{gh} = \dfrac{33 \text{ J}}{(9.80 \text{ m/s}^2)(35.0 \text{ m})} = 0.0962 \text{ kg} = 96.2$ g.

EVALUATE: The Carnot cycle is reversible so $\Delta S = 0$ for the world. However some parts of the world gain entropy while other parts lose it, making the sum equal to zero.

20.39. **IDENTIFY:** The same amount of heat that enters the person's body also leaves the body, but these transfers of heat occur at different temperatures, so the person's entropy changes.

SET UP: 1 food-calorie = 1000 cal = 4186 J. The heat enters the person's body at $37°C = 310$ K and leaves at a temperature of $30°C = 303$ K. $\Delta S = \dfrac{Q}{T}$.

EXECUTE: $|Q| = (0.80)(2.50 \text{ g})(9.3 \text{ food-calorie/g})\left(\dfrac{4186 \text{ J}}{1 \text{ food-calorie}}\right) = 7.79 \times 10^4$ J.

$\Delta S = \dfrac{+7.79 \times 10^4 \text{ J}}{310 \text{ K}} + \dfrac{-7.79 \times 10^4 \text{ J}}{303 \text{ K}} = -5.8$ J/K. Your body's entropy decreases.

EVALUATE: The entropy of your body can decrease without violating the second law of thermodynamics because you are not an isolated system.

20.41. **IDENTIFY:** $pV = nRT$, so pV is constant when T is constant. Use the appropriate expression to calculate Q and W for each process in the cycle. $e = \dfrac{W}{Q_H}$.

SET UP: For an ideal diatomic gas, $C_V = \frac{5}{2}R$ and $C_p = \frac{7}{2}R$.

EXECUTE: **(a)** $p_a V_a = 2.0 \times 10^3$ J. $p_b V_b = 2.0 \times 10^3$ J. $pV = nRT$ so $p_a V_a = p_b V_b$ says $T_a = T_b$.

(b) For an isothermal process, $Q = W = nRT \ln(V_2/V_1)$. ab is a compression, with $V_b < V_a$, so $Q < 0$ and heat is rejected. bc is at constant pressure, so $Q = nC_p\Delta T = \dfrac{C_p}{R}p\Delta V$. ΔV is positive, so $Q > 0$ and heat is absorbed. ca is at constant volume, so $Q = nC_V\Delta T = \dfrac{C_V}{R}V\Delta p$. Δp is negative, so $Q < 0$ and heat is rejected.

(c) $T_a = \dfrac{p_a V_a}{nR} = \dfrac{2.0 \times 10^3 \text{ J}}{(1.00)(8.314 \text{ J/mol} \cdot \text{K})} = 241$ K. $T_b = \dfrac{p_b V_b}{nR} = T_a = 241$ K.

$T_c = \dfrac{p_c V_c}{nR} = \dfrac{4.0 \times 10^3 \text{ J}}{(1.00)(8.314 \text{ J/mol} \cdot \text{K})} = 481$ K.

(d) $Q_{ab} = nRT \ln\left(\dfrac{V_b}{V_a}\right) = (1.00 \text{ mol})(8.314 \text{ J/mol} \cdot \text{K})(241 \text{ K}) \ln\left(\dfrac{0.0050 \text{ m}^3}{0.010 \text{ m}^3}\right) = -1.39 \times 10^3 \text{ J}.$

$Q_{bc} = nC_p\Delta T = (1.00)\left(\dfrac{7}{2}\right)(8.314 \text{ J/mol} \cdot \text{K})(241 \text{ K}) = 7.01 \times 10^3 \text{ J}.$

$Q_{ca} = nC_V\Delta T = (1.00)\left(\dfrac{5}{2}\right)(8.314 \text{ J/mol} \cdot \text{K})(-241 \text{ K}) = -5.01 \times 10^3 \text{ J}. \quad Q_{\text{net}} = Q_{ab} + Q_{bc} + Q_{ca} = 610 \text{ J}.$

$W_{\text{net}} = Q_{\text{net}} = 610 \text{ J}.$

(e) $e = \dfrac{W}{Q_{\text{H}}} = \dfrac{610 \text{ J}}{7.01 \times 10^3 \text{ J}} = 0.087 = 8.7\%$

EVALUATE: We can calculate W for each process in the cycle. $W_{ab} = Q_{ab} = -1.39 \times 10^3 \text{ J}.$

$W_{bc} = p\Delta V = (4.0 \times 10^5 \text{ Pa})(0.0050 \text{ m}^3) = 2.00 \times 10^3 \text{ J}. \quad W_{ca} = 0. \quad W_{\text{net}} = W_{ab} + W_{bc} + W_{ca} = 610 \text{ J},$ which does equal $Q_{\text{net}}.$

20.43.   **IDENTIFY:** $T_b = T_c$ and is equal to the maximum temperature. Use the ideal gas law to calculate T_a. Apply the appropriate expression to calculate Q for each process. $e = \dfrac{W}{Q_{\text{H}}}$. $\Delta U = 0$ for a complete cycle and for an isothermal process of an ideal gas.

SET UP: For helium, $C_V = 3R/2$ and $C_p = 5R/2$. The maximum efficiency is for a Carnot cycle, and $e_{\text{Carnot}} = 1 - T_{\text{C}}/T_{\text{H}}.$

EXECUTE: **(a)** $Q_{\text{in}} = Q_{ab} + Q_{bc}. \quad Q_{\text{out}} = Q_{ca}. \quad T_{\text{max}} = T_b = T_c = 327°\text{C} = 600 \text{ K}.$

$\dfrac{p_a V_a}{T_a} = \dfrac{p_b V_b}{T_b} \rightarrow T_a = \dfrac{p_a}{p_b} T_b = \dfrac{1}{3}(600 \text{ K}) = 200 \text{ K}.$

$p_b V_b = nRT_b \rightarrow V_b = \dfrac{nRT_b}{p_b} = \dfrac{(2 \text{ moles})(8.31 \text{ J/mol} \cdot \text{K})(600 \text{ K})}{3.0 \times 10^5 \text{ Pa}} = 0.0332 \text{ m}^3.$

$\dfrac{p_b V_b}{T_b} = \dfrac{p_c V_c}{T_c} \rightarrow V_c = V_b \dfrac{p_b}{p_c} = (0.0332 \text{ m}^3)\left(\dfrac{3}{1}\right) = 0.0997 \text{ m}^3 = V_a.$

$Q_{ab} = nC_V\Delta T_{ab} = (2 \text{ mol})\left(\dfrac{3}{2}\right)(8.31 \text{ J/mol} \cdot \text{K})(400 \text{ K}) = 9.97 \times 10^3 \text{ J}$

$Q_{bc} = W_{bc} = \int_b^c p\,dV = \int_b^c \dfrac{nRT_b}{V}\,dV = nRT_b \ln\dfrac{V_c}{V_b} = nRT_b \ln 3.$

$Q_{bc} = (2.00 \text{ mol})(8.31 \text{ J/mol} \cdot \text{K})(600 \text{ K})\ln 3 = 1.10 \times 10^4 \text{ J}. \quad Q_{\text{in}} = Q_{ab} + Q_{bc} = 2.10 \times 10^4 \text{ J}.$

$Q_{\text{out}} = Q_{ca} = nC_p\Delta T_{ca} = (2.00 \text{ mol})\left(\dfrac{5}{2}\right)(8.31 \text{ J/mol} \cdot \text{K})(400 \text{ K}) = 1.66 \times 10^4 \text{ J}.$

(b) $Q = \Delta U + W = 0 + W \rightarrow W = Q_{\text{in}} - Q_{\text{out}} = 2.10 \times 10^4 \text{ J} - 1.66 \times 10^4 \text{ J} = 4.4 \times 10^3 \text{ J}.$

$e = W/Q_{\text{in}} = \dfrac{4.4 \times 10^3 \text{ J}}{2.10 \times 10^4 \text{ J}} = 0.21 = 21\%.$

(c) $e_{\text{max}} = e_{\text{Carnot}} = 1 - \dfrac{T_{\text{C}}}{T_{\text{H}}} = 1 - \dfrac{200 \text{ k}}{600 \text{ k}} = 0.67 = 67\%$

EVALUATE: The thermal efficiency of this cycle is about one-third of the efficiency of a Carnot cycle that operates between the same two temperatures.

20.45.   **IDENTIFY:** $e_{\text{max}} = e_{\text{Carnot}} = 1 - T_{\text{C}}/T_{\text{H}}. \quad e = \dfrac{W}{Q_{\text{H}}} = \dfrac{W/t}{Q_{\text{H}}/t}. \quad W = Q_{\text{H}} + Q_{\text{C}}$ so $\dfrac{W}{t} = \dfrac{Q_{\text{C}}}{t} + \dfrac{Q_{\text{H}}}{t}.$ For a temperature change $Q = mc\Delta T.$

SET UP: $T_H = 300.15$ K, $T_C = 279.15$ K. For water, $\rho = 1000$ kg/m³, so a mass of 1 kg has a volume of 1 L. For water, $c = 4190$ J/kg·K.

EXECUTE: **(a)** $e = 1 - \dfrac{279.15\text{K}}{300.15\text{K}} = 7.0\%.$

(b) $\dfrac{Q_H}{t} = \dfrac{P_{out}}{e} = \dfrac{210\text{ kW}}{0.070} = 3.0$ MW. $\dfrac{Q_C}{t} = \dfrac{Q_H}{t} - \dfrac{W}{t} = 3.0$ MW $- 210$ kW $= 2.8$ MW.

(c) $\dfrac{m}{t} = \dfrac{|Q_C|/t}{c\Delta T} = \dfrac{(2.8\times10^6\text{ W})(3600\text{ s/h})}{(4190\text{ J/kg·K})(4\text{ K})} = 6\times10^5$ kg/h $= 6\times10^5$ L/h.

EVALUATE: The efficiency is small since T_C and T_H don't differ greatly.

20.47. **IDENTIFY:** Use $pV = nRT$. Apply the expressions for Q and W that apply to each type of process. $e = \dfrac{W}{Q_H}.$

SET UP: For O_2, $C_V = 20.85$ J/mol·K and $C_p = 29.17$ J/mol·K.

EXECUTE: **(a)** $p_1 = 2.00$ atm, $V_1 = 4.00$ L, $T_1 = 300$ K.

$p_2 = 2.00$ atm. $\dfrac{V_1}{T_1} = \dfrac{V_2}{T_2}$. $V_2 = \left(\dfrac{T_2}{T_1}\right)V_1 = \left(\dfrac{450\text{ K}}{300\text{ K}}\right)(4.00\text{ L}) = 6.00$ L.

$V_3 = 6.00$ L. $\dfrac{p_2}{T_2} = \dfrac{p_3}{T_3}$. $p_3 = \left(\dfrac{T_3}{T_2}\right)p_2 = \left(\dfrac{250\text{ K}}{450\text{ K}}\right)(2.00\text{ atm}) = 1.11$ atm

$V_4 = 4.00$ L. $p_3V_3 = p_4V_4$. $p_4 = p_3\left(\dfrac{V_3}{V_4}\right) = (1.11\text{ atm})\left(\dfrac{6.00\text{ L}}{4.00\text{ L}}\right) = 1.67$ atm.

These processes are shown in Figure 20.47.

(b) $n = \dfrac{p_1V_1}{RT_1} = \dfrac{(2.00\text{ atm})(4.00\text{ L})}{(0.08206\text{ L·atm/mol·K})(300\text{ K})} = 0.325$ mol

process $1 \to 2$: $W = p\Delta V = nR\Delta T = (0.325\text{ mol})(8.315\text{ J/mol·K})(150\text{ K}) = 405$ J.
$Q = nC_p\Delta T = (0.325\text{ mol})(29.17\text{ J/mol·K})(150\text{ K}) = 1422$ J.
process $2 \to 3$: $W = 0$. $Q = nC_V\Delta T = (0.325\text{ mol})(20.85\text{ J/mol·K})(-200\text{ K}) = -1355$ J.
process $3 \to 4$: $\Delta U = 0$ and
$Q = W = nRT_3\ln\left(\dfrac{V_4}{V_3}\right) = (0.325\text{ mol})(8.315\text{ J/mol·K})(250\text{ K})\ln\left(\dfrac{4.00\text{ L}}{6.00\text{ L}}\right) = -274$ J.
process $4 \to 1$: $W = 0$. $Q = nC_V\Delta T = (0.325\text{ mol})(20.85\text{ J/mol·K})(50\text{ K}) = 339$ J.

(c) $W = 405$ J $- 274$ J $= 131$ J

(d) $e = \dfrac{W}{Q_H} = \dfrac{131\text{ J}}{1422\text{ J} + 339\text{ J}} = 0.0744 = 7.44\%.$

$e_{Carnot} = 1 - \dfrac{T_C}{T_H} = 1 - \dfrac{250\text{ K}}{450\text{ K}} = 0.444 = 44.4\%$; e_{Carnot} is much larger.

EVALUATE: $Q_{tot} = +1422$ J $+ (-1355$ J$) + (-274$ J$) + 339$ J $= 132$ J. This is equal to W_{tot}, apart from a slight difference due to rounding. For a cycle, $W_{tot} = Q_{tot}$, since $\Delta U = 0$.

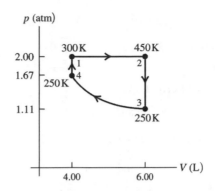

Figure 20.47

20.49. **IDENTIFY:** Since there is temperature difference between the inside and outside of your body, you can use it as a heat engine.

SET UP: For a heat engine $e = \dfrac{W}{Q_{\text{H}}}$. For a Carnot engine $e = 1 - \dfrac{T_{\text{C}}}{T_{\text{H}}}$. Gravitational potential energy is $U_{\text{grav}} = mgh$. 1 food-calorie $= 1000$ cal $= 4186$ J.

EXECUTE: **(a)** $e = 1 - \dfrac{T_{\text{C}}}{T_{\text{H}}} = 1 - \dfrac{303 \text{ K}}{310 \text{ K}} = 0.0226 = 2.26\%$. This engine has a very low thermal efficiency.

(b) $U_{\text{grav}} = mgh = (2.50 \text{ kg})(9.80 \text{ m/s}^2)(1.20 \text{ m}) = 29.4$ J. This equals the work output of the engine.

$e = \dfrac{W}{Q_{\text{H}}}$ so $Q_{\text{H}} = \dfrac{W}{e} = \dfrac{29.4 \text{ J}}{0.0226} = 1.30 \times 10^3$ J.

(c) Since 80% of food energy goes into heat, you must eat food with a food energy of

$\dfrac{1.30 \times 10^3 \text{ J}}{0.80} = 1.63 \times 10^3$ J. Each candy bar gives $(350 \text{ food-calorie})(4186 \text{ J/food-calorie}) = 1.47 \times 10^6$ J.

The number of candy bars required is $\dfrac{1.63 \times 10^3 \text{ J}}{1.47 \times 10^6 \text{ J/candy bar}} = 1.11 \times 10^{-3}$ candy bars.

EVALUATE: A large amount of mechanical work must be done to use up the energy from one candy bar.

20.51. **IDENTIFY:** Use $\Delta U = Q - W$ and the appropriate expressions for Q, W and ΔU for each type of process.

$pV = nRT$ relates ΔT to p and V values. $e = \dfrac{W}{Q_{\text{H}}}$, where Q_{H} is the heat that enters the gas during the cycle.

SET UP: For a monatomic ideal gas, $C_p = \tfrac{5}{2}R$ and $C_V = \tfrac{3}{2}R$.

(a) *ab*: The temperature changes by the same factor as the volume, and so

$Q = nC_p \Delta T = \dfrac{C_p}{R} p_a (V_a - V_b) = (2.5)(3.00 \times 10^5 \text{ Pa})(0.300 \text{ m}^3) = 2.25 \times 10^5$ J.

The work $p\Delta V$ is the same except for the factor of $\tfrac{5}{2}$, so $W = 0.90 \times 10^5$ J.

$\Delta U = Q - W = 1.35 \times 10^5$ J.

bc: The temperature now changes in proportion to the pressure change, and

$Q = \tfrac{3}{2}(p_c - p_b)V_b = (1.5)(-2.00 \times 10^5 \text{ Pa})(0.800 \text{ m}^3) = -2.40 \times 10^5$ J, and the work is zero

$(\Delta V = 0)$. $\Delta U = Q - W = -2.40 \times 10^5$ J.

ca: The easiest way to do this is to find the work done first; W will be the negative of area in the *p-V* plane bounded by the line representing the process *ca* and the verticals from points *a* and *c*. The area of this trapezoid is $\frac{1}{2}(3.00\times10^5$ Pa $+1.00\times10^5$ Pa$)(0.800$ m$^3 - 0.500$ m$^3) = 6.00\times10^4$ J and so the work is

-0.60×10^5 J. ΔU must be 1.05×10^5 J (since $\Delta U = 0$ for the cycle, anticipating part (b)), and so Q must

be $\Delta U + W = 0.45\times10^5$ J.

(b) See above; $Q = W = 0.30\times10^5$ J, $\Delta U = 0$.

(c) The heat added, during process ab and ca, is 2.25×10^5 J $+0.45\times10^5$ J $=2.70\times10^5$ J and the efficiency

is $e = \dfrac{W}{Q_H} = \dfrac{0.30\times10^5}{2.70\times10^5} = 0.111 = 11.1\%$.

EVALUATE: For any cycle, $\Delta U = 0$ and $Q = W$.

20.55.   **(a) IDENTIFY and SET UP:** Calculate e from Eq. (20.6), Q_C from Eq. (20.4) and then W from Eq. (20.2).

EXECUTE: $e = 1 - 1/(r^{\gamma-1}) = 1 - 1/(10.6^{0.4}) = 0.6111$

$e = (Q_H + Q_C)/Q_H$ and we are given $Q_H = 200$ J; calculate Q_C.

$Q_C = (e-1)Q_H = (0.6111 - 1)(200$ J$) = -78$ J (Negative, since corresponds to heat leaving.)

Then $W = Q_C + Q_H = -78$ J $+ 200$ J $= 122$ J. (Positive, in agreement with Figure 20.6.)

EVALUATE: Q_H, $W > 0$, and $Q_C < 0$ for an engine cycle.

(b) IDENTIFY and SET UP: The stoke times the bore equals the change in volume. The initial volume is the final volume V times the compression ratio r. Combining these two expressions gives an equation for V. For each cylinder of area $A = \pi(d/2)^2$ the piston moves 0.864 m and the volume changes from rV to V, as shown in Figure 20.55a.

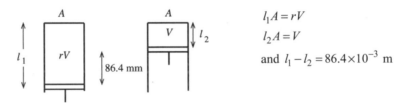

Figure 20.55a

EXECUTE: $l_1 A - l_2 A = rV - V$ and $(l_1 - l_2)A = (r-1)V$

$V = \dfrac{(l_1 - l_2)A}{r-1} = \dfrac{(86.4\times10^{-3}\ \text{m})\pi(41.25\times10^{-3}\ \text{m})^2}{10.6-1} = 4.811\times10^{-5}$ m^3

At point a the volume is $rV = 10.6(4.811\times10^{-5}$ m$^3) = 5.10\times10^{-4}$ m^3.

(c) IDENTIFY and SET UP: The processes in the Otto cycle are either constant volume or adiabatic. Use the Q_H that is given to calculate ΔT for process bc. Use Eq. (19.22) and $pV = nRT$ to relate p, V and T for the adiabatic processes ab and cd.

EXECUTE: point a: $T_a = 300$ K, $p_a = 8.50\times10^4$ Pa and $V_a = 5.10\times10^{-4}$ m^3

point b: $V_b = V_a/r = 4.81\times10^{-5}$ m^3. Process $a \rightarrow b$ is adiabatic, so $T_a V_a^{\gamma-1} = T_b V_b^{\gamma-1}$.

$T_a(rV)^{\gamma-1} = T_b V^{\gamma-1}$

$T_b = T_a r^{\gamma-1} = 300$ K$(10.6)^{0.4} = 771$ K

$pV = nRT$ so $pV/T = nR = $ constant, so $p_a V_a/T_a = p_b V_b/T_b$

$p_b = p_a(V_a/V_b)(T_b/T_a) = (8.50\times10^4$ Pa$)(rV/V)(771$ K$/300$ K$) = 2.32\times10^6$ Pa

point c: Process $b \rightarrow c$ is at constant volume, so $V_c = V_b = 4.81\times10^{-5}$ m^3

$Q_H = nC_V\Delta T = nC_V(T_c - T_b)$. The problem specifies $Q_H = 200$ J; use to calculate T_c. First use the p, V, T values at point a to calculate the number of moles n.